Mathematical Programming

Mathematics and Its Applications (*Japanese Series*)

Mathematical Programming

Recent Developments and Applications

edited by

Masao IRI

Faculty of Engineering
University of Tokyo, Tokyo, Japan

and

Kunio TANABE

Institute of Statistical Mathematics, Tokyo, Japan

KTK Scientific Publishers/Tokyo

Kluwer Academic Publishers

Dordrecht/Boston/London

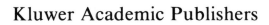

Library of Congress Cataloging in Publication Data

DATA-APPEARS ON SEPARATE CARD

ISBN 0-7923-0490-X

Published by KTK Scientific Publishers (KTK),
302 Jiyugaoka-Komatsu Building, 24-17 Midorigaoka 2-chome, Meguro-ku,
Tokyo 152, Japan,
in co-publication with Kluwer Academic Publishers, Dordrecht, Holland

Sold and distributed in the U.S.A. and Canada
by Kluwer Academic Publishers,
101 Philip Drive, Assinippi Park, Norwell, MA 02061, U.S.A.
in Japan by KTK Scientific Publishers (KTK),
302 Jiyugaoka-Komatsu Building, 24-17 Midorigaoka 2-chome, Meguro-ku,
Tokyo 152, Japan

In all other countries, sold and distributed
by Kluwer Academic Publishers,
P.O. Box 322, 3300 AH Dordrecht, Holland

Printed in Japan

SERIES EDITOR'S PREFACE

Mathematics is a tool for thought. A highly necessary tool in a world where both feedback and non-linearities abound. Similarly, all kinds of parts of mathematics serve as tools for other parts and for other sciences.

Applying a simple rewriting rule to the quote on the right above one finds such statements as: 'One service topology has rendered mathematical physics ...'; 'One service logic has rendered computer science ...'; 'One service category theory has rendered mathematics ...'. All arguably true. And all statements obtainable this way form part of the raison d'être of this series.

This series, *Mathematics and Its Applications*, started in 1977. Now that over one hundred volumes have appeared it seems opportune to reexamine its scope. At the time I wrote

> "Growing specialization and diversification have brought a host of monographs and textbooks on increasingly specialized topics. However, the 'tree' of knowledge of mathematics and related fields does not grow only by putting forth new branches. It also happens, quite often in fact, that branches which were thought to be completely disparate are suddenly seen to be related. Further, the kind and level of sophistication of mathematics applied in various sciences has changed drastically in recent years: measure theory is used (non-trivially) in regional and theoretical economics; algebraic geometry interacts with physics; the Minkowsky lemma, coding theory and the structure of water meet one another in packing and covering theory; quantum fields, crystal defects and mathematical programming profit from homotopy theory; Lie algebras are relevant to filtering; and prediction and electrical engineering can use Stein spaces. And in addition to this there are such new emerging subdisciplines as 'experimental mathematics', 'CFD', 'completely integrable systems', 'chaos, synergetics and large-scale order', which are almost impossible to fit into the existing classification schemes. They draw upon widely different sections of mathematics."

By and large, all this still applies today. It is still true that at first sight mathematics seems rather fragmented and that to find, see, and exploit the deeper underlying interrelations more effort is needed and so are books that can help mathematicians and scientists do so. Accordingly MIA will continue to try to make such books available.

If anything, the description I gave in 1977 is now an understatement. To the examples of interaction areas one should add string theory where Riemann surfaces, algebraic geometry, modular functions, knots, quantum field theory, Kac-Moody algebras, monstrous moonshine (and more) all come together. And to the examples of things which can be usefully applied let me add the topic 'finite geometry'; a combination of words which sounds like it might not even exist, let alone be applicable. And yet it is being applied: to statistics via designs, to radar/sonar detection arrays (via finite projective planes), and to bus connections of VLSI chips (via difference sets). There seems to be no part of (so-called pure) mathematics that is not in immediate danger of being applied. And, accordingly, the applied mathematician needs to be aware of much more. Besides analysis and numerics, the traditional workhorses, he may need all kinds of combinatorics, algebra, probability, and so on.

In addition, the applied scientist needs to cope increasingly with the nonlinear world and the extra mathematical sophistication that this requires. For that is where the rewards are. Linear models are honest and a bit sad and depressing: proportional efforts and results. It is in the non-linear world that infinitesimal inputs may result in macroscopic outputs (or vice versa). To appreciate what I am hinting at: if electronics were linear we would have no fun with transistors and computers; we would have no TV; in fact you would not be reading these lines.

There is also no safety in ignoring such outlandish things as nonstandard analysis, superspace and anticommuting integration, p-adic and ultrametric space. All three have applications in both electrical engineering and physics. Once, complex numbers were equally outlandish, but they frequently proved the shortest path between 'real' results. Similarly, the first two topics named have already provided a number of 'wormhole' paths. There is no telling where all this is leading - fortunately.

Thus the original scope of the series, which for various (sound) reasons now comprises five subseries: white (Japan), yellow (China), red (USSR), blue (Eastern Europe), and green (everything else), still applies. It has been enlarged a bit to include books treating of the tools from one subdiscipline which are used in others. Thus the series still aims at books dealing with:

- a central concept which plays an important role in several different mathematical and/or scientific specialization areas;
- new applications of the results and ideas from one area of scientific endeavour into another;
- influences which the results, problems and concepts of one field of enquiry have, and have had, on the development of another.

In spite of its age of some 50 years 'Mathematical Programming', in the more or less usual meaning of the term, still counts a young member of the family of mathematical specialisms. Viewed as (part of) optimization theory it is a much older tradition, almost as old as mathematics itself. It is, in any case, a most important subfield. An obvious reason for that is its supreme applicability. Some optimization questions are likely to arise in virtually any commercial or industrial set-up; and this is certain to happen just about all the time.

And that means work for the mathematical programming community. Whether this aspect alone could keep the subject alive as a respected part of mathematics is debatable. Fortunately the subject is also good at generating new (theoretical) problems and questions calling for new ideas and techniques. Many of these come directly from   applied practical problems. Examples from the past - and still actual - include, for instance, matters of computational complexity and the probabilistic analysis of efficiency of algorithms.

The present volume consists of most of the special invited lectures at the 13-th Mathematical Programming Symposium, which was held in September 1988 in Tokyo, Japan. Together these 13 lectures give an up-to-date of the state-of-the-art of this important discipline in mathematics and its applications.

The shortest path between two truths in the real domain passes through the complex domain.

J. Hadamard

La physique ne nous donne pas seulement l'occasion de résoudre des problèmes ... elle nous fait pressentir la solution.

H. Poincaré

Never lend books, for no one ever returns them; the only books I have in my library are books that other folk have lent me.

Anatole France

The function of an expert is not to be more right than other people, but to be wrong for more sophisticated reasons.

David Butler

Bussum, July 1989

Michiel Hazewinkel

Preface

Mathematical programming has now nearly half-century long history since the term was coined. It has been developed in various directions, both in theory and applications, and established itself as a vigorous discipline. Since it was started in 1949, the series of international symposia of the Mathematical Programming Society has always been the most important medium among mathematical programmers to exchange ideas and learn the latest developments in the field. The 13th International Symposium on Mathematical Programming was held at Kasuga campus of Chuo University, Tokyo, Japan, on August 29 through September 2, 1988. It was the first in the series that was held outside North America and Europe to enhance the advancement of this important field of science and technology.

Some people argue that Mathematical Programming has been saturated in some research directions. But, every time when it seems saturated, a breakthrough takes place. A breakthrough will be caused usually by those motivated in practical applications and computations of mathematical programming. In view of the strong critical opinion arising within the Mathematical Programming Society against the apparent recent trends of too much sophistication in theory, special emphasis was laid upon "applications" in organizing this symposium in the hope that it makes an epoch for mathematical programming to restore the vigor of its birthdays and to become wider-spread in society and industry, thus acquiring higher estimation both from the theoretical-mathematical circle and from the practical-engineering. This initiative was responded enthusiastically by the participants of the symposium and one third of the presentations turned out to be on applications in diverse fields of Engineering, Medicine, Management, Computer Science and Earth Science.

Besides the usual contributed papers sessions, the scientific program consisted of two Memorial Lectures in memory of Martin Beale and L. V. Kantorovich, one Plenary Lecture, twelve Tutorial-and-Survey Lectures in which distinguished researchers delivered one-hour lectures on topics of current interest in Mathematical Programming, and specially organized sessions organized by recognized experts in the respective fields so that the papers presented at those sessions reflected the latest important developments.

This volume contains the thirteen papers which have materialized from the fifteen invited lectures delivered as plenary, tutorial-and-survey, and memorial lectures at the Symposium. These papers document the state of the art of our field in the year of 1988. There will be no proceedings volume of the invited and contributed papers, besides the book of Abstracts distributed at the Symposium, but a special issue of Mathematical Programming Series B is scheduled to be published consisting of selected applications papers, and many other papers are expected to appear in Mathematical Programming Series A and B in the ordinary course.

In addition to these 15 special one-hour lectures, 542 invited and contributed talks of 30 minutes duration were given in up to fifteen sessions in parallel. Thus, the final statistics of the presentations are as follows:

- 1 Plenary lecture,
- 2 Memorial lectures,
- 12 Tutorial-and-Survey lectures,
- 2 invited talks in a special session for Prize Recipients,
- 8 invited talks in 2 Memorial sessions,
- 279 invited talks in 90 Specially Organized sessions, and
- 253 talks in 84 Contributed Paper sessions.

There were 81 papers cancelled and 5 Contributed Paper sessions cancelled.

Finally we would like to thank the members of the International Program Committee, the organizers of Specially Organized sessions, and all the speakers, as well as the members of the Organizing Committee, for their indispensable contributions to make the Symposium successful. We would also like to thank a number of organizations who sponsored this Symposium, in particular, the Operations Research Society of Japan, the International Federation of Operational Research Societies, and the Association of Asian-Pacific Operational Research Societies within IFORS. We are very much indebted to the various institutions who supported this Symposium financially. In particular, we gratefully acknowledge the financial supports received from the Commemorative Association for the Japan World Exposition and the Telecommunications Advancement Foundation, whose generous supports made the Symposium and the publication of this volume possible. Last but not least, our thanks are due to the authors who kindly cooperated to materialize this volume. We apologize that the publication is delayed far behind the schedule, but would like the participants of the Symposium and the readers of this volume to understand that we have tried to optimize the date of publication with respect to two conflicting objectives — to minimize the delay in publication and to maximize the number of important papers to be included.

Tokyo, July 1989

    IRI Masao
      Chairman of the Organizing Committee
    TANABE Kunio
      Chairman of the Local Program Committee

      The 13th International Symposium on
      Mathematical Programming

Contents

Applications of Combinatorial Optimization

Bernhard KORTE

Research Institute of Discrete Mathematics and Institute of Operations Research, University of Bonn, Nassestrasse 2, Bonn, W. Germany

Abstract

In this paper we survey different applications of combinatorial optimization, especially in production, production planning and production preparation. We report on successful applications of different combinatorial optimization techniques in VLSI-design and -layout. Here we are faced with combinatorial structures of dimension $10^6 - 10^8$. We also discuss very large travelling salesman problems which arise in different production processes, particularly in the production of multi-chip-carrier and printed circuit boards as well as in mask lithography. Another large scale applied problem results in finding perfect matchings in certain graphs, which are constructed to minimize wasted movements of robots and production machines.

1. Introduction

This tutorial lecture reports on successful applications of different combinatorial optimization techniques in design, layout and production processes of computer hardware. About one year ago we were faced for the first time with very interesting combinatorial optimization problems in design and layout of very large scale integrated high performance CMOS-logic-chips. We discuss them in chapter 5.

In connection with these design problems we became acquainted with other interesting applications of combinatorial optimization in design and production

M. Iri and K. Tanabe (eds.), Mathematical Programming, 1–55.
© *1989 by KTK Scientific Publishers, Tokyo.*

of hardware components related to integrated circuits. Lithographic masks or reticles for printed circuit boards, multi-chip modules and VLSI-chips are produced by optical pattern generators using monochromatic light, excimer laser or X-rays. Those pattern generators are extremely precise, but therefore very slow. Some generators need several days for the exposure (writing) of one mask. Thus optimization of the movements of the generator is badly needed. Here the expert might suspect problems like the Chinese postman problem or perfect matching.

The fabrication process of printed circuit boards as well as ceramic multi-chip modules still needs a lot of drilling and punching by classical machine tools. Here combinatorial optimization gets a lot of very large travelling salesman problems and modifications of it. The size ranges from several tens to several hundred-thousands of cities. The largest travelling salesman problem which we got from an exposure machine for via masks of VLSI-chips had about 1.2 million cities.

For inter-chip-wiring one often uses multi-chip-modules which consist of up to 30 to 60 layers of ceramics connected by via holes. Thus a 3-dimensional rectangular grid graph is an appropriate combinatorial model for it. A layout for inter-chip-wiring is then an embedding of edge-disjoint paths or Steiner-trees in this 3-dimensional grid. This embedding can have different objectives. One could be to minimize the number of layers. We have approached this aim with fast heuristics for finding largest cliques in graphs.

Surprisingly, we got quite substantial improvements over the engineering solutions used so far, or we could offer solutions for very large and complex problems, which were not obtained by classical engineering methods. It is not easy to give one unimodal measure for the improvements, since we have dealt with different methods for different problems of different size with even different optimization objectives (size, length, time, homogeneity, wasted movements, yield etc.). However, as a rough estimate we can state that we could improve by means of combinatorial optmization at least 10 %, in the majority of cases between 20 % and 30 %, and in some case over 50 % and even 90 % and more. Some of the underlying production processes, e.g. punching of ceramic layers, have a gross value of several hundred thousand dollars per day. It is easy to figure out that even modest improvements in the production time yield important savings.

In the next chapter we make some general and even philosophical remarks on applications of mathematics or more specifically of mathematical programming.

Chapter 3 reports on large scale Chinese postman problems and perfect matchings derived from mask lithography for VLSI-chips, ceramic multi-chip-modules and printed circuit boards.

Very large scale travelling salesman problems are dealt with in chapter 4. They arise from drilling and punching problems of printed circuit boards and multi-chip modules as well as from via exposures of lithographic masks.

In chapter 5 we give some preliminary results on embedding inter-chip-connections into a 3-dimensional grid, i.e. minimizing the number of layers of multi-chip-modules or printed circuit boards.

Finally, in chapter 6 we report on our experience and results in using combinatorial optimization for VLSI-physical-design. This was the largest and probably the most successful part of our application work so far.

This work was done in the framework of a research contract with IBM Germany. We gratefully acknowledge the most efficient cooperation of many engineers and scientists at the IBM Laboratories in Böblingen and Rüschlikon and at the IBM plants in Sindelfingen and Böblingen.

2. General Remarks on Applications

Since this is a tutorial lecture for a broad audience, I cannot resist on making some general and philosophical remarks on mathematics (or more specific: mathematical programming) and its applications. In my welcoming address at the opening of the XIth International Symposium on Mathematical Programming 1982 in Bonn I quoted in part the famous statement of JOHN VON NEUMANN [1947] on mathematics and real world phenomena. Since I feel that this dictum contains most profound and deep thoughts about mathematics and applications I repeat it here in full length:

"*I think that it is a relatively good approximation to truth - which is much too complicated to allow anything but approximations - that mathematical ideas originate in empirics, although the genealogy is sometimes long and obscure. But, once they are so conceived, the subject begins to live a peculiar life of its own and is better compared to a creative one, governed by almost entirely aesthetical motivations, than to anything else and, in particular, to an empirical science. There is, however, a further point which, I believe, needs stressing. As a mathematical discipline travels far from its empirical source, or still more, if it is a second and third generation only indirectly inspired by ideas coming from "reality", it is beset with very grave dangers. It becomes more and more purely aestheticizing, more and more purely l'art pour l'art. This need not be bad, if the field is surrounded by correlated subjects which still have closer empirical connections, or if the discipline is under the influence of men with an exceptionally well developed taste. But there is a grave danger that the subject will develop along the line of least resistance, that the stream, so far from its source, will separate into a multitude of insignificant branches, and that the discipline will become a disorganized mass of details and complexities. In other words, at a great distance from its empirical source, or after much "abstract" inbreeding, a mathematical subject is in danger of degeneration. At the inception the style is usually classical; when it shows signs of becoming baroque, then the danger signal is up. It would be easy to give examples, to trace specific evolutions into the baroque and the very high baroque, but this, again, would be too technical.*

In any event, whenever this stage is reached, the only remedy seems to me to be the rejuvenating return to the source: the reinjection of more or less directly empirical ideas. I am convinced that this was a necessary condition to conserve the freshness and the vitality of the subject and that this will remain equally true in the future"

In parentheses: Every mathematical programmer should ask himself whether our field is already baroque or even high baroque. In any case, there seems to be some need for mathematical programming and also for combinatorial programming to get involved in new (and exciting) applications.

To elucidate other aspects of mathematics and applications I will again use dicta from other sources. They are more appropriate than my own pedestrian arguments.

There is a long-lasting quarrel about the usefulness of mathematics in general. To this G.M. HARDY [1940] contributed a most elegant but highly controversal saying. He states in his book "A Mathematician's Apology": *"I have never done anything useful. No discovery of mine has made, or is likely to make, directly or indirectly, for good or ill, the least difference to the amenity of the world. I have helped to train other mathematicians, but mathematicians of the same kind as myself, and their work has been, so far at any rate as I have helped them to it, as useless as my own. Judged by all practical standards, the value of my mathematical life is nil; and outside mathematics it is trivial* anyhow. *I have just one chance of escaping a verdict of complete triviality, that I may be judged to have created something worth creating. And that I have created something is undeniable: the question is about its value"*.

I have the impression that mathematics, especially during the last decades, went very much on Hardy's lines. But it is historically provable that great changes and development in mathematics often had its roots in applications. Calculus, the most fruitful source of mathematical development during the last 200 years, was not developed as a l'art pour l'art topic, but for the needs in physics. One of its originators, Newton, was not only a mathematician but even more a physicist. And we should also keep in mind that the other originator of calculus, Leibniz, did not only wonderful pure mathematics but constructed and built one of the first mechanical four-species-calculators. He was elected member of the Royal Academy not for the invention of calculus but for the invention of this machine. Time has changed since then!

However, recently the U.S. National Research Council got concerned about future trends of mathematics and its applicability and therefore established an Ad Hoc Committee. From the report of this committee entitled *"Renewing U.S. Mathematics: Critical Resource for the Future"* [1984] again I quote a very nice statement about mathematical technology: *"The emergence of high technology brought our society into an era of mathematical technology, in which mathematics and engineering interact in new ways. Fifty years ago this was the pattern: mathematics made some tools directly for engineering but basically promoted the development of other sciences, which, in turn, provided the foundations for engineering principles and design. Mathematics and engineering now interact directly, on a broader, deeper scale, greatly to the benefit of both fields, and to technology."*

Indeed in previous centuries natural sciences were the driving force for the development of new mathematics. There is some evidence that today also engineering takes over this part.

I would like to conclude this gallimaufry of quotations by using a last one which emphasises the increasing usefulness of discrete mathematics compared to continuous mathematics. P.R. HALMOS [1981] , evidently not a discrete mathematician, observed that *"in the foreseeable future (as in the present) discrete mathematics will be an increasingly useful tool in the attempt to understand the world, and analysis will therefore play a proportionally smaller role."*

As far as discrete or combinatorial optimization is concerned, I feel that there also is some truth in the statement of Halmos. Probably this lecture will demonstrate this to some extent.

At this point I woud like to list only those combinatorial optimization models and techniques which we have used for application problems in design, layout, and production of chips, modules and boards:

- Travelling salesman problems
- Chinese postman problems
- Mixture of both and modifications
- Matching problems
- Largest-clique-problems
- Graph partitioning problems
- Bin-packing and Knapsack problems
- Shortest path problems and modifications
- Edge-(or vertex-) disjoint path problems
- Multicommodity flow problems
- Steiner tree problems in planar (or rectilinear) graphs

3. Applications of Combinatorial Optimization to Lithographic Systems

In the production process of chips, ceramic multi-chip modules, and printed circuit boards one has to generate masks by a lithographic system. Those masks consist of certain patterns of the chip, module or board. The masks will be used to exposure the patterns on the silicon wafer, the ceramic layer or the printed board layer by an optical stepper system. Since one chip, module or board is an arrangement of layers of many different patterns, mask generation is a substantial, very time consuming and therefore expensive part of the production process. The mask itself is a special glass with a chromium layer covered by a photo resist. The pattern will be exposed on the mask by a pattern generator optically or with electron or ion beams. We will not consider the e-beam pattern generation here. It is done in a scanning technique on a raster and therefore there is not too much to be optimized.

Optical pattern generators use UV-light, X-rays or an excimer laser. They consist of a classical photographic system with a variable aperture and a compound table on which the mask carrier can be moved in the x- and y-direction.

Fig. 3.1 shows the principle of an *excimer laser pattern generator* at the IBM Böblingen Laboratories. For further details see HAFNER [1988].

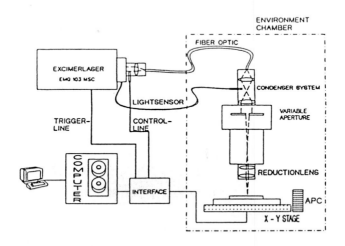

Fig. 3.1. Principle of an excimer laser-pattern generator

Mathematically, pattern generation means to draw a not necessarily connected graph (=write a pattern) by moving the compound table in the x- and y-directions as fast as possible. This again means one has to minimize the wasted movements, i.e. the movements of the compound table which are not used for drawing the graph. In the language of combinatorial optimization this is basically the famous Chinese postman problem.

Since not every reader might be familiar with this problem, I will give here a short introduction by a historical sketch: It is by now a common opinion in combinatorics that LEONHARD EULER (1707-1783) is the founder of graph theory. And even the Chinese postman problem is strongly related to his first and famous graph theoretic problem, the *Königsberg bridge problem*. Fig. 3.2 shows the city of Königsberg at Euler's time. As one can see, the river Pregel subdivides Königsberg and bifurcates for an island, the Kneiphof, in the center. There were seven bridges crossing the river and connecting the different parts of the city. At the time of Euler it had long been an open problem to prove or disprove the existence of a walk crossing each bridge exactly once and returning to the point of origin. Since this problem was spread around the intellectual circles of that time Euler got to know it. He worked on it for several years, first he only got necessary conditions and eventually sufficient conditions. He reported the solution at the St. Petersburg Academy and wrote a 13 page long mathematical paper on it (EULER [1736]). For reasons of curiosity we show in Fig. 3.3 four pages of Euler's original paper. In elegant Latin he did diligent case distinctions and proving arguments. We only mention this because we will see below that today we can give such an easy and elementary argument that a mathematical layman and even a pupil can understand it immediately. As a side-remark, this demonstrates nicely the development of abstract thinking and reasoning of everybody during the last 200 years.

Fig. 3.2. Königsberg with the river Pregel and seven bridges

For the argument see Fig. 3.4. Here we have modelled the problem by a graph. Every region of the city is a vertex and every way across a bridge is an edge of this graph. The Königsberg bridge problem now reads: to traverse all edges of the graph exactly once. But, as we know, this is impossible, since

2. Problema autem hoc, quod mihi satis notum esse perhibebatur, erat sequens: Regiomonti in Borussia esse insulam *A, der Kneiphof* dictam, fluviumque eam cingentem in duos dividi ramos, quemadmodum ex figura (Fig. 1) videre licet; ramos vero huius fluvii septem instructos esse pontibus *a, b, c, d, e, f* et *g*. Circa hos pontes iam ista proponebatur quaestio, num quis cursum ita instituere queat, ut per singulos pontes semel et non plus quam semel transeat. Hocque fieri posse, mihi dictum est, alios negare alios dubitare; neminem vero affirmare. Ego ex hoc mihi sequens maxime generale formavi problema: quaecunque sit fluvii figura et distributio in ramos atque quicunque fuerit numerus pontium, invenire, utrum per singulos pontes semel tantum transiri queat an vero secus.

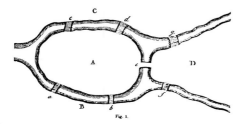

Fig. 1.

3. Quod quidem ad problema Regiomontanum de septem pontibus attinet, id resolvi posset facienda perfecta enumeratione omnium cursuum, qui institui possunt; ex his enim innotesceret, num quis cursus satisfaceret an vero nullus. Hic vero solvendi modus propter tantum combinationum numerum et nimis esset difficilis atque operosus et in aliis quaestionibus de multo pluribus pontibus ne quidem adhiberi posset. Hoc porro modo si operatio ad finem perducatur, multa inveniuntur, quae non erant in quaestione; in quo procul dubio tantae difficultatis causa consistit. Quamobrem missa hac me-

thodo in aliam inquisivi, quae plus non largiatur, quam ostendat, utrum talis cursus institui queat an secus; talem enim methodum multo simpliciorem fore sum suspicatus.

4. Innititur autem tota mea methodus idoneo modo singulos pontium transitus designandi, in quo utor litteris maiusculis *A, B, C, D* singulis regionibus adscriptis, quae flumine sunt separatae. Ita, si quis ex regione *A* in regionem *B* transmigrat per pontem *a* sive *b*, hunc transitum denoto litteris *AB*, quarum prior praebet regionem, ex qua exierat viator, posterior vero dat regionem, in quam pontem transgressus pervenit. Si deinceps viator ex regione *B* abeat in regionem *D* per pontem *f*, hic transitus repraesentabitur litteris *BD*; duos autem hos transitus successive institutos *AB* et *BD* denoto tantum tribus litteris *ABD*, quia media *B* designat tam regionem, in quam primo transitu pervenit, quam regionem, ex qua altero transitu exit.

5. Simili modo si viator ex regione *D* progrediatur in regionem *C* per pontem *g*, hos tres transitus successive factos quatuor litteris *ABDC* denotabo. Ex his enim quatuor litteris *ABDC* intelligetur viatorem primo in regione *A* existentem transiisse in regionem *B*, hinc esse progressum in regionem *D* atque ex hac esse profectum in *C*; cum vero hae regiones fluviis sint a se invicem separatae, necesse est, ut viator tres pontes transierit. Sic transitus per quatuor pontes successive instituti quinque litteris denotabuntur; et si viator trans quotcunque pontes eat, eius migratio per litterarum numerum, qui unitate est maior quam numerus pontium, denotabitur. Quare transitus per septem pontes ad designandum octo requirit litteras.

6. In hoc designandi modo non respicio, per quos pontes transitus sit factus, sed si idem transitus ex una regione in aliam per plures pontes fieri potest, perinde est, per quemnam transeat, modo in designatam regionem perveniat. Ex quo intelligitur, si cursus per septem figurae pontes ita institui posset, ut per singulos semel ideoque per nullum bis transeatur, hunc cursum octo litteris repraesentari posse easque litteras ita esse debere dispositas, ut immediata litterarum *A* et *B* successio bis occurrat, quia sunt duo pontes *a* et *b* has regiones *A* et *B* iungentes; simili modo successio litterarum *A* et *C* quoque debet bis occurrere in illa octo litterarum serie; deinde successio litterarum *A* et *D* semel occurret similiterque successio litterarum *B* et *D* iterumque *C* et *D* semel occurrat necesse est.

1*

summam numeri pontium unitate aucti dimidium, si numerus pontium fuerit impar; ipsius vero numeri pontium medietatem, si fuerit par. Deinde si numerus omnium vicium adaequet numerum pontium unitate auctum, tum transitus desideratus succedit, at initium ex regione, in quam impar pontium numerus ducit, capi debet. Sin autem numerus omnium vicium fuerit unitate minor quam pontium numerus unitate auctus, tum transitus succedet incipiendo ex regione, in quam par pontium numerus ducit, quia hoc modo vicium numerus unitate est augendus.

14. Proposita ergo quacunque aquae pontiumque figura ad investigandum, num quis per singulos semel transire queat, sequenti modo operationem instituo. Primo singulas regiones aqua a se invicem diremptas litteris *A, B, C* etc. designo. Secundo sumo omnium pontium numerum eumque unitate augeo atque sequenti operationi praefigo. Tertio singulis litteris *A, B, C* etc. sibi subscriptis cuilibet adscribo numerum pontium ad eam regionem deducentium. Quarto eas litteras, quae pares adscriptos habent numeros, signo asterisco. Quinto singulorum horum numerorum parium dimidia adiicio, imparium vero unitate auctorum dimidia ipsis adscribo. Sexto hos numeros ultimo scriptos in unam summam coniicio; quae summa si vel unitate minor fuerit vel aequalis numero supra praefixo, qui est numerus pontium unitate auctus, tum concludo transitum desideratum perfici posse. Hoc vero est tenendum, si summa inventa fuerit unitate minor quam numerus supra positus, tum initium ambulationis ex regione asterisco notata fieri debere; contra vero ex regione non signata, si summa fuerit aequalis numero praescripto. Ita ergo pro casu Regiomontano operationem instituo, ut sequitur:

Numerus pontium 7, habetur ergo 8.

Pontes		
A,	5	3
B,	3	2
C,	3	2
D,	3	2

Quia ergo plus prodit quam 8, huiusmodi transitus nequaquam fieri potest.

15. Sint duae insulae *A* et *B* aqua circumdatae, qua cum aqua communicent quatuor fluvii, quemadmodum figura (Fig. 3) repraesentat. Traiecto porro sint super aquam insulas circumdantem et fluvios quindecim pontes *a, b, c, d* etc. et quaeritur, num quis cursum ita instituere queat, ut per

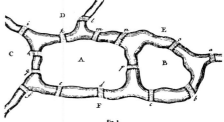

Fig. 3.

omnes pontes transeat, per nullum autem plus quam semel. Designo ergo primum omnes regiones, quae aqua a se invicem sunt separatae, litteris *A, B, C, D, E, F*, cuiusmodi ergo sunt sex regiones. Dein numerum pontium 15 unitate augeo et summam 16 sequenti operationi praefigo.

	16	
A,	8	4
B,	4	2
C,	4	2
D,	3	2
E,	5	3
F,	6	3
		16

Fig. 3.3. Four pages of the original paper by Euler on the Königsberg bridge problem

every vertex of the graph has *odd degree* (=odd number of edges incident at the vertex). Euler's theorem says that an arbitrary connected graph has an *eulerian tour*, i.e. is traversable, if and only if every vertex has even degree. For traversing the edges we enter each vertex and leave it by using exactly two edges. Thus if one vertex has odd degree, this odd number minus two remains odd. Hence we will eventually reduce the number of non-traversed edges at this node to one. This means we will enter the node finally but we cannot leave.

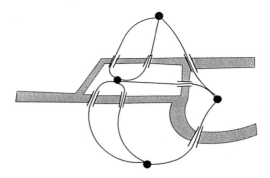

Fig. 3.4. Königsberg bridge problem as a graph model

But now we are very close to the problem of minimizing the wasted movement of a machine which has to traverse a graph (e.g. draw a picture or expose a line pattern). We modify the original question of the bridge problem and ask, how many additional bridges (=wasted movements) would have to be built by the people of Königsberg in order to make the walk possible. Fig. 3.5 shows a solution. Two additional bridges suffice. For an arbitrary connected graph we take all vertices with odd degree and we consider the complete graph on them. Every edge in this graph is a possible wasted movement. If these wasted movements have different cost (length) we may put weights on the edges. Minimizing the wasted movements means to find a *minimum weight perfect matching* in this graph. A matching is a subset of edges of the graph such that no two are adjacent, i.e. meet at the same vertex. A matching is called perfect if every vertex of the graph is covered by one edge of the matching.

This problem got the name *Chinese postman problem*, since the Chinese math-

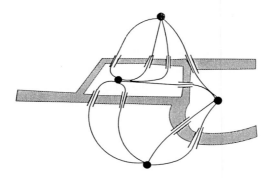

Fig. 3.5. Two additional bridges suffice to solve the Königsberg bridge problem

ematician GUAN MEI-GU had to work during the period of cultural revolution at a post office. There he was faced with the problem of minimizing the route of a postman who has to traverse all streets of his district (cf. GUAN MEI-GU [1962]). If the graph which has to be traversed is not connected one has to connect its components first in a minimal way before constructing the perfect matching.

The matching problem belongs to the class of "good problems". Thus it is solvable by a polynomial algorithm. However, exact matching algorithms have the complexity $0(n^3)$. This is far too slow for very large practical problems. For real world problems of the size we are dealing with only linear time algorithms are applicable. Even approximation algorithms of complexity $0(n^2)$ or $0(nlogn)$ are far too slow. Reports in the literature on matching heuristics with guaranteed performances are not too optimistic (cf. REINGOLD and TARJAN [1981] and SUPOSIT, PLAISTED and REINGOLD [1980]). They normally require the triangle inequality, i.e. euclidean problems, which is not the case in our problems.

However, our Japanese colleagues have developed very fast (linear time) approximate algorithms for perfect matchings which exploit the geometrical structure of the problem. (cf. IRI, MUROTA and MATSUI [1981], [1982] and ASANO, EDAHIRO, IRI and MUROTA [1985]). We cannot go into details here and therefore we have to refer the interested reader to the above-mentioned papers. The main idea is a decomposition approach using bucketing techniques and geometrical procedures called serpentine rack algorithms. They have used this approach for minimizing the wasted movements of a pen plotter. Mathematically this is the same problem as our problem in mask lithography. We have borrowed substantially from the ideas of these papers in our approach.

By courtesy of Professor Iri, the chairman of this Symposium, we demonstrate in Fig. 3.6 some of his applications. Fig. 3.6 (a) shows the road map of Kanto district in Japan, which had to be drawn with a classical pen plotter. This is a graph with 20726 nodes and total $(L_\infty-)$ length of 44.47 meters (on the plot). In (b) we see the total wasted movement when the data are shuffled, as it happens often in practical applications. Note that the total wasted movement is about 100 times longer than the road map. Fig. (c) shows the wasted movement when the data are manually sorted and searched, while Fig. (d) shows the effect of the approximate matching algorithm. The total wasted movement is only 13.62 meters. The computing time for the matching algorithm is relatively small, but the saving in plotting (=production) time is 94 %.

Before we report on our applications we would like to comment on the objective functions under consideration. As mentioned, we have to find a minimum-weight perfect matching. But what are the weights on the edges? Certainly not the euclidean distances. The plotter or the compound table of a pattern generator moves only in two orthogonal directions, namely the x- and y-axis. If the device has only one numerically controlled processor, the movements are done sequentially, say first x- then y-axis. Thus we have no L_2- but L_1-norm. If the device has two independently controlled processors we are faced with the L_∞-norm, i.e. the maximum of the two directions matters.

But not only the correct norm is important. The velocity and acceleration of the device is different in both directions. Normally, one direction is preferred and faster. In one direction only the object is moved, while in the other direction the whole compound table has to be moved. Therefore, stop and adjustment times are also different in both directions. Moreover, the device might have different modes, like a fast "flash on the fly"-mode and a slow, but very precise "stop, expose and go"-mode. The speed and the acceleration depends on the direction but also on the total length. Thus we are confronted with a highly *non-linear* objective function.

A very good knowledge or estimate of the objective function is most essential for practical results of the optimization approach. Very often it is not sufficient to

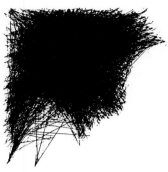

(a)

20726 nodes (including
        degree-2 nodes)
25336 links (straight
        line segments)
25 connected components

Total L_∞-length of lines
(on 60cm×60cm sheet): 44.47m

Fig.1. Road map of Kanto
        district of Japan

(b)

Total length of wasted
    movement $\fallingdotseq$ 4702.53m ($\fallingdotseq$ 4.7km)

Total plotting time $\fallingdotseq$ 9h 50min

CPU time of the host computer
(HITAC M-200H)
to input data   : 7.9s
to output plotter file:14.0s

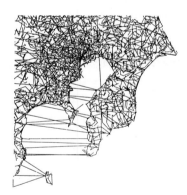

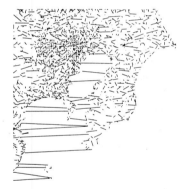

(c)

Total length of wasted movement
            $\fallingdotseq$ 57.80m

Total plotting time $\fallingdotseq$ 51min

CPU time of the host computer
(HITAC M-200H)
to input data   :7.9s
to output plotter file:5.1s

(d)

Total length of wasted movement
            $\fallingdotseq$ 13.62m

Total plotting time $\fallingdotseq$ 36min 45s

CPU time of the host computer
(HITAC M-200H)
to input data   :7.9s
to output plotter file:3.9s
to preprocess data   :0.85s

    0.28s for COMPONENT
    0.27s for CONNECT
    0.07s for MATCH
    0.23s for EULER

Fig. 3.6. Road map of Kanto district in Japan drawn with a pen plotter

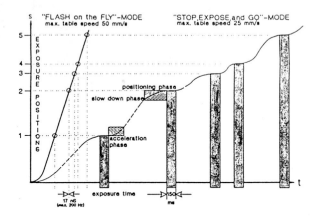

Fig. 3.7. Velocity diagram of an excimer laser pattern generator

use for the objective function only the given parameters (velocity, acceleration, slow down) of the device. The objective function has to be empirically evaluated by special test patterns and by measuring the real time at the device. The velocity diagram of an excimer laser pattern generator in Fig. 3.7 demonstrates the non-linearity of the objective function. In order to have an impression of the structure of the objective function, we give here one for a device with L_∞-norm and distance-independent acceleration:

$$DIST = max\{FCONST, ACONST * ACHANGE,$$
$$DCONST * DCHANGE,$$
$$XCONST * XCHANGE + XFACT * XDIST$$
$$+ SCONST * SCHANGE, YCONST * YCHANGE +$$
$$YFACT * YDIST + XCHANGE\}$$

with

$XDIST, YDIST$ = distances in x- resp. y-direction

$XFACT, YFACT$ = time (in ms) for x-resp. y-move

$FCONST$ = loading time of the exposure device

$DCONST$ = time for changing the aperture

$ACONST$ = time for changing the shutter angle

$XCONST, YCONST$ = constant time (partitioning)

in x- resp. y-direction

$SCONST$ = stammering time (peculiarity of the device)

in x-direction

$ACHANGE, SCHANGE, DCHANGE, XCHANGE, YCHANGE \in \{0,1\}$.

Our method uses bucketing techniques together with nearest neighbour heuristics and local search. It is a linear time algorithm. The computing time on an IBM 3081 KX is several minutes only. Even for large problems with several hundred thousand edges it goes up to at most 15 minutes.

Tables 3.1, 3.2 and 3.3 demonstrate that the production time for lithographic masks can be reduced substantially by using combinatorial optimization. The three tables represent three different devices. NCW and NCT are excimer laser lithographic systems, while BRA is an optical pattern generator with UV-light.

Mask ID number	number of flashes	Time before after optimization hours (decimal)		Reduction %
NCW32	6,684	1.14	0.27	76.3
NCW31	9,484	1.51	0,59	60.9
NCW17	12,675	1.64	0.66	59.8
NCW04	116,173	9.95	0.24	97.6
NCW16	136,112	9.33	3.60	61.4
NCW15	168,325	4.10	1.54	62.4
NCW02	173,397	7.23	2.78	61.5
NCW09	202,946	5.73	2.42	57.8
NCW07	292,768	12.02	5.52	54.1
NCW10	497,709	22.92	9.92	56.5

Table 3.1. Time Reduction with Combinatorial Optimization for Lithographic System NCW

Mask ID number	number of flashes	Time before optimization hours (decimal)		Reduction %
		before	after	
NCT37	2,891	0.56	0.14	75.0
NCT42	10,094	0.79	0,48	39.2
NCT41	19,042	5.86	1.08	81.6
NCT38	75,945	6.19	3.27	47.2
NCT35	170,403	6.48	2.71	58.2
NCT34	171,741	6.50	2.72	58.2
NCT33	215,196	4.32	2.22	48.6
NCT36	1,032,493	5.19	5.16	0.6

Table 3.2. Time Reduction with Combinatorial Optimization for Lithographic System NCT

Mask ID number	number of flashes	Time before optimization hours (decimal)		Reduction %
		before	after	
BRA03	1,754	0.25	0.11	56.0
BRA04	1,763	0.29	0.14	51.7
BRA08	3,861	0.40	0.18	55.0
BRA06	4,870	0.75	0.42	44.0
BRA07	5,667	1.14	0.51	55.3

Table 3.3. Time Reduction with Combinatorial Optimization for Optical Pattern Generator BRA

The first column gives the mask ID number. The second column shows the number of flashes (edges). This is a measure for the size of the pattern, but not necessarily for its complexity. Note that frequent changes of the aperture and angle of the shutter makes a pattern more difficult. Column 3 shows the real (gross) production time at the device (including set-up time etc.) before optimization, while column 4 shows the same time - again measured at the device - after solving the associated perfect matching problem heuristically. In column 5 we have displayed the relative reduction (one minus quotient of columns 4 and 3). The average reduction lies between 50 and 60 % and depends on the structure of the pattern. To demonstrate the reduction graphically, we show in Fig. 3.8 the pattern of mask NCW04.

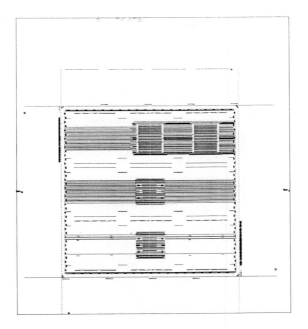

Fig. 3.8. Mask pattern NCW04

Fig. 3.9 demonstrates the pattern together with the wasted movement before optimization. Fig. 3.10 shows the same picture after optimization. This is the example with the largest reduction (97.6 %). Figures 3.11, 3.12 and 3.13 demonstrate the same for mask NCW15 (reduction 62.4 %).

The figure in column 5 of the last line of table 3.2 is not a misprint. Here the reduction was only 0.6 %. But this is clear if we look at the pattern of mask NCT36 (Fig. 3.14). A complete square had to be drawn. It is evident that there is no wasted movement and therefore nothing to be optimized.

We have very briefly described what kind of heuristics we used to minimize wasted movements. But we have not yet mentioned how the original data of the engineering solution ("before optimization", column 3 of the tables) was generated. The microprocessors which control the devices normally sort the coordinates lexicographically. The faster coordinate gets higher lexicographic

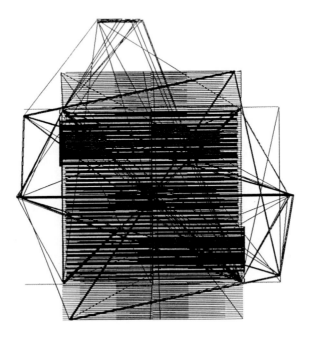

Fig. 3.9. Mask pattern NCW04 and wasted movement before optimization

priority. Since the patterns are relatively dense there seems to be some evidence for this rule. Actually, in a handbook for programming of the NC-controllers we found the remark that "this rule is optimal in the average". Of course this is not the case, as we have demonstrated.

Finally, we should mention that lithographic systems are very expensive, maintenance dependent, and always a bottleneck in the total production process. Thus savings of 50 % and more are important figures.

4. Applications of Combinatorial Optimization to Drilling and Punching of Printed Circuit Boards and Multi-Chip Modules

The previous chapter used mainly the Chinese postman problem as a combinatorial optimization model. Here we will use mainly the travelling salesman

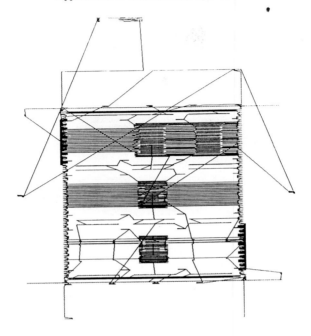

Fig. 3.10. Mask pattern NCW04 and wasted movement after optimization

problem as an appropriate model for applications. Thus in the previous application problems one had to traverse the edge of a certain graphical pattern, here the vertices.

First let us briefly describe the technological problems involved here. Multi-chip modules are ceramic carriers of size 9×9 centimeters which can carry up to 133 chips. Each carrier consists of 30 to 60 layers, each 0.2 mm thick. Thus, the whole carrier has a width of 0.5 to 1.0 cm. Each layer has only horizontal or only vertical "wires" for interchip connections. The layers can be connected by vias. This are holes punched into the layer, which are filled with molybdenum to make the electrical connections. Fig. 4.1 shows the principal arrangement of a multi-chip module. To demonstrate the construction principle some layers are partially removed. The diameter of the vias is 0.12 mm. A multi-chip module can have up to 15960 vias (holes). It carries the chips on top and it has 1800 pins on the bottom for further connections.

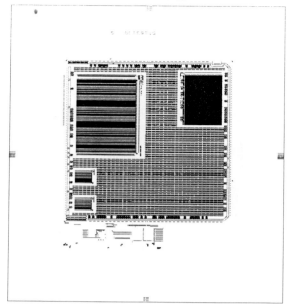

Fig. 3.11. Mask pattern NCW15

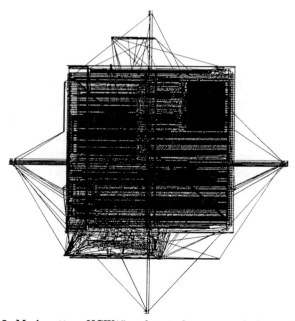

Fig. 3.12. Mask pattern NCW15 and wasted movement before optimization

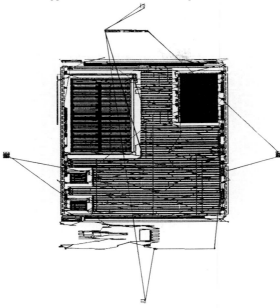

Fig. 3.13. Mask pattern NCW15 and wasted movement after optimization

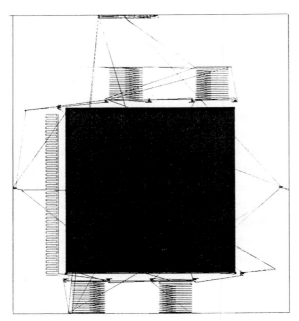

Fig. 3.14. Mask pattern NCT36

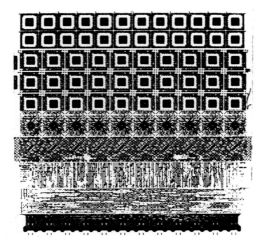

Fig. 4.1. Principal arrangement of a multi-chip module

Multi-chip modules are connected on printed circuit boards. These boards are made of fiber glass and copper. They have basically the same structure as the multi-chip modules, but are larger. They can have a size up to 1 square meter with up to 20 copper-layers for signal and power transmission. They can carry many multi-chip modules. One printed board can have up to 1 kilometer of wires.

We will deal here with the drilling and punching of vias or connection holes into the layers of a multi-chip module or boards. The wiring on the layers is done by optical exposure. Actually, some of the lithographic masks discussed in the previous chapter are masks for multi-chip modules or board. The drilling or punching is performed with devices similar to those of lithographic systems (cf. Fig. 3.1). The layer is moved on a compound table in two directions only, while the drilling or punching device is fixed. Unlike the lithographic problem, drilling and punching is done mechanically. Therefore it is more time consuming. Even if the drilling is done by laser it takes more time than just the exposure time. Each drilling or punching machine can have several drills or punchers with fixed or changeable distances.

The goal is to minimize the total drilling or punching procedure. Of course this is nothing but the famous travelling salesman problem and we are back to combinatorial optimization.

Unlike the previous chapter I do not give a historical account of the problem here. Instead, I refer to the interesting historical investigations of HOFFMAN and WOLFE [1985]. I feel that the travelling salesman problem is so widely known among mathematical programmers that it needs no additional introduction. For further details see LAWLER et al. [1985].

There has been great progress in the attempt to solve the travelling salesman problem optimally. About 20 years ago problems with at most 50 cities could be solved, basically by branch-and-bound methods. Due to the development of polyhedral combinatorics we are now able to solve problems with several thousands of cities. Partial knowledge of the linear description of the polyhedron of the travelling salesman problem has improved the bound of solvable problem sizes substantially. The paper by PADBERG and GRÖTSCHEL [1985] gives a good introduction into this branch of combinatorial optimization. At the moment PADBERG and RINALDI [1988] are the record holder for large real world travelling salesman problems. They can solve up to optimality problems with about 2000 cities. They got their large problems by duplicating the famous 532 cities drilling problem, which was originally also solved by PADBERG and RINALDI [1987]. The largest geographical problem with 666 cities was solved optimally by O. HOLLAND [1987], a student of mine. This optimal tour through 666 cities of the world is shown in Fig. 4.2. For curiosity reasons the total length of this tour is 294358 kilometers. And of course, one cannot do it better!

The travelling salesman problems we have to deal with in our applications are much too large to be solved optimally. Thus we have to apply heuristics. Although we know only little about guaranteed performance (worst case bounds) of heuristics for the travelling salesman problem, we have by now a very good empirical judgement about their practical performance. GOLDEN and STEWART [1985] give a good account of the empirical analysis of heuristics. We should also refer to the experiences reported by BLAND and SHALLCROSS [1987] and JOHNSON, McGEOCH and ROTHBERG [1988].

Our empirical experience coincides completely with the results of the above-mentioned papers. They can be summarized as follows: The *nearest*

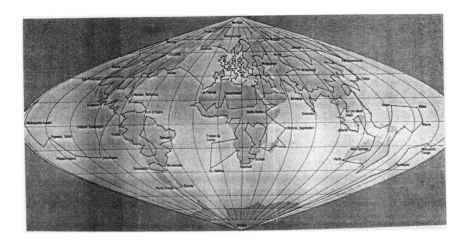

Fig. 4.2. Optimal tour through 666 cities of the world

neighbour heuristic is a fast (linear time) and always useful algorithm for geometrical drilling and punching problems. It gets solutions which are about 20–
25 % off the optimum. The best heuristic for the travelling salesman problem is
evidently the *k-opt-algorithm*, first proposed by LIN and KERNIGHAN [1973].
Other investigations and also our results corroborate that its solution is almost
at most 5 % off the optimum. In very many cases its relative error is 2 % or less.
Thus its empirical performance guarantee is more than sufficient for practical
problems. Since *k*-opt is an exchange heuristic its quality may depend on the
starting solution. We have got best experience with a starting solution generated by the CHRISTOFIDES heuristic (cf. CHRISTOFIDES [1976]), where
the matching subproblem to make the graph eulerian was solved by a (linear)
matching heuristic. The *convex hull heuristic* and the *nearest neighbour heuristic* were also used, but with somewhat less success. Unfortunately the *k*-opt
algorithm has a running time of order $O(n^k)$. Thus it cannot be used for very
large problems. In these cases we used the simpler exchange heuristics, *2-opt* or

3-opt with running times $O(n^2)$ resp. $O(n^3)$ or a variant of these two which is known in the literature under the name *Or-opt* (cf. GOLDEN and STEWART [1985]). The computing time for the application of k-opt to our problems very often went up to several ten hours. In one case we used more than 100 CPU-hours on an IBM 3081 KX. But this was a production problem which was used about 10^4 times on a machine and k-opt gained a saving of 15 % of the total production time. Thus the huge computing time was justified.

We would like to report on one additional empirical observation: The k-opt algorithm very often gets its best tour already during the first 10 % of its running time. The remaining computing time is used to prove that relative to the exchanges of the algorithm no further improvement can be obtained.

One final word on how to get the relative error of approximation algorithms, since we do not know the optimal solution (the problems are far too large to be solved optimally): We compute as a lower bound for the optimal solution the HELD and KARP solution (cf. HELD and KARP [1970]). However, to compute this lower bound was often more expensive than the heuristic itself.

Before we report on our results, we would like to make some additional remarks on the objective function. All that we stated about the objective function for the perfect matching problem in the previous chapter is equally true for the travelling salesman problem. The practical success depends heavily on a very good knowledge of the real world objective function of the equipment. Again, only L_∞- or L_1-norms should be considered and different velocities, accelerations, slow-downs, adjustments, position times and even drilling and punching times play a very crucial role. In all cases we had to measure the (highly nonlinear) objective function empirically with test problems at the device.

However, we should emphasize one point which is quite different here compared to the previous chapter: the fixed time for visiting a city (drilling or punching time) can be very high compared to the travelling time. Thus, even a remarkable improvement of the tour length (=tour time) will have littel influence on the total production time. We will distinguish in the following between *net improvement*, i.e. the improvement of the tour traversing time, and *gross improvement*, which includes all fixed times for drilling, set-up and starting the machine. Of course, the production engineer is only interested in over-all improvements, i.e. in the gross figures.

The device for punching via holes in ceramic layers for multi-chip modules has only one microprocessor for numerical control. Thus the L_1-norm is appropriate here. Fig. 4.3 shows one punching problem with 644 holes and the engineering solution of it. Again the microprocessor control used a lexicograhic ordering of the coordinates of the points. This seems to be reasonable, since all vertical lines are more or less equally packed with points. As a surprise, the solution generated by the k-opt algorithm shown in Fig. 4.4 looks completely different.

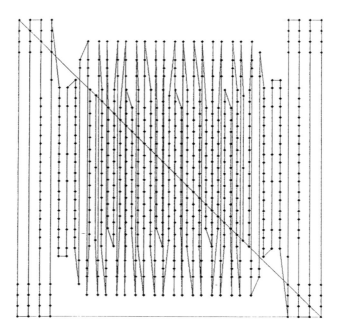

Fig. 4.3. Engineering solution of a puncher for ceramic layers (644 holes, L_1-norm)

Note that crossing of lines is allowed, since the problem is not euclidean and the triangle inequality does not hold. The net improvement by k-opt was about 30 %, the gross improvement is about 15 %. These improvement figures were also obtained by larger samples of test problems. Thus, even considering the relatively big punching and set up times, the improvements gained by

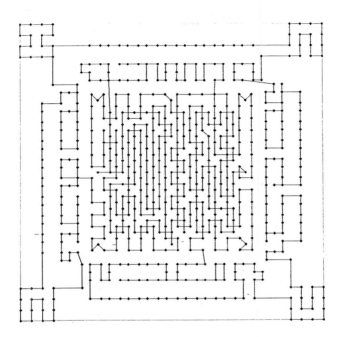

Fig. 4.4. k-opt solution of a puncher for ceramic layers (644 holes, L_1-norm)

combinatorial optimization were substantial.

We also did some comparisons of devices with one or two controllers (L_1- or L_∞-norm). The results for three test samples are shown in the tables 4.1 and 4.2.

Pattern number	Net			Gross		
	time of engineering solution	time of k-opt solution	improve-ment	time of engineering solution	time of k-opt solution	improve-ment
GS1	16,769ms	12,305ms	26.6%	30,116ms	25,625ms	14.8%
GS2	38,166ms	29,093ms	23.8%	64,130ms	55,057ms	14.2%
GS3	11,453ms	10,637ms	7.1%	25,327ms	24,511ms	3.2%

Table 4.1. Net and Gross Times for One-Controller-Puncher (L_1-norm) (all times in milliseconds)

Pattern	Net			Gross		
number	time of engineering solution	time of k-opt solution	improve-ment	time of engineering solution	time of k-opt solution	improve-ment
GS1	15,455ms	11,831ms	23.5%	28,802ms	25,178ms	12.6%
GS2	36,089ms	26,415ms	26.8%	62,053ms	52,379ms	15.6%
GS3	10,703ms	10,364ms	3.2%	24,577ms	24,238ms	1.4%

Table 4.2. Net and Gross Times for Two-Controller-Puncher (L_∞-norm) (all times in milliseconds)

These groups of samples show that the gross improvement is about one half of the net improvement. The net improvements for the L_∞-norm are higher than those for the L_1-norm. But this is not the case for the gross improvements because of the highly non-linear objective function. The pattern $GS3$ was very special. It was almost a complete two-dimensional grid, where all grid points had to be punched. These are the top-most or bottom-most layers of the module. They have to get all connections to the chip or they have to be connected with all pins on the bottom. Of course lexicographic ordering of the coordinates must be almost optimal for this pattern. Thus the improvements by k-opt must be marginal. In this very special case the maximum norm is even worse than the L_1-norm due to different speeds in different directions. One interesting result of this investigation was that sometimes machines with two controllers get smaller gross improvements than those with one, again due to nonlinear objective functions. We should also note, that we are faced here with all the (unpleasant) details of the real world, e.g. two machines which are equal by their parameters are not equal in practice. And even the same machine might run differently on Monday compared to Friday. This is reality not only for workers but also for machines.

The production process of multi-chip modules is very costly. Thus over-all improvements of about 10-15 % are remarkable indeed.

In table 4.3 we show some gross improvements for drilling problems of printed circuit boards. All drilling machines have two controllers. We distinguish here between the improvement of the nearest neighbour heuristic over the

engineering solution (lexicographic ordering) and the additional improvement
of the k-opt algorithm over the nearest neighbour solution. The improvement
by using combinatorial optimization techniques is substantial, the additional
gross improvement by k-opt is marginal. The different magnitudes of the tour
length (time) are due to the fact that all problems were run on different drilling
equipment.

Number of holes	Engineering solution length (time)	nearest neighbour solution length (time)	improve- ment %	k-opt length(time)	additional of k-opt over nearest neighbour
2723	2129	931	56.3 %	893	1.8 %
4040	865121	450728	52.1 %	368882	5.3 %
5133	3340	1534	54.0 %	1451	2.5 %
8064	6298	2687	57.3 %	2553	2.2 %
15991	857	110	87.2 %	91	2.2 %

Table 4.3. Gross improvements of nearest neighbour solution and additional gross
improvement of k-opt solution for drilling problems of different size (L_∞-norm)

The largest problem with almost 16,000 holes demonstrates a substantial improvement of the nearest neighbour solution. Therefore we demonstrate its
graphical solution. Fig. 4.5 shows the original data of the holes. They are so
grid-like that a lexicographic ordering of the points seems to make sense. Fig.
4.6 shows the tour according to this order. However Fig. 4.7 with the next
neighbour solution shows a much sparser picture. Fig. 4.8 shows the k-opt
solution.

   Those solutions generated by combinatorial optimization are not only
much faster as demonstrated, but even aesthetically nicer.

   We have mentioned above the empirical observation with other real world
travelling salesman problems that nearest neighbour is about 20 % off the
optimum, while k-opt about 2-5 % off the optimum. Thus one should gain
another 10-15 % by applying k-opt over nearest neighbour. This is only the
case for net times (tour length) but not for gross times.

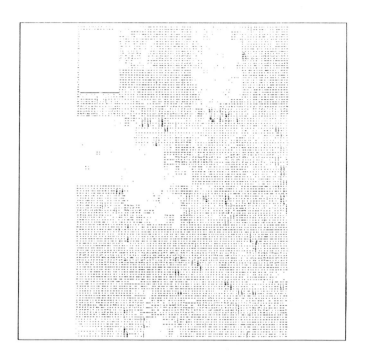

Fig. 4.5. Hole-pattern of a drilling problem with 15991 holes

Problem ID	Net production time (sec)		Net Improvement of	Gross production time (sec)		Gross Improvement of
	nearest neighbour solution	k-opt solution	k-opt over nearest neighbour	nearest neighbour solution	k-opt solution	nearest neighbour
F8517/18	344	281	18.3 %	3,942	3,897	1.1 %
F8517/43	103	89	13.6 %	803	791	1.5 %
F6762/18	236	207	12.3 %	2,494	2,465	1.2 %
F6762/43	149	138	7.4 %	1.293	1,282	0.9 %
F4086/18	298	263	11.7 %	3,168	3,133	1.1 %
F4086/43	119	110	7.6 %	939	930	1.0 %

Table 4.4. Net and gross production time improvements of *k*-opt solutions over nearest neighbour solutions

Fig. 4.6. Engineering solution of a drilling problem with 15991 holes (L_∞-norm)

We show this in table 4.4 where we compose additional improvements of k-opt over nearest neighbour gross and net.

We summarize that applying sophisticated (and therefore very time consuming) algorithms does not pay off with drilling problems, since fixed costs for drilling etc. are very high. However, fast heuristics like nearest neighbour again give a very substantial improvement even of the over-all production times.

Finally, we should mention that we are now working on additional improvements by using *multi-drill devices* with fixed or changeable distances. The problem with variable distances of the drills leads to very interesting combina-

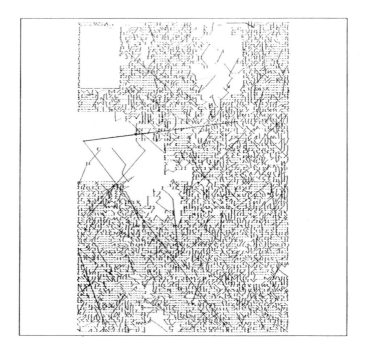

Fig. 4.7. Nearest neighbour solution of a drilling problem with 15991 holes (L_∞-norm)

torial optimization approaches. We were able to demonstrate that optimizing the tours of a multi-drill device yields another 20-30 % of gross improvements.

5. Application of Combinatorial Optimization to Carrier-Layout

In this short chapter we will report on work which is still much under progress. The main idea is just to mention another interesting application of combinatorial optimization which seems to be very useful and important.

As mentioned earlier multi-chip-modules as well as printed circuit boards can be considered as 3-dimensional grid-graphs usable for inter-chip- or inter-module-wiring. Thus the combinatorial optimization problem here is to find a packing of edge-disjoint Steiner-trees which uses a minimum number of layers

Fig. 4.8. k-opt solution of a drilling problem with 15991 holes (L_∞-norm)

or which has minimum total length. Minimizing the number of layers saves production costs. Minimizing the net length or some critical bottleneck nets saves
time for inter-chip connections. Especially for multi-chip-modules carrying fast
bipolar chips the time for signal transfer in the carrier is crucial.

In order to prove that combinatorial optimization can be applied here,
too, we have redesigned so far only a few multi-chip-modules for the IBM 4300
series, which have up to 8 layers. To minimize the number of layers we used
some heuristics for finding the maximum stable set (or maximum clique) in a
graph. In figures 5.1 and 5.2 we show one example. Fig. 5.1 shows the original
engineering solution with 8 layers. Fig. 5.2 shows a redesign by combinatorial
optimization which uses only 6 layers. The illustrations are different since they
were plotted on different devices. Structure and size of the layers are the same,

of course. Again the practical embedding of Steiner trees in 3-dimensional grids
has to follow many technical ground rules, which we do not mention here.

Our preliminary experience shows savings of 25 % to more than 50 % of
the layers, which results in much shorter nets and smaller production cost. We
feel that this problem has great potentials for application.

6. Applications of Combinatorial Optimization to VLSI-Design

The development of highly integrated electronic circuits during the last two
decades was breathtaking. Very large scale integrated chips became one of
the driving forces of high-tech. The technological growth of microelectronics
has shown to be overexponential. The four letters "VLSI" have become an
imaginary symbol for progress, future, and superiority. Thus it is quite natural
that almost everybody wants to "jump on the band-waggon". Sports teams
used to greet each other by cheering "hip-hip-hurrah". In variation of this one
could scoff modern high-tech society by calling out "chip-chip-hurrah".

Thus, does combinatorial optimization also want to jump on the band-
wagon before the train leaves the station? This is, at least so far as our observation is concerned, not the case. As long as integrated circuits were relatively
small, say several hundred components, sophisticated computer-aided design
tools were sufficient to get efficient layouts of chips. Nowadays highly integrated chips have one million and more components. Thus it is evident that they
cannot any more be designed by men interacting with a graphical computer
terminal. Since the underlying structures of chip-design are of course discrete,
it was more or less natural that combinatorial optimization was asked by electronic engineers to provide help. And indeed, very many techniques and tools
of combinatorial optimization developed in the past and independently of VLSI
turned out to be extremely well applicable here.

Before we report on this we would like to give at least some very general
explanation of the trends of modern computer electronics. Fig. 6.1 is a diagram
of the degree of integration versus time. It shows the development during the
last 25 years and it gives a forecast for the next 30 years. We deduce from
this diagram that the number of components per chip grows by the factor of
100 every decade. Today we are in the range of 10^6. The forecast shows a
pessimistic and an optimistic curve. All experts agree that the technological
development of chips has a research-and-development potential of at least two

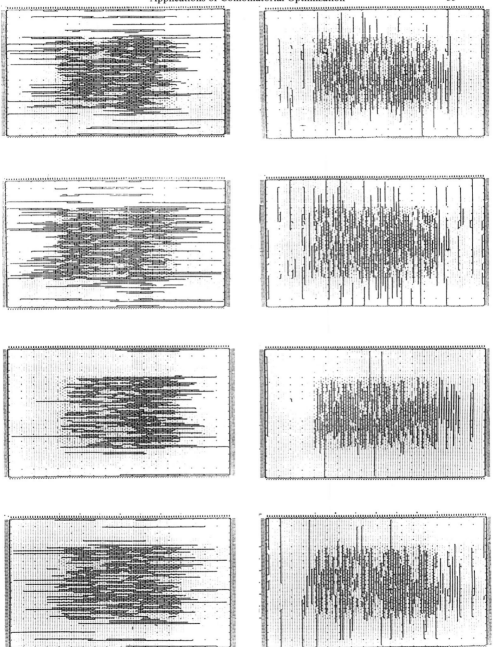

Fig. 5.1. Engineering design of a multi-chip-module using 8 layers

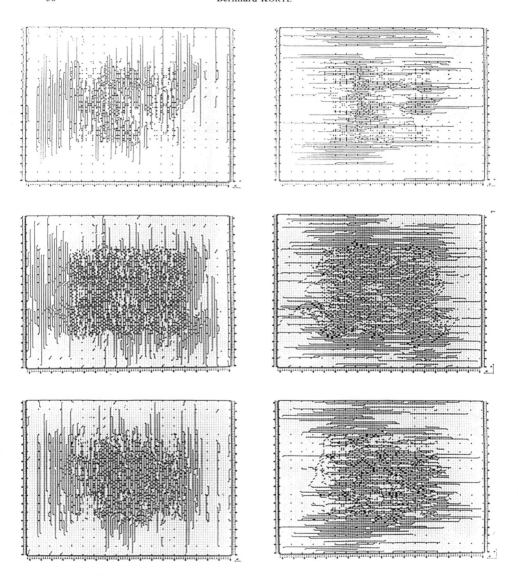

Fig. 5.2. Redesign of a multi-chip-module by combinatorial optimization techniques using 6 layers

further decades. Integration densities of 10^9 components per chip will be reached. Abbreviations on the vertical axis of Fig. 6.1 mean: MSI=Medium scale integration, LSI=Large scale integration, VLSI=Very large scale integration; ULSI=Ultra large scale integration, GSI=Giga scale integration.

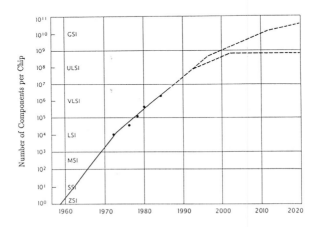

Fig. 6.1. Degree of integration: number of components (transistors) per chip

Fig. 6.2 shows a similar development diagram, again degree of integration versus time. But we distinguish here between memory and logic chips. *Memory chips* are very uniform. They have only one elementary electronic unit (gate), which is placed million times on a chip. The package density of a memory chip is technologically impressive. However, its design is trivial and there is nothing to be optimized by mathematical methods in the layout. The structure of *logic chips* is far more complex and irregular. Layout and physical design of today's logic chips can be done only by mathematical methods. In the following we will only deal with logic chips and combinatorial optimization methods for them.

On the trendline of Fig. 6.2 we have indicated the size of memory chips, e.g. 256k-bit, 1 Megabit, 4 Megabit etc.. Today the megabit-chip is produced in large scale production. Batch production of the 4-megabit-chip will start 1989. Prototype masks of a 16-megabit chip have already been shown at a conference in February 1988. There are rumors coming from Japan that 64-megabit-chips

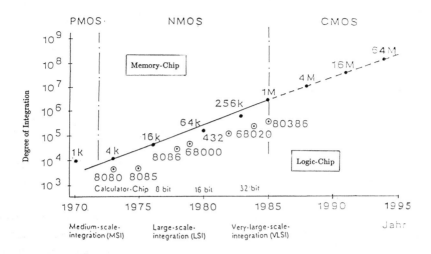

Fig. 6.2. Degree of integration for memory- and logic-chips

and even larger ones will be technologically possible very soon.

The 4- or 5-digit-numbers in the lower part of Fig. 6.2 are type-numbers of complete microprocessors, i.e. logic chips. PC-freaks can associate with these numbers certain micros of Intel, Motorola or Texas Instruments. To the right of the dotted vertical line at the year 1985 we have no further type numbers of micros. Here is the area of *mega-logic-chips* (1 million transistors) which are under development at the present time. Our experience reported below deals with mega-logic-chips.

Since I have given some historical sketches in the previous chapters, I cannot resist to doing it here too. Fig. 6.3 gives a historical synopsis of the development of integrated circuits during the last three decades. It is highly impressive to follow this line from the very first planar transistor in 1959 to a complete microprocessor with several hundred thousand transistors. Since Fig. 6.3 ends with 1985 I will add one picture for 1988. In Fig. 6.4 we show a complex memory – management-chip (MMU) for IBM S/370 architecture of the Capitol chip series of IBM Böblingen laboratories. This chip has 800,000 transistors. We have made several designs for it using combinatorial optimization techniques.

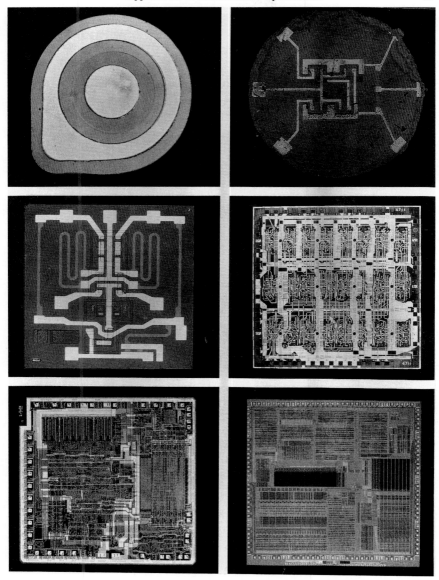

Fig. 6.3. History of integrated circuits 1959–1985 demonstrated with 6 chips of the Fairchild Semiconductor Corporation: Top left: first planar transistor, 1959; Top right: first planar integrated circuit on a planar chip with four transistors, 1961; Middle left: first integrated circuit with five transistors for batch production, 1964; Middle right: bipolar logic chip with 180 transistors, 1968; Bottom left: first 16-bit microprocessor on a chip with 20,000 transistors, 1978; Bottom right: microprocessor with 132,000 transistors, 1985

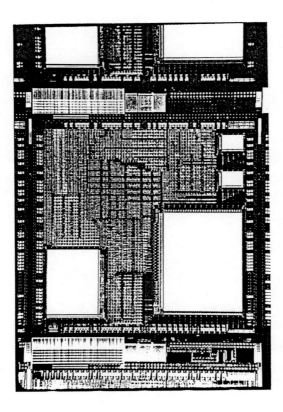

Fig. 6.4. Memory-management-chip for IBM S/370 architecture of the Capitol chip
series of IBM Böblingen laboratories, 1988 (by courtesy of G. Koetzle)

Development and design of logic chips have the following major steps:

– System specification
– Logic design
– Logic simulation and verification
– Physical design
– Physical layout
– Simulation and testing
– Mask construction

- Simulation and testing
- Mask construction
-   $\vdots$
- etc.

This list does not claim to be complete, the steps are not disjunctive and some of them are done iteratively or in a loop. We will restrict ourselves here only to the steps of *physical design* and *layout*. Nevertheless, the other major steps also contain some very interesting combinatorial optimization problems, e.g. in logic synthesis or timing analysis.

For the purpose of combinatorial optimization we may view a chip basically as a two-dimensional grid graph as shown in the principal illustration of Fig. 6.5. The horizontal and vertical lines of the grid have the same distance, but they are located on different layers separated by an insulating layer. Wiring of the chip can be done by using only these horizontal and vertical lines. Connections of horizontal and vertical lines can be done only at a grid point. A little hole, called a *via*, will be made in the insulating layer in order to connect the horizontal and vertical layer at this point. Vias are indicated in Fig. 6.5 by little circles. If a horizontal and a vertical line cross outside a circle at a grid point, then they are disjoint. Those parts of the grid graph of Fig. 6.5 which are used for wiring are drawn in heavy lines.

What has to be routed or wired on a chip? We show in Fig. 6.5 some hatched rectangles or compositions of rectangles. Those objects are called *(standard) cells*. They are the actual elementary electronic units of the chip. They are produced in a very complicated chemical and diffusion process on the silicon. There is no need to go into technical details. It is sufficient for our purposes to think of these cells as being placed on (more exactly: under) the grid graph. Each chip generation has only a relatively small collection of different cells, say 20 to 40. They are sometimes called the "books" of the cell-library. All complicated circuits of a chip have to be constructed in the logic design phase with those few books. In the interior of the cells we have *pins* for inputs and outputs. Pins are indicated in Fig. 6.5 by little squares. The cells have to be placed so that the pins coincide with grid points. According to the rules of logic design pins of different cells have to be connected using vertical and horizontal lines of the grid graph. But not only two pins have to be connected by a path in the grid. Very often many pins (20 and more) have to be interconnected, since one output signal of a cell can be the input signal for many other cells.

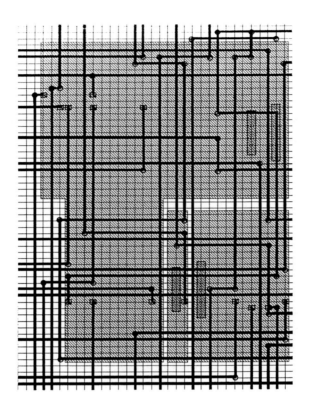

Fig. 6.5. Principal illustration of wiring and placement of a chip

For the experts we should mention that this chip principle is different from classical approaches like *standard cell, gate array* or *custom macro design*. It is sometimes called *master image design*. This design principle could also be named *channels standard cell design* or *sea-of-cell-design*. For further details of this design principle and of the underlying technology we refer to SPRUTH [1989] and KOETZLE [1987], [1987a]. For our purposes we should remark that this design concept can use not only channels for wiring but more or less the whole grid graph. Only some small parts of the grid under certain cells are forbidden for wiring. Note, that power supply is done on separate lines of the grid which are not indicated in Fig. 6.5.

The combinatorial optimization problem of the chip-layout problem is to place the cells on the grid (as densely as possible) so that pins are placed on grid points and moreover that the required connections (nets) can be wired *edge-disjoint* by using vertical and horizontal lines of the grid. As long as chips were relatively small this process was done manually or in a man-machine-interactive way using CAD tools. For chips with 10^6 components this is not possible anymore. Sophisticated optimization techniques are needed.

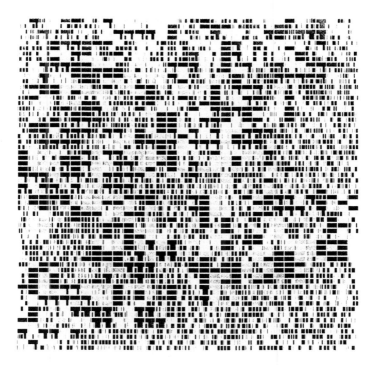

Fig. 6.6. Complete placement of a part of a chip with about 5000 cells (density over 95 %)

Before we mention some of these techniques we would like to discuss briefly the objective function. Since this is a very complex design process there is no unimodal objective function. Very often it is extremely difficult to find a feasible solution or even to decide whether there a feasible solution exists. Thus optimization approaches which predict wireability, i.e. the existence of a feasible solution, are highly welcome at an early stage of logic design. One major goal is to get the chip as fast as possible. This can be obtained by having the total length of all nets (=all wires) small or the length of some critical nets as small as possible or under a certain threshold (bottleneck problem). We should note that the wire length gives only a rough estimate for the timing. The total delay of a signal on a net depends on rather complicated delay equations for cells, fan outs and wire length. We are now working on combinatorial optimization techniques for *performance driven placement* and *wiring*. The preliminary results are promising.

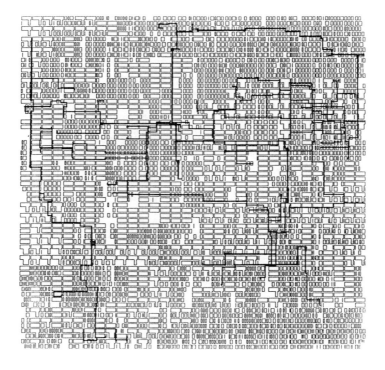

Fig. 6.7. Complete placement of a part of a chip with one embedded Steiner tree

Another goal of the optimization approach should be the uniformity of wires on the grid. Cumulation areas should be avoided since they may be sources of noise. We found out that engineers have more than 30 different goals when evaluating the quality of a design. We will mention some below when we discuss the results.

We are now ready to discuss some details of the combinatorial optimization techniques in particular. But since this tutorial lecture is basically a survey, we only mention these problems here and refer to some technical papers of ours in which parts of the design approach are described in detail (cf. FROLEYKS, KORTE and PRÖMEL [1987], KORTE, PRÖMEL, SCHWIETZKE and STE-GER [1988] and KORTE, PRÖMEL and STEGER [1989]).

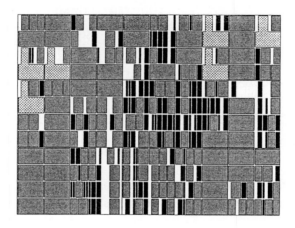

Fig. 6.8. Zoom part of a placement

We do the optimization approach for chip layout hierarchically. First the cells are placed on the grid, then they are connected by a routing or wiring approach. Of course, a decomposition of a problem always has drawbacks. A very good and dense placement may create problems for the routing phase or may be not wireable at all. The optimization criteria of a hierarchical approach can be controversial. The optimization attempt basically consists of four steps:
-   Global placement (=partitioning)
-   Global routing
-   Local placement
-   Local routing

However, these four steps interact at several stages. For instance, a global

routing is also done after the local placement. Moreover, after each step of
the iterative partitioning approach a global routing is done, which specifies the
objective function for the next partitioning step. Thus the first and the second
steps are done in a loop.

Fig. 6.9. Complete wiring of the zoom part

Partitioning is mainly done using algorithms of *graph-partitioning*. The
problem is to partition the node set of a graph such that the edges of the
cut have minimal weight sum. Since graph-partitioning is NP-hard, we have to
use fast heuristics. The exchange heuristic of KERNIGHAN and LIN [1970] is
known to be very efficient. For our purposes the nodes of the graph are the cells,
while the edges are connections. Since the cells have different sizes we have to
put weights on the nodes too and the node partition has to give subsets of equal
size or of prescribed size. Moreover, the nets between the cells normally have
more than two elements. Thus instead of edges we have to consider hyperedges
for the partitioning algorithm. We had to do further modifications of the basic
ideas of KERNIGHAN and LIN, but they are too technical to be explained
here. We do the graph partitioning sequentially. First we partition the set of
all cells of the chip into equally sized sets (or sets of prescribed size), say the
upper and the lower part. Then we take one of the two parts and subdivide
it again, and so on. One of the main features of our algorithm is that we run
a complete global routing after each partitioning step. This goes as follows:

Each partition is considered as a "supernode" in a global routing graph. The edges have capacities according to the number of vertical or horizontal lines in the grid connecting the supernodes. Of course, a net which lies completely in one class cannot be considered for the global routing. But all other nets will be globally routed according to the capacities of the edges. This global routing information is essential for the objective function for the next steps of the partitioning phase. By combining partitioning and global routing we get much better placements of the chip. For further details see KORTE, PRÖMEL, SCHWIETZKE and STEGER [1988].

Fig. 6.10. Complete design of the CPU-chip of the Capitol series

The partitioning algorithm has a very local criterion, namely to minimize the edge weight of a very specific cut. This might cause global wiring problems in other parts of the chip. However, the combination with a global routing at every step corrects this. By this we guarantee the wireablity of the chip through the whole phase of partitioning.

Fig. 6.11. Complete design of the MMU-chip of the Capitol series

Since the cut criterion is local we have to stop the partitioning procedure at a certain point. Two adjacent cells have to have maximal possible connections since they are closest. They have to leave a maximum weighted cut, but not a minimum weighted one. We decided by empirical tests to stop the partitioning at a *page size* which can contain 80 inverters. These are the smallest cells. Measured in the underlying grid structure a page consists of 108 horizontal and 60 vertical lines. The bus structure for the power supply of the cells also plays a role in specifying this size.

After the complete partitioning step is done, we again do a global routing. This specifies all intra-page nets. The whole topology of the chip is now specified. In particular, the chip-wireability is now decidable. The local placement step arranges the cells within a page. At this stage we can use the information of the global routing, namely whether a net leaves the page on top or on bottom or to the right or to the left. Moreover local placement tries to get short in-page-nets, again by graph partitioning or by bin-packing approaches.

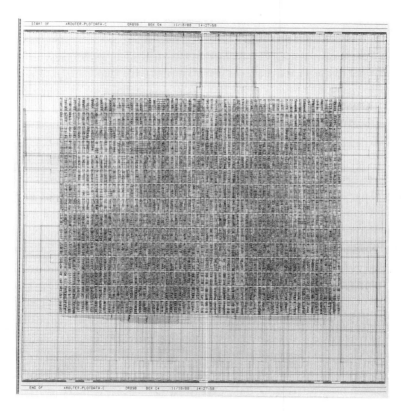

Fig. 6.12. Complete redesign of the STC-chip of the Capitol series using only 47,1 % of the previously used space

We did not mention how the global routing is done. But since the algorithmic principles of global routing and local routing are one and the same, we now briefly mention some basic principles of *local routing*.

After all cells and therefore all pins have been placed, the routing problem consists of placing all nets disjointly into the grid. Each net can be modelled by a *Steiner tree*. Thus we have to pack edge-disjoint Steiner trees into the grid graph, connecting the prescribed and fixed pins. Again, we cannot mention all details, e.g. if two Steiner trees bend at the same grid point, they cannot use the same via. However, from a graph theoretic point of view they are edge-disjoint. This situation is called "knock-kne". It has to be handled specifically.

We used two different types of Steiner tree heuristics. One is based on a shortest path approach. First we construct a *minimum spanning tree* for a net using the Manhattan distance of the pins. Then we shell this tree which gives a vertex sequence. The Steiner tree is then built sequentially in the inverse order of the vertex sequence using shortest path algorithms, but not from point to point, but from a point to a set of points. This set of points consists of all vertices connected so far and the Steiner points generated so far. For the shortest path subroutine we used modifications of α-β-router (cf. HU and SHING [1985]) which have additional penalties for via and for highly saturated cuts. This should guarantee a homogeneous distribution of the wiring. Another heuristic for Steiner trees which we tried successfully is the algorithm of WU, WIDMAYER and WONG [1986] which builds up Steiner trees not from the interior but from its leaves. We have no final judgement about the empirical performance of these different approaches. For the global routing we used modifications of WU, WIDMAYER and WONG. Our local router still has a strong shortest path component.

We feel quite uncomfortable with the description above. It does not give enough details for the expert and even the layman could not gain too much. But we hope that we have given at least some flavour of the problem to the reader. However, we have sketched only a very small portion of the mathematical and technological problems involved. To give a vague idea of the complexity of chip design, we list here some of the major criteria which engineers use to evaluate a design: cell usage, data flow, floor plan, total net length, number of vias, cycle time, input delay, output timing, slow rise times, load limit violations, clock loads, clock stops, clock driver distributions, cache signals, shift paths, spare books. We had to deal with all these criteria in our algorithms.

In closing this chapter I would like to demonstrate some results. First some illustrations: Fig. 6.6 shows the final placement of a part of a chip with about 5000 cells of different shape and size. The density is more than 95 %. One can prove that this is maximal, no denser packing is possible or wireable. Fig. 6.7 shows the same part of the chip where only the boundary of the cells is drawn and one very large Steiner tree is embedded. Fig. 6.8 is again a zoom part of Fig. 6.6. Different hatchings of the cells mean different functions. Fig. 6.9 shows the complete wiring of the zoom part. We had to take such a small part in order to really show all details of the routing.

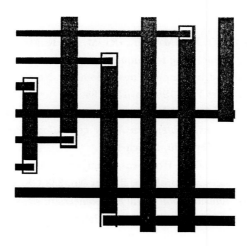

Fig. 6.13. Enlargement of wiring detail or modern art?

We have designed or redesigned so far about one dozen very complex mega-logic-chips and their modifications. Some of our designs are already being produced. Compared to the earlier engineering approach we can confirm substantial improvements. In some cases we could reduce the space on silicon by 40 – 60 %. The major criteria like total net length, timing, number of vias could be improved between 20 % – 30 %. Moreover the total turn-around-time for the complete design of one chip could be drastically reduced to 10 – 15 days. Note that one continuous week (7 days of 24 hours) is expended on computing time only. The wireability of a chip can be predicted (up to 5 % of the net length) within a few hours.

Fig. 6.10, 6.11 and 6.12 show three of our complete designs of chips of the S1370 architecture Capitol series (CPU, MMU and STC). The storage controller has been redesigned using only 47.1 % of the space of the engineering solution. Therefore its boundary is empty.

With the very final Fig. 6.13 I would like to demonstrate that combinatorial optimization is not only very useful for chip design but can also compete with modern art.

References

ASANO, T., EDAHIRO, M., IMAI, H., IRI, M. and K. MUROTA [1985]:*Practical use of bucketing techniques in computational geometry*, in: G.T. Toussaint (ed.): Computational Geometry. North Holland Publ. Amsterdam/New York 1985, p. 153–195.

BLAND, R. and D. SHALLCROSS [1987]: *Large Traveling Salesman Probems arising from experiments in X-ray crystallography*, SORIE Cornell Univ. Tech. Rep. 730, March 1987.

CHRISTOFIDES, N. [1976]: *Worst-case analysis of a new heuristic for the traveling salesman problem*, Report No. 338 GSIA Pittsburgh 1976.

EULER, L. [1936]: *Solutio problematis ad geometriam situs pertinentis.* Commentarii academiae scientiarum Petropolitanae 8 (1936), 1741, p. 128–140, reprinted in: F. Rudio, A. Krazer, A. Speiser, L.G. du Pasquier (eds.): Leonardi Euleri Opera omnia. Ser. I, Vol. 7, p. 1–10, Teubner Liepzig/Berlin 1923.

FROLEYKS, B., KORTE B. and H.J. PRÖMEL [1987]: *Routing in VLSI-Layout.* Report No. 87494–OR, Discrete Mathematics/Operations Research University of Bonn 1987.

GOLDEN, B.L. and W.R. STEWART [1985]: *Empirical analysis of heuristics*, in: LAWLER et al. [1985], p. 207–249.

GUAN, Mei-Gu [1962]: *Graphic programming using odd and even points.* Chinese Math. 1 (1962), p. 273–277.

HAFNER, B.F. [1988]: *Optical Pattern Generator using Excimer Laser.* Proceedings of SPIE. The International Society of Optical Engineering Vol. 922 Optical/Laser Microlithography, Santa Clarce 1988, p. 417–423.

HALMOS, P.R. [1981]: *Applied Mathematics Is Bad Mathematics*, in: L.A. Steen (ed.): Mathematics Tomorrow, Springer Verlag Berlin/Heidelberg/New York 1981, p. 9–20, also reprinted in P.R. Halmos: Selecta-Expositary Writing, Springer Verlag Berlin/Heidelberg/New York 1983, p. 279–290.

HARDY, G.H. [1940]: *A Mathematician's Apology*. Cambridge University Press 1940 (latest reprint 1987).

HELD, M. and R.M. KARP [1970]: *The traveling-salesman problem and minimum spanning trees*. Operations Reseach 18 (1970), p. 1138–1162, part II: Mathematical Programming 1 (1971) p. 6–25.

HOFFMAN, A.J. and P. WOLFE [1985]: *History*, in: LAWLER et al. (eds.) [1985], p. 1–15.

HOLLAND, O.A. [1987]: *Schnittebenenverfahren für travelling-Salesman- und verwandte Probleme*. Ph. D. dissertation. Report 87479–OR Discrete Mathematics/Operations Research, University of Bonn 1987.

HU, T.C. and M.T. SHING [1985]: *The α-β-routing*, in: Hu, T.C. and E.S. Kuh: VLSI circuit layout: theory and design. IEEE Press, New York 1985, p. 139–152.

IRI, M., MUROTA, K. and S. MATSUI [1981]: *Linear-time approximation algorithms for finding the minimum-weight perfect matching on a plane*. Information Processing Letters 12, 4 (1981) p. 206–209.

IRI, M., MUROTA, K. and S. MATSUI [1982]: *An Approximate Solution for the Problem of Optimizing the Plotter Pen Movement*, in: Lecture Notes in Control and Information Science, Vol. 38, Springer Verlag Berlin/Heidelberg/New York 1982, p. 572–580.

JOHNSON, D.S., MCGEOCH, L.A. and E.E. ROTHBERG [1988]: *Near optimal solutions of very large scale traveling salesman problems*, (invited lecture this meeting).

KERNIGHAN, B.W. and S. LIN [1970]: *An efficient heuristic procedure for partitioning graphs*. Bell Syst. Tech. J. 49, no. 2 (1970), p. 291–307.

KOETZLE, G. [1987]: *Hierarchisches Designkonzept für anwendungsspezifische VLSI Chips mit über 1000000 Transistoren*, in: ITG-Fachberichte 98: Großintegration vde-verlag Berlin 1987, 55–58.

KOETZLE, G. [1987a]: *System implementation on a highly structured VLSI master image*, in W.E. Proebster and H. Reiner (eds.): Proceedings VLSI and Computers, IEEE, 1987, 604–609.

KORTE, B., PRÖMEL, H.J., SCHWIETZKE, E. and A. STEGER [1988]: *Partitioning and placement in VLSI-layout*. Report No. 88500–OR, Discrete Mathematics/Operations Research, University of Bonn 1988.

KORTE, B., PRÖMEL, H.J. and A. STEGER [1989]: *Steiner trees in VLSI-layout*. Report No. 89567, Discrete Mathematics/Operations Research, University of Bonn, April 1989. To appear in: Korte, Lovász, Prömel, Schrijver (eds.): Paths, Flows and VLSI-Layout. Algorithms and Combinatorics, Vol. 7, Springer Verlag Heidelberg/Berlin/New York 1989.

LAWLER, E.L., LENSTRA, J.K., RINNOOY KAN, A.H.G. and D.S. SHMOYS (eds.) [1985]: *The travelling Salesman Problem. A Guided Tour of Combinatorial Optimization*. Wiley New York 1985.

LIN, S. and B.W. KERNIGHAN [1973]: *An effective heuristic for the traveling salesman problem*. Operations Research 21 (1973), p. 972–989.

NATIONAL RESEARCH COUNCIL [1984]: *Renewing U.S. Mathematics: Critical Resource for the Future*. Report of the Ad Hoc Committee on Resources for the Mathematical Sciences. National Academy Press, Washington 1984.

VON NEUMANN, J. [1947]: *The Mathematician*, in R.B. Heywood (ed.): The Works of the Mind. Vol. I, no. I, University of Chicago Press 1947, p. 180–196, also reprinted in: John von Neumann: Collected Works, Vol. I, Pergamon Press 1961, p. 1–9.

PADBERG, M.W. and M. GRÖTSCHEL [1985]: *Polyhedral computations*, in LAWLER et al. (eds.) [1985], p. 307–360.

PADBERG, M.W. and G. RINALDI [1987]: *Optimization of a 532-city symmetric travelling salesman problem by branch and cut*. Operations Research Letters 6 (1), 1987, p. 1–7.

PADBERG, M.W. and G. RINALDI [1988]: *An LP-based algorithm for the resolution of large-scale traveling salesman problems*. Preprint. New York University 1988.

REINGOLD, E.M. and R.E. TARJAN [1981]: *On a greedy-heuristic for complete matching*. SIAM J. Comp. 10,4, p. 676–681.

SPRUTH, W. (ed.) [1989]: *The design of a microprocessor*, to appear: Springer Verlag Berlin/Heidelberg/New York 1989.

SUPOWIT, K.J., PLAISTED, D.A. and E.M. REINGOLD [1980]: *Heuristics for weighted perfect matchings*. Proc. 12th ACM Symposium on Theory of Computing 1980, p. 398–419.

WU, Y.F., WIDMAYER, P. and C.K. WONG [1986]: *A faster approximation algorithm for the Steiner problem in graphs*. Acta Informatica 23, 1986, p. 223–229.

Mathematical Programming Applications in National, Regional and Urban Planning

Michael FLORIAN

Department of Computer Science and Operations Research and Transportation Research Center, Université de Montréal, Montréal, Québec, H3C 3J7, Canada

Abstract

This paper presents the models used in the analysis of urban and regional multi-modal networks for passengers transportation, which include route choice models on road and transit networks, and models for the analysis of national and regional multi-modal networks for freight transportation.

1. Introduction

The prediction of the flows of passengers and freight by quantitative aproaches naturally leads to the formulation of network models which represent the spatial characteristics of the transport infrastructure. The issues that motivate the need to formulate, analyse, solve and apply such network models may be addressed by determining the choices that people, vehicles and goods make or should make in using an existing or proposed transportation infrastructure. Such models may be subdivided into two classes: descriptive models which attempt to reproduce observed behavior patterns or normative models, which attempt to prescribe how to make use of the transportation infrastructure efficiently.

The practical contexts which require the use of such models are varied and the stimuli for developing them arise from urban and regional contexts, as well as national and inter-national situations. When congestion phenomena are present, the cost functions that are appropriate to model such situations are non linear; in most applications they are convex or monotone. Wherever economies of scale phenomena are present, the cost functions are naturally concave. In many cases, the resulting models may be formulated as nonlinear cost optimization models over an appropiate network. More general versions of these problems may be formulated as nonlinear complementarity problems or variational inequalities with imbedded network structures.

The purpose of this paper is to present a partial survey of these models and to point out the principal methods used for their solution in practice. The progress that has been made in the past twenty years in the applications of such models, was stimulated in part

M. Iri and K. Tanabe (eds.), Mathematical Programming, 57–81.
© *1989 by KTK Scientific Publishers, Tokyo.*

by the attention that researchers with mathematical programming interest have given to these problems and by the advent of digital computers, which has made the solution of these models practical. The study of models that are used in transportation planning applications continues to pose significant challenges in the formulation of models and in the analysis and development of solution algorithms.

Since the application of these models requires usually the handling of large amounts of data, over the past 10 years several codes have been developed that use interactive-graphic user interface, in order to build and update the necessary data bases, as well as producing the results of the models as plots of flows and service levels on the analysed networks.

The paper is organized in the following way. First we present models used in the analysis of urban and regional multi-modal networks for passenger transportation, which include route choice models on road and transit networks. Then we present models for the analysis of national and regional multi-modal networks for freight transportation. As the body of knowledge in this field is extensive, no attempt has been made to describe all the solution algorithms in detail, rather the emphasys is on model formulation and applications of these models in the contexts mentioned above. References to solution algorithms are given in the bibliography.

2. Urban and Regional Multi-Modal Passenger Networks

The mathematical programming models that are used in the planning of urban and regional multi-modal passenger networks are aimed to predict the route choice on road networks by automobile traffic and individuals route choice on transit network by individuals travelling on public transit. Simultaneous models of road and transit route choice may be specified and applied if econometric demand models (mode choice) have been estimated. In the following we discuss the network equilibrium model for route choice on road networks, transit route choice models and models which combine mode choice with route choice on multi-modal networks.

2.1. The Network Equilibrium Model-Route Choice on Road Networks

Traffic equilibria models are descriptive models that aim to predict link flows and travel times that result from the way in which drivers choose routes from origins to destinations on a tranportation network. The behavioral assumption used is stated by Wardrop (1951), who postulates that "journey times on all routes actually used are equal and less than those which would be experienced by a single vehicle on any unused route". This characterization of route choice, on traffic equilibrium, which is identical to ideas formulated by Knight (1924), became accepted over the past 30 years as a sound behavioral principle for the static simulation of road traffic under congested traffic conditions and their perception by the user. The traffic flows that satisfy this principle are usually referred to as 'user-optimized' flows, since each user chooses the route that he perceives to be the best. Beckmann, McGuire and Winsten (1956) were the first to formulate and analyse a model of traffic equilibrium on a network. Their seminal work motivated the other contributions to follow in this area, which were achieved mostly within the past twenty years. The presentation of the model below is based on the work by Smith (1979), Aashtiani and Magnanti (1981) and Dafermos (1980, 1982b).

Consider a transportation network which permits the flow of one type of traffic (vehicles) on its links. The nodes n represent origins, destinations and intersections of links; the links a represent the transportation infrastructure. The flow of trips on link a is v_a and the cost of travelling on a link is given by a user cost function $s_a(v)$, where v is the vector of the link flows over the entire network. This cost function may model the time delay for travel on that arc, in which case it is commonly referred to as a volume/delay function; however, it may model other costs, such as fuel consumption. The vector user cost function $s(v)$ is assumed defferentiable and monotone:

$$(s(v') - s(v''))^T (v' - v'') \geq 0 \ \forall \ v', \ v'' \text{ feasible.} \tag{1}$$

The origin to destination demand between O/D pair d, g_d may use directed paths k. The flow h_k on path k satisfies conservation of flow and nonnegativity constraints

$$\sum_{k \in K_d} h_k = g_d, \ \ d \in D,$$

$$h_k \geq 0, \ \ k \in K_d, \ d \in D, \tag{2}$$

where D is the set of origin/destination pairs and K_d is the set of paths for the O/D pair d.

The corresponding link flows v_a are given by

$$v_a = \sum_{d \in D} \sum_{k \in K_d} \delta_{ak} h_k, \ \ a \in A \tag{3}$$

where

$$\delta_{ak} = \begin{cases} 1 & \text{if link } a \text{ belong to path } k, \\ 0 & \text{otherwise,} \end{cases}$$

and A is the set of links of the network. The cost $s_k (= s_k(v))$ of each path k is the sum of the user costs on the links of k:

$$s_k = \sum_a \delta_{ak} s_a(v), \ \ k \in K_d, \ d \in D. \tag{4}$$

Let $u_d (= u_d(v))$ be the cost of the least cost path for any O/D pair d:

$$u_d = \min_{k \in K} s_k, \ \ d \in D. \tag{5}$$

For each $d \in D$ the demand g_d is given by a function $G_d(u)$, where u is the vector of least cost travel times for all the O/D pairs of the network

$$g_d = G_d(u), \ \ d \in D. \tag{6}$$

The vector demand function $G(u)$ is assumed bounded from above, differentiable and strictly monotone decreasing $(-G(u)$ is strictly monotone):

$$(G(u')-G(u''))^T(u'-u'') < 0, \quad u' \neq u''. \tag{7}$$

We suppose that for every O/D pair Wardop's user optimal principle is satisfied:

$$s_k^* - u_d^* \begin{cases} = 0 & \text{if } h_k^* > 0, \\ \geq 0 & \text{if } h_k^* = 0, \end{cases} \quad k \in K_d, \ d \in D, \tag{8}$$

over the feasible set (2)-(6).

It is relatively straightforward to show that (8) may be restated in the 'complementarity' form

$$u_d^* \leq s_k^* \text{ and } (s_k^* - u_d^*)\, h_k^* = 0, \ k \in K_d, \ d \in D, \tag{9}$$

and that (8) and (5) are equivalent to

$$s_{k_1}^* \leq s_{k_2}^*, \ \text{if } h_{k_1}^* > 0, \ k_1, \ k_2 \in K_d, \ d \in D. \tag{10}$$

Thus, Wardrop's principle may be stated mathematically in several forms.

The 'complementarity form' (9) has been used by Aashtiani and Magnanti to formulate the network equilibrium problems as a nonlinear complementarity problem by letting

$$F(x) = \left\{ (s_k - u_d), \ k \in K_d, \ d \in D; \left(\sum_{k \in K_d} h_k - G_d(u) \right), \ d \in D \right\} \tag{11}$$

and $x = (h, \ u)$. Then the network equilibrium problem is equivalent to solving

$$F(x)x = 0, \ F(x) \geq 0, \ x \geq 0. \tag{12}$$

The condition $\left(\sum_{k \in K_d} h_k - G_d(u)\right) u_d = 0, d \in D$, does not correspond directly to the original statement of the problem; however, under the reasonable assumptions that $G_d(u) > 0, \ d \in D$ and $s_k > 0, \ k \in K_d, \ d \in D$, it is equivalent to (2).

This formulation may be used to derive existence results. A solution $(h, \ u)$ exists if the cost functions $s_a(v), \ a \in A$ are continuous and positive and if $G_d(u), d \in D$ are continuous nonnegative functions bounded from above. These conditions are relatively weak and are certainly satisfied by most cost and demand functions which arise in practice (see Aashtiani and Magnanti, 1981).

Another, more useful, restatement of the network equilibrium problem is in the form of a variational inequality. This may be done by direct derivation or by applying the result (see Kinderlehrer and Stampacchia, 1980) that $x^* \in R^+$ is a solution to the nonlinear complementarity problem (12) if and only if x^* is a solution of the variational inequality

$F(x^*)(x - x^*) \geq 0, \quad x \in R^+$. The variational inequality that results by applying this equivalence is

$$\sum_{d \in D} \sum_{k \in K_d} (s_k^* - u_d^*)(h_k - h_k^*) - \sum_{d \in D} \left(G_d(u^*) - \sum_{k \in K_d} h_k^* \right) (u_d - u_d^*) \geq 0. \tag{13}$$

(see also Fisk and Boyce, 1983; Magnanti, 1984).

If the demand function is invertible, that is $W_d(g) = G_d^{-1}(g) = u_d, d \in D$, or in vector notation $W(g) = u$, then (13), which is equivalent to

$$\sum_{d \in D} \sum_{k \in K_d} s_k^*(h_k - h_k^*) - \sum_{d \in D} \sum_{k \in K_d} u_d^* h_k + \sum_{d \in D} \sum_{k \in K_d} h_k^* u_d - \sum_{d \in D} G_d(u^*)(u_d - u_d^*) \geq 0 \tag{14}$$

reduces to

$$s(v^*)^T(v - v^*) - W(g^*)(g - g^*) \geq 0 \tag{15}$$

which is the variational inequality formulation derived by Dafermos (1982b). When the demand is constant, that is $G_d(u) = \bar{g}_d, d \in D$, (13) reduces to:

$$s(v^*)^T(v - v^*) \geq 0 \tag{16}$$

which is the variational inequality formulation derived by Smith (1979).

The usefulness of (15) and (16) is that large scale models of the network equilibrium problems formulated in the space of link flows may be easier to solve than such problems formulated in the space of path flows.

For a network model that is used to simulate traffic flows for different future network configurations, it is important that, for a given scenario, the flows and origin to destination costs be unique. Without uniqueness, a comparison of different future situations would be difficult to carry out; differences between scenarios would depend on which solution was found for each case as well as on the different network configurations. Fortunately, for many situations of interest, the network equilibrium model has unique flows , demands and origin to destination costs. This is ensured when the user cost function is strictly monotone and the demand function (and its inverse) is strictly monotone decreasing, as assumed in the model formulation.

For the details of the uniqueness proofs see Asmuth (1978), Smith (1979) and Aashtiani and Magnanti (1981).

Most of the application of the network equilibrium in practice have been achieved for simpler versions of the model (14) subject to (2)-(3). The user cost functions that have been calibrated succesfully are separable; that is $s_a(v) = s_a(v_a), a \in A$. The demand functions depend only on the travel cost between the origin and the destination of the node pair d, that is $G_d(u) = G_d(u_d)$. In this case (15) may be restated as:

$$\sum_{a \in A} s_a(v_a^*)((v_a - v_a^*) - \sum_{d \in D} W_d(g_d^*)(g_d - g_d^*) \geq 0. \tag{17}$$

With these separability conditions, (17) is equivalent to the convex cost optimization problem

$$MinZ = \sum_{a \in A} \int_0^{v_a} s_a(x)\mathrm{d}x - \sum_{d \in D} \int_0^{g_d} W_d(y)\mathrm{d}y \qquad (18)$$

subject to (2)-(3). Since the user cost functions are increasing and the demand functions are strictly decreasing, the objective functon (18) is convex and (17) may be interpreted as the first order optimality condition, that the directional derivative Z be nonnegative. Kinderlehrer and Stampacchia (1980) give a rigourous proof of this equivalence. The Kuhn-Tucker conditions of (18) subject to (2)-(3) are equivalent to the statement of Wardrop's user optimal principle in the form (8), which is the classical result found by Beckmann, McGuire and Winsten (1956). The existence of the equivalent optimization problem opened the way for practical applications, since a number of efficient algorithms are available for solving such problems.

The version of the problem which found most application in practice is the fixed demand version, that is, when $g_d = \bar{g}_d$, $d \in D$ and turn penalties are present at selected nodes of the network. The formulation of this problem is

$$Min \sum_{a \in A} \int_0^{v_a} s_a(x)\mathrm{d}x + \sum_{i \in P} \sum_{a_i \in A_i^-} \sum_{a_i \in A_i^+} \int_0^{v_{a_1 a_2}} p_{a_1 a_2}(x)\mathrm{d}x \qquad (19)$$

Subject to (2), (3) and

$$v_{a_1 a_2} = \sum_{d \in D} \sum_{k \in K_d} \delta_{a_1 a_2} \cdot h_k, \quad a_1 a_2 \in P \qquad (20)$$

where A_i^+, (A_i^-) denotes the set of arcs that have node i as tail (head) node, $(a_1 a_2)$ is a penalized turn that belongs to the set of turns P, $P \subset N$, $p_{a_1 a_2}(v_{a_1 a_2})$ is the cost function associated with the turn $a_1 a_2$ and

$$\delta_{a_1 a_2 k} = \begin{cases} 1 & \text{if } a_1 a_2 \in K \\ 0 & \text{otherwise} \end{cases} \qquad (21)$$

From the large repertory of methods that are available for solving convex cost minimization problems subject to linear constraints, the adaptation of the linear approximation method of Frank and Wolfe (1956) and some of its variants are the commonly used algorithms. Bruynooghe, Gibert and Sakarovitch (1968) were the first to propose the method, however the subsequent work of Le Blanc, Morlok and Pierskalla (1975), Golden (1975), Nguyen (1976), Florian and Nguyen (1976), Dow and Van Vliet (1979) made this method popular in practice. In the context of 'packet switching' networks, the same algorithm was also proposed by Fratta, Gerla and Kleinrock (1973).

Starting from an initial feasible solution, the linear approximation method obtains a feasible direction by linearizing the objective function, solving a linear programming subproblem and then finding an improved solution on the line segment between the current solution and the solution of the subproblem. For the network equilibrium problem, the linearized problem is solved by a sequence of shortest path computations, which take into account the turn penalties present. An iteration is completed by performing a one dimensional line search. Since the adaptation of the method has been well documented, the details are not given here. A disadvantage of the algorithm is its slow convergence in the vicinity of the optimal solution (see Wolfe, 1970; Guélat and Marcotte, 1986). This has stimulated interest in improving its rate of convergence in way that still take advantage of its simplicity.

Variants of the linear approximation method, that exibit empirically a better rate of convergence have been developed by Dembo and Tulowitzki (1983), Fukushima (1983), Florian, Guélat and Speiss (1987). Of particular interest are the variants based on the restricted simplicial decomposition methods of Lawphongpanich and Hearn (1984), Pang and Yu (1984) which were motivated by the earlier work of Holloway (1974) and Cantor and Gerla (1974). Florian (1977b) and then Guélat and Marcotte (1986) proposed adaptations of Wolfe's (1970) 'away' step within a restricted simplicial decomposition approach. Unfortunately, few of those methods have been implemented in commercially available codes, probably since the solutions obtained with the linear approximation method are acceptable in practice.

Other algorithmic approaches have been developed and tested computationally for this problem. Nguyen (1974) has adapted the convex simplex method by using a decomposition of the flows v_a by origin and keeping a 'copy' of the network with flows v_a^p for each origin p. The storage requirements of this method and the input and output manipulations required each time an origin is considered may impede its application, although empirically the computational results obtained are very good. Another approach to solving this problem when the flows are decomposed by origin is to apply an algorithm based on finding negative cycles, as proposed by Weintraub and Gonzales (1980). Alternatively the subproblem may be converted into a circulation problem which may be solved with an algorithm based on finding negative cycles (Ferland, 1974). Dembo and Klincewicz (1981) have proposed second-order algorithms for solving the problems for each origin. Conclusive computational comparisons are still to be performed.

When the number of paths for an origin/destination pair is relatively small, such as in an airline or a railway network and path information is necessary, it may be advantageous to decompose the problem by origin/destination pairs and cyclically solve smaller problems in the space of path flows. The solution strategy used may be stated as follows.

Step 1.   Given an initial solution, set $j = 0$, $m = 0$ and continue.

Step 2.   if $m = |I|$, the total number of origin/destination pairs, terminate; otherwise set $j = j$ modulo $|I| + 1$ and continue.

Step 3.   if the current flow is optimal for the jth origin/destination pair, set $j = j+1$, $m = m+1$ and return to step 2; otherwise, solve the jth subproblem, update the current flow, set $m = 1$, and return to step 2.

Such solution approaches were suggested by Gibert (1968), Dafermos and Sparrow (1969), Dafermos (1972), Leventhal, Nemhauser and Trotter (1973), Nguyen (1974) and Soumis (1978). The convergence of this decomposition approach is ensured by the fact that the objective function is strictly convex and it may be shown to be strictly decreasing after each solution of a subproblem in Step 3. Dafermos and Sparrow (1969) and Auslander (1976, pp.94-96) give rigorous convergences proofs.

Bertsekas (1982) proposed the adaptation of the Goldstein-Levitan-Polyak gradient projection method for solving the problem in the space of path flows.

The interested reader is referred to the summary papers of Fernadez and Friesz (1983), Florian (1986) and Boyce, Le Blanc and Chang (1988) for both a historical prospective and more detailed descriptions of the algorithms mentioned above.

2.2. Optimal Strategies – Route Choice on Transit Networks

Route choice of transit riders is fundamentally different than the same problem on a road network, due to the phenomenon of waiting for the arrival of the arrival of a transit vehicle. In the case of the private car, the computation of shortest paths is symmetric with respect to an origin/destination pair. This symmetry does not hold in the case of transit route choice.

The transit route choice problem has been studied by several authors in the past, either as a separate problem (e.g. see Dial, 1967; Chriqui, 1974; Chriqui and Robillard, 1975; Rapp et al., 1976; Andreasson, 1976) or as a subproblem of more complex models, such as transit nework design (e.g. see Lampkin and Saalmans, 1967; Scheele, 1977; Mandl, 1979; Hasselström, 1981), or multimodal network equilibrium (Florian, 1977a; Florian and Spiess, 1983). Computer programs, such as IGTDS (General Motors, 1978), TRANSCOM (Chapleau, 1974), TRANSEPT (Last and Leak, 1976), NOPTS (Rapp et al., 1976), the VOLVO system (Andreasson, 1976; Hasselström, 1981) and ULOAD (UMTA/FHWA, 1977), are each based on one of these methods and are widely used by transit companies and government agencies for planning purposes.

The presentation of the model of route choice in this section is based on the paper by Spiess and Florian (1988).

Let T denote a transit network consisting of a set of nodes, a set of lines, each defined as a sequence of nodes at which passengers amy board and alight, and set of walk links, each connecting two nodes. The times (or costs) that are associated with each walk link and each segment of a transit line are constant and known. At each node that is served by a line, the distribution of interarrival times of the passengers at each node, one can construct the distribution of the waiting time for the arrival of the first vehicle, among any given set of lines passing at the same node, as well as the probality of each line arriving first.

Each walk link may be replaced by a transit line of one link with a zero waiting time (very high frequency). In the following, the walk links will therefore not always be distinguished from the segments of the transit lines. In order to simplify the formulation, it is also assumed that the transit network T is strongly connected.

Consider the transit traveler whose trip originates at the node A and is destined to node B. It is assumed that he behaves in a way that minimizes his total expected travel time, which is the weighted sum of expected walk, wait, and in-vehicle time. This assumption seems natural and it also parallels the behavior assumptions made for car traffic route choice. Since the weights may always be normalized to 1 by scaling the times used in the definition of the network, one need not consider the weights separately. Thus, the term time will be used, even if these "times" might actually include weight factors and should then more appropriately be called "generalized costs".

Any element of the traveler's choice set is called a *strategy*. A strategy is a set of rules that, when applied, allows the traveler to reach his destination. The number and the kind of different strategies that the traveler may choose from depend on the information that is available to him during his trip. If no additional information becomes available during the trip, a strategy simply defines a path. For the transit network given in Figure 1, an exemple of such strategy would be:

- Take line 2 to node Y; tranfer to line 3 and exit at node B.

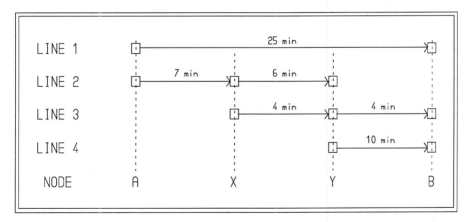

Figure 1

If the traveler, while waiting at a node, knows the line which is to be served next, a strategy may include subsets of lines and would then be:

- Take the next vehicle among lines 1 and 2; if line 1 was taken, exit at B; if line 2 was taken, transfer at node Y to any of the lines 3 or 4 and exit at B

Of course, the more information available to the traveler (such as elapsed waiting time, arrival times of vehicles seen while waiting at a stop, information on other vehicles while riding in a vehicle by looking out of the window, ...), the more complex the strategies may become; for example:

- Wait up to five minutes for a vehicle of line 1; otherwise take line 2; if at node X you see a vehicle of line 3 (express bus), transfer to line 3, otherwise continue to node Y and transfer there to any of the lines 3 or 4.

The assumption is made that the only information that is available to the traveler during his trip is that he finds out, while waiting at a node, which line is about to be served next; i.e., he may at that time decide whether or not to board this vehicle. Consequently, only strategies that correspond to the second example given above are considered. It is then possible to define a strategy of this type by specifying at every node of the network, except at the destination node, a non-empty set of attractive lines, and for each of these lines the (upstream) node at which the traveler alights. Note that when walk links are included in the network, a strategy may contain at a node an outgoing walk link instead of the set of attractive lines. Given a strategy, an actual trip is then carried out according to a mechanism that may be stated systematically as:

0:    Set NODE to origin node.
1:    Board vehicle that arrives first among the vehicles of the set of attractive lines at NODE.
2:    Alight at the predetermined node.
3:    If not yet at destination, set NODE to current node and return to step 1. Otherwise the trip is completed.

A strategy is said to be *feasible* if the graph defined by the attractive lines of the strategy does not contain cycles.

A transit trip consists in general of several trip components that may include some or all of the following:
-    access from origin to transit stop,
-    waiting for a vehicle,
-    riding in a vehicle,
-    alighting a vehicle,
-    walking between two transit stops,
-    egress from transit stop to destination.

A non-negative time (or cost) is usually used to quantify each of these trip components, with the exception of "waiting for a vehicle", which is quantified by using the statistical distribution of waiting times for the arrival of the first vehicle of a given transit line at a given stop.

In order to simplify the presentation, every trip component includes a constant time as well as a distribution of waiting times. The trip components are represented by links $a \in A$ of a network $T = (I, A)$ with nodes $i \in I$. Let $A_i^+(A_i^-)$ denote the set of outgoing (incoming) links at node $i \in I$. Each link $a \in A$ is characterized by the pair (c_a, G_a) where c_a is a non-negative link travel time and G_a the distribution function for the waiting time

$$G_a(x) = Prob\{\text{waiting time on a link } a \leq x\}.$$

The functions $G_a(x)$ can be obtained from the distribution of interarrival times (headways) and the distribution of passenger arrival times.

Using this generalized formulation one represent any transit network T. The mechanics that are involved in constructing the links and the nodes of the generalized network depend very much on the particularities of the transit network and the degree of aggregation considered.

Since this generalized formulation no longer explicitly identifies transit lines but relies only on links, one can no longer use the terminology associated with the transit lines, such as waiting for a vehicle of a transit line. Instead, in the following, the convention used is that each link is *served* by vehicles and that the passenger, instead of waiting for a vehicle of a certain line, waits for a *link to be served*.

For the transit route choice problem in this generalized form, a strategy to reach destination node r is defined by a partial network $T_r = (I, \bar{A})$ that contains only those links that will be used as a consequence of this strategy. One can therefore denote a strategy by the corresponding subset of Links $\bar{A} \subseteq A$, or alternatively, by the partition of these links according to their i-node $\{\bar{A}_i^+ = A_i^+ \cap \bar{A}, \ i \in I\}$. links that are included in the strategy $\bar{A}$ are never used. Among the links that are included in the strategy $\bar{A}$, at each node $i \in I$, a traveler boards the first vehicle that serves any of the links $a \in \bar{A}_i^+$. Hence, $\bar{A}_i^+$ corresponds to the set of attractive lines that was discussed in Section 2 and, of course $\bar{A}_i^+ \neq \varnothing$ for $i \neq r$. For a strategy to be *feasible* it must contain at least one path from each node to the destination node r and it must not contain any cycle.

Let $W(\bar{A}_i^+)$ denote the expected waiting time for the arrival of the first vehicle serving any of the links $a \in \bar{A}_i^+$. $W(\bar{A}_i^+)$ is called the *combined waiting time* of link $a \in \bar{A}_i^+$. Let further $P_a(\bar{A}_i^+)$ be the probability that link a is served first among the links $\bar{A}_i^+$. For convenience we define $P_a(\bar{A}_i^+) = 0$ for all $a \notin \bar{A}_i^+$.

The model that is presented here is based on the assignment of the trips from all nodes toward a single node r, the *destination* node. This is different from most other route choice models, transit or highway, which always consider the trips from one origin node to all destination nodes. Let $g_i, i \in I - \{r\}$, be the demand (number of trips) from node i to the destination node r. In order to simplify the notation, let

$$g_r = -\sum_{i \neq r} g_i.$$

Given a strategy $\bar{A}$, the demand g_i, $i \in I$, is assigned to the network yielding link volumes v_a, $a \in A$. The volume at a node, which is denoted V_i, $i \in I$, is the sum of the volumes of all incoming links and the demand at that node

$$V_i = \sum_{a \in A_i^-} v_a + g_i, \quad i \in I. \tag{22}$$

The node volume V_i is distributed on the outgoing links according to their link probabilities under strategy $\bar{A}$

$$v_a = P_a(\bar{A}_i^+)V_i, \quad a \in A_i^+, \quad i \in I. \tag{23}$$

Since $\sum_{a \in A_i^+} P_a(\bar{A}_i^+) = 1$, (22) and (23) ensure that the conservation of flow equations

$$\sum_{a \in A_i^+} v_a - \sum_{a \in A_i^-} v_a = g_i, \quad i \in I$$

are satisfied.

The optimal strategy $\bar{A}^*$ is the strategy which minimizes the expected total travel time including waiting time. The most general formulation of the transit model is described by Spiess (1984).

In the following, the model is specialized to the case where the waiting time distribution of each link a (or transit line, in the original form of the problem) is quantified by a positive parameter f_a, which is called the *frequency* of a link. The expected combined waiting time and the link probabilities are derived from the frequencies in the following way:

$$W(\bar{A}_i^+) = \frac{\alpha}{\sum_{a \in A_i^+} f_a}, \alpha > 0, \tag{24}$$

$$P_a(\bar{A}_i^+) = \frac{f_a}{\sum_{a' \in \bar{A}_i^+} f_{a'}}, \quad a \in \bar{A}_i^+. \tag{25}$$

The case $\alpha = 1$ corresponds to an exponential distribution of interarrival times of the vehicles with mean $1/f_a$ and a uniform passenger arrival rate at the nodes.

The case $\alpha = 1/2$ is an approximation of a constant interarrival time $1/f_a$ for the vehicles on link a. This measure of waiting time is the most widely used approach in practice (Chapleau, 1974; Dial, 1967; Le Clercq, 1972; Rapp et al., 1976; UMTA/FHWA, 1977), in spite of the fact that it is based on a rough approximation.

The factor α may, of course, also model the effect of different perceptions of waiting time and in-vehicle time. Since the link probabilities (25) are independent of the units in which f_a is specified, it is always possible to scale the frequencies by a factor $1/\alpha$. Without loss of generality we can therefore assume $\alpha = 1$ in the following.

Based on the assumptions it is possible to show (see Spiess and Florian, 1988) that the transit route choice model is equivalent to the following linear programming problem.

$$\text{Min} \sum_{a \in A} c_a v_a + \sum_{i \in I} w_i \tag{26}$$

subject to

$$\sum_{a \in A_i^+} v_a - \sum_{a \in A_i^-} v_a = g_i, \ i \in I \tag{27}$$

$$v_a \leq f_a w_i, \ a \in A_i^+, \ i \in I \tag{28}$$

$$v_a \geq 0, \ a \in A \tag{29}$$

where v_a are the link volumes and w_i are the total expected times at node i. This model may be solved by a polynomial time algorithm. The algorithm is composed of two parts. In a first pass, from the destination node to all origins, the optimal strategy $\bar{A}^*$ and the expected total travel times u_i^* from each node $i \in I$ to the destination node r are computed. In a second pass, from all origins to the destination, the demand is assigned to the network according to the optimal strategy.

Part 1: Find Optimal Strategy

1.1 (Initialization) $u_i := \infty, \ i \in I - \{r\}; \ u_r := 0;$

$f_i := 0, \ i \in I;$

$S := A; \bar{A} := \oslash$

1.2 (Get next link)   If $S = \oslash$ then STOP,

else find $a = (i, j) \in S$ which satisfies

$u_j + c_a \leq u_j + c_{a'}, \ a' = (i', j') \in S;$

$S := S - \{a\}.$

1.3 (Update node label)   If $u_i \geq u_j + c_a$ then

$u_i := \frac{f_i u_i + f_a (u_j + c_a)}{f_i + f_a},$

$f_i := f_i + f_a, \bar{A} := \bar{A} + \{a\};$

go to step 1.2.

Part 2: Assign Demand According to Optimal Strategy

2.1 (Initialization) $V_i := g_i, \ i \in I;$

2.2 (Loading)   Do for every link $a \in A$, in decreasing order of $(u_j + c_a)$:

if $a \in \bar{A}$ then $v_a := \frac{f_a}{f_i} V_i,$

$V_j := V_j + v_a,$

else $v_a := 0.$

In step 1.1 the node labels u_i, i.e. the expected travel times to reach the destination, are set to infinity for all nodes except the destination, for which it is initialized to zero. The auxiliary variables f_i, $i \in I$, that contain the combined frequencies of all selected links at node i, are initialized to zero. The set S is used to identify the links that have not yet been examined, and the set $\bar{A}$ is used to identify the optimal strategy.

In step 1.2 the link nearest to the destination is selected among the links not yet examined. The time considered is $u_j + c_a$, that is, the time from the i-node of the link to the destination not including the waiting time at node i. If this time is smaller than the current time associated with node i, u_i, link a is included in the optimal strategy and u_i and f_i are updated according to the formulas given in step 1.3. Note that when the label of a node i, u_i, is improved for the first time, we have $f_i u_i = 0 \cdot \infty$. In order to keep the algorithm as compact as possible we adapt the convention $0 \cdot \infty = 1$ whenever this case occurs. (When $\alpha \neq 1$ is used in (24), this convention becomes $0 \cdot \infty = \alpha$). Part 1 of the algorithm terminates when all links have been examined.

In the second part of the algorithm the demand from node i to the destination, g_i, is assigned to the network according to the strategy $\bar{A}$. This is done by assigning to each link $a \in \bar{A}$ the proportion of the node volume v_i that corresponds to its frequency. Since the links are processed in reverse topological order (decreasing $u_j + c_a$), the node volumes may be updated in parallel, hence making it possible to examine every link only once.

This model has been applied in many urban and regional contexts for analysing bus, metro and commuter railway networks.

2.3. Multi-modal network equilibrium problems

The formulation of network equilibrium models which considers travel by several modes and choices are made by travellers both for mode choice and route choice result in network equilibrium models which do not have equivalent optimization problems, but may be formulated as nonlinear complementary problems or variational inequality problems. Recent developments have made it possible to tackle succesfully certain classes of these problems and some practical applications have been achieved.

We reconsider now the variational inequality formulation of the network equilibrium problem in the form

$$s(v^*)^T(v-v^*)-W(g^*)^T(g-g^*) \geq 0.$$

There are many variants of the network equilibrium problem which fit within this framework. When flows may occur by several modes (car, bus, train) on some or all of the links of the network, the user costs for travel on a link a depends on the flow of all modes that link, that is

$$s(v) = (s_a^m(v), \ a \in A, \ \forall m) \tag{30}$$

and the demand for travel by mode m depends on the vector of travel costs by all modes, that is

$$G(u) = (G_d^m(u), \ d \in D, \ \forall m) \tag{31}$$

where u is the vector of travel costs for origin/destination pair i by all modes. If the vector function $G(u)$ admits a smooth inverse $W(g)$, the model can be formulated as the variational inequality (15).

A similar formulation results for situations where the total demand for travel by all modes is constant and the demand function predicts the proportion of trips that occur by each mode. When two modes are considered, say private car and bus, then $g_d^1 + g_d^2 = \bar{g}_d, \ d \in D$ and the model may be reformulated by eliminating the demand of one of the modes, say mode 2. In that case, Florian and Spiess (1983) showed that the resulting variational inequality is

$$s(v^*)^T(v-v^*) - W(g^1 - g^{1*}) \geq 0 \tag{32}$$

where $W(g^1)$ is the mapping of inverse mode choice functions, which are taken to be strictly decreasing functions of g^1, the demand for travel by mode 1. This variational inequality may be derived from (15) by the appropriate substitutions and some algebraic manipulations. This model was also studied by Fisk and Nguyen (1982).

Network equilibrium problems may be much larger than problems that may be solved by general purpose algorithms for nonlinear complementarity or variational inequality problems. Nevertheless, recent progress in this area has led to the succesful solution of large problems by using methods that amount to solving a sequence of network equilibrium problems which have equivalent optimization formulations. The two most natural approaches are the adaptation of the Jacobi and Gauss-Seidel methods for solving systems of linear and nonlinear equations. These methods are quite intuitive and their use was suggested even before rigorous convergence proofs were found.

In the Jacobi method, given a feasible solution $(v,g)^l$ at iteration l, the off-diagonal elements of the user cost and inverse demand functions are fixed by using the current solution elements. This results in user cost functions $s(v,v^l)$ and inverse demand functions $W(g,g^1)$, as follows:

$$v^l \rightarrow s(v,v^l) \text{ where } s_a(v,v^l) = s_a(v_1^l, ..., v_{a-1}^l, v_a, v_{a+1}^l, ..., v_{|A|}^l), \tag{33}$$

$$g^l \rightarrow W(g,g^l) \text{ where } W_d(g,g^l) = W_d(g_1^l, ..., g_{d-1}^l, g_d, g_{d+1}^l, ..., g_{|D|}^l), \qquad .$$

The corresponding variational inequality

$$s(v^*,v^l)^T(v-v^*) - W(g^*,g^l)(g-g^*) \geq 0 \tag{34}$$

has an equivalent optimization formulation and may be solved by any of the methods discussed above. Depending on the nature of the problem, this solution strategy may result in a decomposition by origin/destination pairs, by origin, by mode or by origin and mode.

$(v,g)^{l+1}$ is obtained by solving (34) by an appropriate algorithm and repeatig the process of redefining the cost and demand functions. This typical iteration is repeated until a convergence criterion is satisfied.

The Gauss-Seidel method is very similar, with the exception that, when the variational inequality decomposes into more than one subproblem, the most recent information is used; that is, given that the variational inequality decomposes into M subproblems with

$$(v,g)^l = \{(v,g)^l_1, (v,g)^l_2, ..., (v,g)^l_M\} \tag{35}$$

then $(v,g)^{l+1}_m$ is obtained for $m = 1, ..., M$ by solving the variational inequality which depends on the vector

$$\{v, g^{l+1}_1, ..., (v,g)^{l+1}_{m-1}, (v,g)^l_{m+1}, ..., (v,g)^l_M\}. \tag{36}$$

Thus the variational inequality solved as an equivalent optimization problem in the mth component of the original vector functions, where for each subproblem m the off-diagonal elements are fixed with the solution of subproblems $1, ..., m-1$ at iteration $l+1$ and with the solutions of the subproblems $m+1$ to M at iteration l. An iteration is completed by solving M subproblems.

Another approach which has been suggested is a Newton type linearization scheme which determines the vectors $(v,g)^l_m$, $l = 1, 2, ...$ iteratively by obtaining $(v,g)^{l+1}$, from $(v,g)^l$ by replacing the cost functions with their first order Taylor approximation at $(v,g)^l$. $(v,g)^{l+1}$ is the solution of the affine cost problem

$$s_{lin}(v^*, v^l)^T (v-v^*) - W_{lin}(g^*, g^l)^T (g-g^*) \geq 0 \tag{37}$$

when $s_{lin}(v^*, v^l) = s(v^l) + \nabla s(v^l)(v - v^l)$ is the first order of $s(v)$ at v^l and $W_{lin}(g, g^l)$ is similarly defined. This approach requires the solution of a sequence of variational inequalities with affine costs. this problem may be converted into a linear complementarity problem (see Aashtiani and Magnanti, 1982), however the pivoting methods available for its solution are not effective for large scale problems. Alternatively, one could solve each linearized subproblem by an appropriate decomposition scheme, such as the Jacobi or Gauss-Seidel methods.

Another approach to solve the general network equilibrium problem includes projection methods (Dafermos, 1980; Bertsekas and Gafni, 1982). In Particular Bertsekas and Gafni (1982) proposed the use of a projection method to solve an asymmetric network equilibrium in the space of path flows, by embedding the projection approach in a Gauss-Seidel decomposition by origin/destination pair. Lawphongpanich and Hearn (1984) suggested the use of a similar projection method with in a simplicial decomposition scheme which uses a restriction approach. Yet another method proposed by Nguyen and Dupuis (1984) is a cutting plane algorithm applied to a dual formulation of the variational inequality formulation of the network equilibrium problem with a nondifferentiable objective (see Hearn,

1982; Hearn and Nguyen, 1982).

The recent interest in solving the general network equilibrium problem has stimulated convergence studies of the solution approaches mentioned above. The convergence of the Jacobi method, proposed in this context by Dafermos (1972), Florian (1977a) and Abdulaal and Le Blanc (1979), Hall, Van Vliet and Willumsen (1980) has been studied by Pang and Chan (1982) and Florian and Spiess (1982) by using ideas similar to those proposed by Ahn (1979) for showing the convergence of the Jacobi method for an economic equilibrium model. Josephy (1979) and Pang and Chan (1982) have established local convergence criteria for Newton type linearization algorithms. Dafermos (1982a) has first studied global convergence properties of the Jacobi method and then Dafermos (1983), has provided a unified way for proving the convergenceof the Jacobi, Gauss-Seidel and projection methods for variational inequalities. Soumis (1978) and Pang (1985) have provided proofs for the convergence of the Gauss-Seidel method. The essence of the convergence results, which are obtained by showing that the algorithmic map are contractive, is that the strong monotonicy of the cost and demand functions, which ensures the uniqueness of the solution, is a necessary condition for convergence. Furthermore, The Jacobian matrices of the cost functions must satisfy additional conditions for which the diagonal dominance (or block diagonal dominance) of these matrices is a sufficient condition. To date, the Jacobi method (see Florian et al., 1979) and the Gauss-Seidel method (see Soumis, 1978) are the only methods which have been applied in practice to solve general network equilibrium problems. However, the rapid evolution of the solution approaches in this field may lead to the emergence of other, more efficient methods.

3.   National and Regional Multi-Modal Freight Networks

The prediction of multicommodity freight flows over a multimodal network has attracted much interest in recent years. In contrast to urban transportation, where the prediction of passenger flows over multimodal networks has been studied extensively and many of the research results have been transferred to practice, the study of freight flows at the national or regional level, perhaps due to the inherent difficulties and complexities of such problems, recieved less attention.

The class of models that was well studied in the past for prediction of interregional flows is the *spatial price equilibrium model* and its variants. The models, stated initially by Samuelson (1952) and extended by Takayama and Judge (1964, 1970) then by Florian and Los (1982), Friesz, Tobin, and Harker (1983) has been used extensively for analyzing interregional commodity flows. This class of models determines simultanouesly the flows between *producing* and *consuming* regions as well as the *selling* and *buying* prices. The transportation network is usually modeled in a simplitic way and these models rely to a large extent on the *supply* and *demand* functions of the producers and consumers respectively. The calibration of these functions is essential to the application of these models and the transportation costs are unit costs or may be functions of the flow on the network. There have been so far few multicommodity applications of this class of models, with the majority of applications having been carried out in the agricultural and energy sectors in an international or interregional setting. It is not this class of models which is presented in this paper.

Another class of models of interest are network models wich enable the prediction of multicommodity flows over a multimode network, where the physical network is modeled at a level of detail appropiate for a nation or a large region and represent the physical facilities with relatively little abstraction. The demand for the transportation services is exogenous and may originate from an input-output model, if one is available, or from other sources, such as observed demand or scaling of observed past demand. The choice of mode, or subsets of modes used is exogenous and intermodal shipments are permitted. In this sense, these models may be integrated with econometric demand models as well. The emphasis is on the network representation and the proper representation of congestion effects in a static model aimed to served for comparative static analyses or for discrete time multiperiod analyses.

The Harvard Model (Kresge and Roberts, 1971), which is probably the first published freight network model of the type that interests us, resorted to a fairly simple "direct link" representation of the physical network and congestion effect were not considered. Later, the Multi-State Transportation Corridor Model (McGinnis et al., 1981; Jones and Sharp, 1979; and Sharp, 1979) went a step further in representing an explicit multimodal network, but without any consideration of congestion. The Transportation Network Model (Bronzini, 1980) does not consider congestion effects either. The first model that considers congestion phenomena in this field, is the Freight Network Equilibrium Model (Friesz, Gottfried and Morlok, 1986). This is a sequential model which uses two network representations: an aggregate network that is perceived by the users, which serves to determine the *carriers* chosen by the *shippers* and then more detailed separate networks for each carrier, where commodities are transported at least total cost.

The model that is presented in detail in this paper is a multi-product multi-mode formulation based on the normative assumption that total costs are minimized. The simplifying assumption is made that the goods are shipped at least cost, even though the shipment of freight is governed by a variety of micro-circumstances that prevent the achievement of total cost minimization. The justification for this assumption is that, at a strategic level, predicting freight flows based on this assumption, when reliable modal demand is available, is satisfactory for the purpose of scenario comparisons when the investements considered are of large magnitude.

The *multi-modal* aspects of a national transportation system are accounted for in the network representation chosen. A link of the multi-modal network is defined by its origin and destination nodes and a single mode. Parallel links are allowed between two adjacent nodes, one for each mode available to transport goods between them. The intermodal transfers at a node of network are modeled as link to link permitted movements. Appropriate cost functions may be associated with links and intermodal transfers.

The *multi-product* aspects of a national transportation system are accounted for in the formulation of the predictive model and is taken advantage of in the solution procedure. The algorithm developed for this problem exploits the natural decomposition by product and results in a Gauss-Seidel like procedure.

The multi-mode multi-product model is formulated in the most general way, permitting in principle nonconvex and asymmetric cost functions. Nevertheless, certain simplifying assumptions made on the structure of the cost functions may simplify the problem and permit the solution of large size problems in reasonable computational times.

The physical network infrastructure represented by the network model chosen supports the transportation of several products on several modes. A *product* is any commodity (collection of similar products), goods or passengers, that generates a link flow specifically associated with it. A *mode* is a means of transportation that has its own characteristics, such as vehicle type and capacity, as well as a specific cost function.

The *base network* is the network that consists of the nodes, links and modes that represent all the physical movements possible on the available infrastructure. The model that we have chosen defines a link as a triplet (i, j, m), where i is the origin node, $i \in N$, where N is the set of nodes of the network, j is the destination node, $j \in N$, and m is the mode allowed on the arc, $m \in M$, where M is the set of modes available on the network. Parallel links are used to represent the situation where more than one mode is available for transporting goods between two adjacent nodes.

The network that considered consists of a set of nodes N, a set of arcs A, $A \subseteq N \times N$, a set of modes M and a set of transfers T, $T \subseteq A \times A$. We denote their cardinality, n_N, n_A, n_M and n_T respectively. With each arc a, $a \in A$, a *cost function* $s_a(.)$ is associated which depends on the volume of goods on the arc, or possibly, on the volume of goods on the other arcs of the network. Similarly, a cost function $s_t(.)$ is associated with each transfer $t \in T$.

The products transported over the multi-modal network are denoted by index p, $p \in P$, where P is the set of all products considered, which is of cardinality n_P. Each product is shipped from origins o, $o \in O \subseteq N$, to destinations d, $d \in D \subseteq N$, of the network. The demand for each product and for all origin/destination (O/D) pairs is specified by a set of O/D matrices for the corresponding subsets of modes. It is assumed that the mode choice is determined exogenously. Let $g^{m(p)}$ be the demand matrix associated with product $p \in P$, where $m(p)$ is a subset of modes that belong to $M(p)$, the set of all subsets of modes that are used to transport product p.

The flows of product p on the multimodal network is denoted by v^p and consists of the induced flows of this products on links and transfers :

$$v^p = \left(\begin{array}{ll} (v_a^p), & a \in A \\ (v_t^p), & t \in T \end{array} \right)$$

The flow of all the products on the multimodal network is denoted by $v = (v^p)$, $p \in P$, and is a vector of dimension $n_P(n_A + n_T)$.

The average cost functions $s_a^p(v)$, on links, and $s_t^p(v)$, on transfers, correspond to a given flow vector v. The average cost functions for product p are denoted, similar to the notation used for the flow v^p, s^p, $p \in P$, where

$$s^p = \begin{pmatrix} (s_a^p), & a \in A \\ (s_t^p), & t \in T \end{pmatrix}$$

and $s = (s^p)$, $p \in P$, is the vector of average cost functions of dimension $n_P(n_A \times n_T)$.

The total cost of the flow on arc a, $a \in A$, for the product p, $p \in P$, is the product $s_a^p(v)v_a^p$; the total cost of the flow on transfer t, $t \in T$, is $s_t^p(v)v_t^p$. The total cost of the flows of all products over the multi-modal network is the function F that we seek to minimize

$$F = \sum_{p \in P} \left(\sum_{a \in A} s_a^p(v)v_a^p + \sum_{t \in T} s_t^p(v)v_t^p \right) \qquad (= s(v)^T v) \tag{38}$$

over the set of flows which satisfy the conservation of the flow and nonnegativity constraints. In order to write these constraints for the multi-product multi-mode network defined above, the following notation is used. Let $K_{od}^{m(p)}$ denote the set of paths that lead from origin o, $o \in O$, to destination d, $d \in D$, by using only modes of $m(p) \in M(p)$, $p \in P$. The conservation of flow equations are then

$$\sum_{k \in K_{od}^{m(p)}} h_k = g_{od}^{m(p)}, \qquad o \in O, d \in D, m(p) \in M(p), p \in P \tag{39}$$

where h_k is the flow on path k. The nonnegativity constraints are

$$h_k \geq 0, \qquad k \in K_{od}^{m(p)}, o \in O, d \in D, m(p) \in M(p), p \in P \tag{40}$$

Let Ω be the set of flows v that satisfy (39) and (40). Since the conservation of flow equations are stated in the space of path flows, for notational convenience, the specification of Ω requires the relation between arc flows and path flows, which is

$$v_a^p = \sum_{k \in K^p} \delta_{ak} h_k, \qquad a \in A, p \in P \tag{41}$$

where

$$K^p = \bigcup_{m(p) \in M(p)} \bigcup_{o \in O} \bigcup_{d \in D} K_{od}^{m(p)}$$

is the set of all paths that may be used by product p, and

$$\delta_{ak} = \begin{cases} 1 & \text{if } a \in k \\ 0 & \text{otherwise} \end{cases}$$

is the indicator function which identifies the arcs of a particular path. Similarly, the flows on tranfers are

$$v_t^p = \sum_{k \in K^p} \delta_{tk} h_k, \qquad t \in T, p \in P \tag{42}$$

where
$$\delta_{ak} = \begin{cases} 1 & \text{if } t \in k \\ 0 & \text{otherwise .} \end{cases}$$

The transfer t belongs to the path k if the two arcs that define the transfer belong to it.

In conclusion, the system optimal multi-product, multi-mode assignment model consists of minimizing (38) subject to (39)-(40) with the definitional constraints (41)-(42).

The model is sufficiently general to be adapted for different ways of specifying the demand. Even though typical applications of the model are likely to be carried out after an "a priori" mode choice calculation, which would allocate the demand for a product g^p to a set of mode subsets, it is equally possible to permit the demand for a product to be transported over all the allowed modes, that is $m(p)$ is the set of all modes of the network and $M(p)$ has a single component, which in this case is $m(p)$. Also, the model is flexible in the specification of intermodal movements. The mode to mode transfers may be restricted to occur only at specific nodes of the network, and only between specific modes.

The structure of this model suggests a natural decomposition by product Ω is the direct product $\prod_p \Omega^p$, where Ω^p is the set of feasible flows of product p on the subnetwork $m(p) \in M(p)$. The algorithm for solving this problem, as described in Guélat, Florian and Crainic (1987) may be characterized as a Gauss-Seidel-Linear Approximation algorithm, which is convergent. This model and solution have been used for analysing strategic planning situations in Brazil and more recently in Europe. The Brazil freight transportation studied consist of 211 origins and destinations, 1243 regular nodes, 4957 links and 5718 transfers. Problems involving up to six products and 10 modes were solved in that application.

References

Aashtiani H. and Magnanti T., "A linearization and decomposition algorithm for computing urban traffic equilibrium", *Proceeding of the 1982 IEEE large scale systems symposium* (October, 1982).

Aashtiani H. and Magnanti T., "Equilibria on a congested transportation network", *SIAM Journal on Algebraic and Discrete Methods* 2 (1981) 213-226.

Abdullal M. and Leblanc L., "Methods for combining model split equilibrium assignment models", *Transportation Science* 13 (1979) 213-314.

Ahn B., *Computation of market equilibria for policy analysis: The Project Independence Evaluation System (PIES) approach* (Garland, New York, 1979).

Andreasson I., "A method for the analysis of transit networks", 2nd *European Congress on Operations Research*, edited by Mark Roubens (North Holland, 1976).

Asmuth R., "Traffic network equilibria", Technical Report SOL 78-2, Department of Operations Research, Stanford University (Stanford, CA, 1978).

Auslender A., *Optimization, méthodes numériques* (Masson, Paris, 1976).

Beckmann M.J., McGuire C.B. and Winsten C.B., *Studies in the economics of transportation* (Yale University Press, New Haven, 1956).

Bertsekas D.P., "Optimal routing and flow control methods for communications networks", in: Bensoussan A. and Lions J.L., eds., *Analysis and optimization of system* (Springer-Verlag, Berlin, New York, 1892) pp.615-643.

Bertsekas D.P. and Gafni E.M., "Projection methods for variational inequalities with application to the traffic assignment problem", *Mathematical Programming Study* 17 (1982) 139-159.

Boyce D., Le Blanc L. and Chang K., "Network equilibrium models of urban location and travel choices: a retrospective survey", *Journal of Regional Science* 28 (1988) 159-183.

Bronzini M.S., "Evolution of a Multimodal Freight Transportation Network Model", Proceeding of the 21st Annual Meeting Transportation Research Forum (1980) 475-485.

Bruynooghe A., Gibert A. and Sakarovitch M., "Une méthode d'affecttion du traffic", *Proceedings of fourth symposium on the theory of traffic flow* (Karlsruhe, 1968).

Cantor D. and Gerla M., "Optimal routing in a packet switched computer network", *IEEE Transactions on Computers* 10 (1974) 1062-1068.

Chapleau R., "Réseaux de Transport en Commun: Structure Informatique et Affectation", Ph.D *Thesis*, Département d'informatique et de recherche opérationnelle, Université de Montréal (1974).

Chriqui C., "Réseaux de Transport en Commun: Les problèmes de cheminement et d'Accès", Ph.D *Thesis*, Département d'informatique et de recherche opérationnelle, Université de Montréal (1974).

Chriqui C. and Robillard P., "Common Bus Lines", *Transportation Science* 9 (1975) 115-121.

Dafermos S., "The traffic assignment problem for multiclass-user transportation networks", *Transportation Science* 6 (1972) 73-87.

Dafermos S., "Traffic equilibrium and variational inequalities", *Transportation Science* 14 (1980) 42-54.

Dafermos S., "Relaxation algorithms for the general asymmetric traffic equilibrium problem", *Transportation Science* 16 (1982a) 231-240.

Dafermos S., "The general multimodal network equilibrium problem with elastic demand", *Networks* 12 (1982b) 57-72.

Dafermos S., "An iterative scheme for variational inequalities", *Mathematical Programming* 26 (1983) 40-47.

Dafermos S. and Sparrow T., "The traffic assignment problem for a general network", *Journal of Research of the National Bureau of Standards* B 73B (1969) 91-117.

Dembo R.S. and Klincewicz J.G, "A scale reduced gradient algorithm for network flow problems with convex separable costs", *Mathematical Programming Study* 15 (1981) 125-147.

Dembo R.S. and Tulowitzki U., "Successive inexact quadratic programming and its application to traffic problems", Working Paper Series B, No. 65, School of Organization and Management, Yale University (New Haven, 1983).

Dial R.B., "Transit Path Finder Algorithm", *Highway Research Record* 205 (1967) 67-85.

Dow P. and Van Vliet D., "Capacity restrained road assignment", *Traffic Engeneering and Control* 20 (1979) 296-305.

Ferland J.A., "Minimum cost multicommodity circulation problems with convex arc-costs", *Transportation Science* 8 (1974) 355-360.

Fernandez J.E. and Friesz T.L., "Travel market equilibrium: the state of the art", *Transportation Research* 17B (1983) 155-172.

Fisk C. and Boyce D., "Alternative variational inequality formulations of the network equilibrium - Travel Choice Problem", *Transportation Science* 17 (1983) 454-463.

Fisk C. and Nguyen S., "Existence and uniqueness properties of an asymetric two-mode equilibrium model", *Transportation Science* 16 (1982) 318-328.

Florian M., "A traffic equilibrium model of travel by car and public transit modes", *Transportation Science* 8 (1977a) 166-179.

Florian M., "An improved linear approximation algorithm for the network equilibrium (packet switching) problem", *IEEE Proceedings, Decision and Control* (1977b) 812-818.

Florian M., "Nonlinear cost networks models in transportation analysis", *Mathematical Programming Study* 26 (1986) 167-196.

Florian M., Chapleau R., Nguyen S., Achim C., James-Lefebvre L. and Fisk C., "Validation and application of EMME: An equilibrium based two-mode urban transportation planning method", *Transportation Research Record* 728 (1979) 14-22.

Florian M. and Los M., "A New Look at Static Price Equilibrium Models", *Regional Science and Urban Economics* 12 (1982) 579-597.

Florian M. and Nguyen S., "An application and validation of equilibrium trip assignment methods", *Transportation Science* 10 (1976) 374-389.

Florian M., Guélat J. and Spiess H., "An efficient implementation of the Partan Variant of the linear approximation method for the network equilibrium problem", *Network* 17 (1987) 319-339.

Florian M. and Spiess H., "The convergence of diagonalization algorithms for the asymmetric equilibrium problem", *Transportation Research* B 16B (1982) 447-483.

Florian M. and Spiess H., "On binary mode choice/assignment models", *Transportation Science* 17 (1983) 32-47.

Frank M. and Wolfe P., "An algorithm for quadratic programming", *Naval Research Logistics Quarterly* 3 (1956) 95-110.

Fratta L., Gerla M. and Kleinrock L., "The flow deviation method: An approach to store and forward communication networking design", *Networks* 3 (1973) 97-133.

Friesz T.L., Tobin R.L. and Harker P.T., "Predictive Intercity Freight Network Models", *Transportation Research* 17A (1983) 409-417.

Friesz T.L., Gottfried J.A. and Morlok E.K., "A Sequential Shipper-Carrier Network Model for Predicting Freight Flows", *Transportation Science* 20 (1986) 80-91.

Fukushima M., "A modified Frank-Wolfe algorithm for solving the traffic assignment problem", *Transportation Research* B 18B(2) (1983) 169-177.

General Motors Transportation System Center, "Interactive Graphic Transit Design System (IGTDS): User's Manual", General Motors Technical Center, Warren, MI (1978).

Gibert A., "A method for the traffic assignment problem when demand is elastic", unpublished report LBS TNT-B5 (1968).

Golden B., "A miunimum cost multi-commodity network flow problem concerning imports and exports", *Network* 5 (1975) 331-356.

Guélat J., Florian M. and Crainic T.G., "A multimode multiproduct network assignment model for strategic planning of freight flows", Publication #549, Centre de recherche ur les transports, Université de Montréal (March 1987) to appear in *Transportation Science*.

Guélat J. and Marcotte P., "Some comments on Wolfes's 'away step' ". *Mathematical Programming* 35 (1986) 110-119.

Hall M.D., Van Viet D. and Willumsen L.G., "SATURN–A simulation assignment model for the evaluation of traffic manegement schemes", *Traffic Engineering and Control* (1980) 168-176.

Hasselström D., "Public Transportation Planning - A Mathematical Programming Approach", Ph.D. *Thesis*, Department of Business Administration, University of Gothenburg, Sweden (1981).

Hearn D.W., "The gap function of a convex program", *Operations Research Letters* 1 (1982) 67-71.

Hearn D.W. and Nguyen S., "Dual and saddle functions related to the gap function", Research Report 82-4, Department of Industrial and Systems Engeneering, University of Florida (Gainesville, 1982).

Holloway C.A., "An Extension of the Frank and Wolfe method of feasible directions", *Mathematical Programming* 6 (1974) 14-27.

Jones P.S. and Sharp G.P., "Multi-Mode Itercity Freight Transportation Planning for Underdeveloped Regions", Proceedings of the 20th Annual Meeting Transportation Research Forum (1979).

Josephy N., "Newton's method for generalized equations", Technical Report 1965, Mathematics Research Center, University of Wisconsin (Madison, (1979).

Kinderlehrer D. and Stampacchia, *An Introduction to Variational Inequalities and Applications* (Academic Press, New York, 1980).

Knight F.H., "Some fallacies in the interpretation of social cost", *Quarterly Journal of Economics* 38 (1924) 582-606.

Kresge D.T. and Roberts P.O., "System Analysis and Simulation Models", *Techniques of Transportation Planning*, ed: Meyers Jr., Vol. 2, Brookings Institute, Washington D.C. (1971).

Lampkin W. and Saalmans P.D., "The Design of Routes, Services Frequencies and Schedules for a Municipal Bus Undertaking: A Case Study", *Operations Research Quarterly* Vol. 18, (1967) 375-397.

Last A. and Leak S.E., "Transept: a Bus Model", *Traffic Engineering and Control* (January 1976) 14-20.

Lawphongpanich S. and Hearn D.W., "Simplicial decomposition of the asymmetric traffic assignment problem", *Transportation Research* B 18B (1984) 123-133.

Le Blanc L.J., Morlok E.K. and Pierskalla W.P., "An efficient approach to solving the road network equilibrium traffic assignment problem", *Transportation Research* 5 (1975) 309-318.

Le Clercq F., "A Public Transport Assignment Method", *Traffic Engineering and Control* (June 1972) 91-96.

Leventhal T., Nemhauser G. and Trotter L.E., Jr., "A column generation algorithm for optimal traffic assignment", *Transportation Science* 7 (1973) 168-176.

Magnanti T.L., "Models and algorithms for predicting urban traffic equilibrium", in: Florian M., ed., *Transportation Planning Models* (North-Holland, Amsterdam, 1984) pp. 153-186.

Mandl C.E., "Evaluation and Optimization of Urban Public Transportation Networks", presented at the 3rd European Congress on Operations Research (Amsterdam, Netherlands, 1979).

McGinnis L.F., Sharp G.P. and Yu D.H.C., "Procedures for Multi-State, Multi-Mode analysis: Vol. IV", Transportation Modeling and Analysis, U.S., D.O.T report DOT-OST-80050-17/V.N., (1981).

Nguyen S., "An algorithm for the traffic assignment problem", *Transportation Science* 8 (1974) 203-216.

Nguyen S., "A unified approach to equilibrium methods for traffic assignment", in: Florian M., ed., *Traffic equilibrium methods*, Lecture Notes in Economics and Mathematical Systems 118 (Springer-Verlag, Berlin, 1976) pp. 148-182.

Nguyen S. and Dupuis C., "An efficient method for computing traffic equlibria in network with asymmetric transportation costs", *Transportation Science* 18 (1984) 185-232.

Pang J.-S., "Asymmetric variational inequality problems over product sets: applications and iterative methods", *Mathematical Programming* 31 (1985) 206-219.

Pang J.-S. and Chan D., "Iterative methods for variational and complementary problems", *Mathematical Programming* 24 (1982) 284-313.

Pang J.-S. and Yu C.S., "Linearized simplicial decomposition methods for computing traffic equilibrium on networks", *Networks* 14(3) (1984) 427-438.

Rapp M.H., Mattenberger P., Piguet S. and Robert-Grandpierre A., "Interactive Graphic System for Transit Route Optimization", *Transportation Research Record* 619 (1976).

Samuelson P.A.,"Spatial Price Equilibrium and Linear Programming", *American Economic Review* 42 (1952) 283-303.

Scheele C.E, "A Mathematical Programming Algorithm for Optimal Bus Frequencies", Institute of Technology, University of Linköping, Sweden (1977).

Sharp G.P., "A Multi-Commodity Intermodal Transportation Model", Proceedings of the 20th Annual Meeting Transportation Research Forum (1979).

Smith M.J., "Existence, uniqueness and stability of traffic equilibria", *Transportation Research* B 1B (1979) 295-304.

Soumis F., "Planification d'une flotte d'avions", Publication #133, Centre de Recherche sur les Transports, Université de Montréal (1978).

Spiess H., "Contributions a la théorie et aux outils de planification des réseaux de transport urbain", Ph.D. *thesis*, Département d'informatique et de recherche opérationnelle, Université de Montréal (1984).

Spiess H. and Florian M., "Optimal strategies: a new algorithm for transit assignment", To appear in Transportation Research B (1988).

Takayama T.and Judge G.G, "Alternative Spatial Price Equilibrium Models", *Journal of Regional Science* 10 (1970) 1-12.

Takayama T.and Judge G.G, "Equilibrium among Spatially Separated Markets: A Reformulation", *Econometrica* 32 (1964) 510-524.

UMTA/FHWA, UTPS *Reference Manual*, U.S. Department of Transportation (1977).

Wardrop J., "Some theoretical aspects of road traffic research", in: *Proceeding of the Institute of Civil Engineers*, part II, Vol. 1 (1951) 325-378.

Weintraub A. and Gonzales J., "An algorithm for the traffic assignment problem", *Networks* 10 (1980) 191-210.

Wolfe P., "Convergence theory in nonlinear programming", in Abadie J., ed., *Nonlinear and integer programming* (North-Holland, Amsterdam, 1970) pp. 1-36.

On Automatic Differentiation[1]

Andreas GRIEWANK

Mathematics and Computer Science Division, Argonne National Laboratory, Argonne, IL 60439, U.S.A.

Abstract

In comparison to symbolic differentiation and numerical differencing, the chain rule based technique of automatic differentiation is shown to evaluate partial derivatives accurately and cheaply. In particular it is demonstrated that the reverse mode of automatic differentiation yields any gradient vector at no more than five times the cost of evaluating the underlying scalar function. After developing the basic mathematics we describe several software implementations and briefly discuss the ramifications for optimization theory and applications.

1 Introduction

In 1982 Phil Wolfe [31] made the following observation regarding the ratio between the cost of evaluating a gradient with n components and the cost of evaluating the underlying scalar function.

If care is taken in handling quantities which are common to the function and derivatives, the ratio is usually around 1.5, not $n+1$. [31]

The main purpose of this article is to demonstrate that Phil Wolfe's observation is in fact a theorem (with the average 1.5 replaced by an upper bound 5) and that *care* can be taken automatically. This remarkable result is achieved by one variant of *automatic differentiation* [25], which simply implements the chain rule in a suitable fashion. The same approach can be used to compute second and higher derivatives. At least since the fifties these techniques have been developed by computational scientists in various fields, and several software implementations are now available. Although a theorem confirming Wolfe's assertion for rationals was published in 1983 by Baur and Strassen [2], the optimization community took little notice of these developments. This can be partly explained by a lack of clarity in the customary terminology.

Automatic differentiation is often confused with symbolic differentiation or even with the approximation of derivatives by divided differences. For algebraically rather simple functions, the explicit derivative expressions obtained by symbolic differentiation may be readable to an experienced user and thus provide an extremely useful extension of research with pencil and paper. However, for functions of any complexity in more than three variables, the analytic expressions for gradient or Hessian tend to take up several pages and are unlikely to facilitate any insights.

[1]This work was supported by the Applied Mathematical Sciences subprogram of the Office of Energy Research, U.S. Department of Energy, under contracts W-31-109-Eng-38.

M. Iri and K. Tanabe (eds.), Mathematical Programming, 83–107.
© *1989 by KTK Scientific Publishers, Tokyo.*

In this article we will concentrate on the goal of obtaining numerical derivative values at given arguments. The need for efficient and accurate derivative evaluations arises in particular during the iterative solution of nonlinear problems and the subsequent sensitivity analysis. Following several other authors, notably Iri [15], we will argue that **for these numerical purposes the reverse mode of automatic differentiation is far superior to symbolic differentiation or divided difference approximations.** The latter technique is always less accurate and about as costly as the forward form of automatic differentiation.

The paper is organized as follows. The remainder of this Section we briefly discuss the historical development and applications of automatic differentiation. In Section 2 we utilize two simple example functions to illustrate the characteristic properties of various techniques for evaluating gradients. In Section 3 we develop the two modes of automatic differentiation for the general case and conclude that the cost of evaluating gradients in the reverse mode is additive with respect to function composition. As a corollary we obtain Wolfe's assertion with 1.5 replaced by the uniform bound 5. Section 4 describes several implementations of automatic differentiation that require the user to do little more than provide a subroutine for the evaluation of the underlying function. In the final Section 5 we briefly discuss the implications of automatic differentiation on the design and selection of optimization algorithms.

The literature relating to automatic differentiation is extensive and very diverse. The main stream of research and implementation has been concerned with the automatic evaluation of gradients (or more generally truncated Taylor series) in the forward mode. This effort goes back at least to Beda et al [3] in the Soviet Union and Wengert [30] in the United States. Numerous other references are contained in the paper by Kedem [21], the books by Rall [25] and Kagiwada et al [19], and the recent report by Fischer [11]. In general the researchers in this main stream were unaware of the reverse mode or continued to consider it as a somewhat obscure approach of a rather theoretical nature.

Mathematically the reverse mode is closely related to adjoint differential equations. Nuclear engineers have long used *adjoint sensitivity analysis* [4], [5] to evaluate the partial derivatives of certain system responses (e.g. the reactor temperature) with respect to thousands of design parameters. This approach yields all sensitivities simultaneously at a cost comparable to only a few reactor simulations. In contrast, thousands of these lengthy calculations would be needed to approximate all sensitivities by divided differences. For a recent survey on the software and applications in this field see the paper by Worley [32]. Similarly, in atmospheric and oceanographic research, adjoints of the governing partial differential equations have been used to obtain the gradients of residual norms with respect to initial conditions and other unknown quantities [29]. Here the residuals represent discrepancies between observed and predicted conditions in the atmosphere or ocean. Even though these 3D calculations may involve millions of variables, the gradient of the sum of squares can be obtained at essentially the same cost as an evaluation of the residual vector itself. In order to avoid any storage and manipulation of matrices the gradient is then utilized in a conjugate gradient like minimization routine.

Apparently the first general purpose implementation of the reverse mode was the precompiler JAKE due to Speelpenning. In his unpublished thesis [28] Speelpenning showed that Wolfe's assertion is true, but did not state it formally. His original intention was to optimize the gradient code generated in the forward mode by sharing common expressions.

During this attempt he realized that the optimal gradient code can be obtained directly without any optimization by (what we call here) the reverse mode of automatic differentiation. Several other papers proposing the reverse or *top down* mode are referenced in the survey [17]. This excellent article discusses also the closely related issue of estimating evaluation errors. Now let us examine various techniques for evaluating gradients on a couple of simple problems.

2   Comparisons on two Examples

The use of a cubic equation of state [24] yields the Helmholtz energy of a mixed fluid in a unit volume at the absolute temperature T as

$$f(x) = RT \sum_{i=1}^{n} x_i \log \frac{x_i}{1 - b^T x} - \frac{x^T A x}{\sqrt{8 b^T x}} \log \frac{1 + (1 + \sqrt{2}) b^T x}{1 + (1 - \sqrt{2}) b^T x} \quad ,$$

where R is the universal gas constant and

$$0 \leq x, b \in \mathbf{R}^n \quad , \quad A = A^T \in \mathbf{R}^{n \times n}.$$

During the simulation of an oil reservoir this function and its gradient have to be evaluated at thousands of points in space and time. Typically the number of fluid components n is restricted to less than 20, but we will include larger values in our comparative timings.

2.1   MACSYMAl Results on the Helmholtz Energy

First let us examine the results of symbolic differentiation with MACSYMA, version 309, distributed by Symbolics Inc. After entering $f(x)$ and computing its gradient with the *diff* command one may translate the symbolic representations into FORTRAN using the *fortran* command. On the following page we list the resulting code for the evaluation of $f(x)$ and the first component of its gradient when $n = 5$. Actually the original code had to be modified, mainly because it contained more than the maximum of 19 continuation lines allowed in FORTRAN 77. Due to our familiarity with the function we could break the expression for the first gradient component $g(1)$ in the middle, but in general that would be a rather challenging task. Even after this problem and some type conflicts in the original code were overcome the results are clearly unimpressive. Just imagine this code segment had been inserted into a subroutine and subsequently the programmer made a trivial editing error. Then it would be quite difficult to determine by inspection whether the segment had been corrupted and nearly impossible to correct it. In other words the code is not only inefficient but unmaintainable.

While some aspects of MACSYMA's FORTRAN interface are annoying, they are by no means the root of our problems. The main culprit is the wrong-headed idea of generating separate expressions for the function and each gradient component, directly in terms of the independent variables. By definition this approach eliminates any possibility of utilizing common expressions during the evaluation. **Instead one should write a program for evaluating the function efficiently and then generate an extended program that evaluates the function and gradient simultaneously.** As we will see later the extended program can be generated automatically.

MACSYMA generated FORTRAN code for Helmholtz energy and first partial

```
RUTU=DSQRT(2.D0)
    F=0.0013564*(-(x(5)+x(4)+x(3)+x(2)+x(1))*DLOG(-b(5)*x(5)-b(4)*x(
 1   4)-b(3)*x(3)-b(2)*x(2)-b(1)*x(1)+1)+x(5)*DLOG(x(5))+x(4)*DLOG
 2   (x(4))+x(3)*DLOG(x(3))+x(2)*DLOG(x(2))+x(1)*DLOG(x(1)))-(x(5
 3   )*(x(5)*a(5,5)+x(4)*a(5,4)+x(3)*a(5,3)+x(2)*a(5,2)+x(1)*a(
 4   5,1))+x(4)*(a(4,5)*x(5)+x(4)*a(4,4)+x(3)*a(4,3)+x(2)*a(4,2
 5   )+x(1)*a(4,1))+x(3)*(a(3,5)*x(5)+a(3,4)*x(4)+x(3)*a(3,3)+x
 6   (2)*a(3,2)+x(1)*a(3,1))+x(2)*(a(2,5)*x(5)+a(2,4)*x(4)+a(2,
 7   3)*x(3)+x(2)*a(2,2)+x(1)*a(2,1))+x(1)*(a(1,5)*x(5)+a(1,4)*
 8   x(4)+a(1,3)*x(3)+a(1,2)*x(2)+x(1)*a(1,1)))*DLOG(((RUTU+1
 9   )*(b(5)*x(5)+b(4)*x(4)+b(3)*x(3)+b(2)*x(2)+b(1)*x(1))+1)/(
 :   (1-RUTU)*(b(5)*x(5)+b(4)*x(4)+b(3)*x(3)+b(2)*x(2)+b(1)*
 ;   x(1))+1))/(b(5)*x(5)+b(4)*x(4)+b(3)*x(3)+b(2)*x(2)+b(1)*x( 1))
  g(1)=b(1)*(x(5)*(x(5)*a(5,5)+x(4)*a(5,4)+x(3)*a(5,3)+x(2)*a(5,2)
 1   +x1(1)*a(5,1))+x(4)*(a(4,5)*x(5)+x(4)*a(4,4)+x(3)*a(4,3)+x(2)
 2   *a(4,2)+x(1)*a(4,1))+x(3)*(a(3,5)*x(5)+a(3,4)*x(4)+x(3)
 3   *a(3,3)+x(2)*a(3,2)+x(1)*a(3,1))+x(2)*(a(2,5)*x(5)+a(2,4)*x(4
 4   )+a(2,3)*x(3)+x(2)*a(2,2)+x(1)*a(2,1))+x(1)*(a(1,5)*x(5)+a
 5   (1,4)*x(4)+a(1,3)*x(3)+a(1,2)*x(2)+x(1)*a(1,1)))*DLOG(((RUTU
 6   +1)*(b(5)*x(5)+b(4)*x(4)+b(3)*x(3)+b(2)*x(2)+b(1)*x(1)
 7   )+1)/((1-RUTU)*(b(5)*x(5)+b(4)*x(4)+b(3)*x(3)+b(2)*x(2)
 8   +b(1)*x(1))+1))/(b(5)*x(5)+b(4)*x(4)+b(3)*x(3)+b(2)*x(2)+b
 9   (1)*x(1))**2-(x(5)*a(5,1)+a(1,5)*x(5)+x(4)*a(4,1)+a(1,4)*x
 7   (4)+x(3)*a(3,1)+a(1,3)*x(3)+x(2)*a(2,1)+a(1,2)*x(2)+2*x(1)
 7   *a(1,1))*DLOG(((RUTU+1)*(b(5)*x(5)+b(4)*x(4)+b(3)*x(3)+b
 7   (2)*x(2)+b(1)*x(1))+1)/((1-RUTU)*(b(5)*x(5)+b(4)*x(4)+b
 1   (3)*x(3)+b(2)*x(2)+b(1)*x(1))+1))/(b(5)*x(5)+b(4)*x(4)+b(3
 6   )*x(3)+b(2)*x(2)+b(1)*x(1))
  g(1)=g(1)+0.0013625*(-DLOG(-b(5)*x(5)-b(4
 7   )*x(4)-b(3)*x(3)-b(2)*x(2)-b(1)*x(1)+1)+DLOG(x(1))+b(1)*(x(
 7   5)+x(4)+x(3)+x(2)+x(1))/(-b(5)*x(5)-b(4)*x(4)-b(3)*x(3)-b(
 7   2)*x(2)-b(1)*x(1)+1)+1)-((1-RUTU)*(b(5)*x(5)+b(4)*x(4)+
 7   b(3)*x(3)+b(2)*x(2)+b(1)*x(1))+1)*((RUTU+1)*b(1)/((1-RUTU
 7   )*(b(5)*x(5)+b(4)*x(4)+b(3)*x(3)+b(2)*x(2)+b(1)*x(1))
 7   +1)-(1-RUTU)*b(1)*((RUTU+1)*(b(5)*x(5)+b(4)*x(4)+b(3
 7   )*x(3)+b(2)*x(2)+b(1)*x(1))+1)/((1-RUTU)*(b(5)*x(5)+b(4
 7   )*x(4)+b(3)*x(3)+b(2)*x(2)+b(1)*x(1))+1)**2)*(x(5)*(x(5)*a
 7   (5,5)+x(4)*a(5,4)+x(3)*a(5,3)+x(2)*a(5,2)+x(1)*a(5,1))+x(4
 7   )*(a(4,5)*x(5)+x(4)*a(4,4)+x(3)*a(4,3)+x(2)*a(4,2)+x(1)*a(
 7   4,1))+x(3)*(a(3,5)*x(5)+a(3,4)*x(4)+x(3)*a(3,3)+x(2)*a(3,2
 7   )+x(1)*a(3,1))+x(2)*(a(2,5)*x(5)+a(2,4)*x(4)+a(2,3)*x(3)+x
 7   (2)*a(2,2)+x(1)*a(2,1))+x(1)*(a(1,5)*x(5)+a(1,4)*x(4)+a(1,
 7   3)*x(3)+a(1,2)*x(2)+x(1)*a(1,1)))/(b(5)*x(5)+b(4)*x(4)+b(
 7   3)*x(3)+b(2)*x(2)+b(1)*x(1))*((RUTU+1)*(b(5)*x(5)+b(4)*
 7   x(4)+b(3)*x(3)+b(2)*x(2)+b(1)*x(1))+1))
```

Everything may be done *by hand* on our second example

$$f(x) \equiv \prod_{i=1}^{n} x_i = x_1 \cdot x_2 \cdots x_{n-1} \cdot x_n$$

which was already used by Speelpenning [28]. Obviously the $i - th$ component of the gradient $\nabla f(x)$ is given by

$$\partial f / \partial x_i = \prod_{j \neq i} x_j = x_1 \cdots x_{j-1} \cdot x_{j+1} \cdots x_n$$

If calculated in this form each gradient component involves $n - 1$ multiplications and is thus almost as expensive to evaluate as the function f itself. Since symbolic differentiators generate separate algebraic expressions for each component of $\nabla f(x)$ they require exactly n times as many arithmetic operations for evaluating function and gradient jointly as for evaluating the function by itself. Formally we may write $q\{f\} = n$, where

$$q\{f\} \equiv work\{f, \nabla f\} / work\{f\} \quad .$$

Since the work ratio $q\{f\}$ is even slightly larger for divided differences this may at first seem a fair price to pay. However, according to Wolfe's assertion we should be able to do a lot better, namely to bound $q\{f\}$ by a constant independent of n.

2.2   Automatic Differentiation of the Product Example

In order to obtain the gradient cheaply one could use the identity

$$\partial f(x) / \partial x_i = f(x) / x_i \quad if \quad x_i \neq 0 \quad .$$

Unfortunately, this 'solution' suggests that the efficient evaluation of gradients involves some special cancellations, which have to be detected by human inspection and require numerical exception handling when certain denominators are zero or small. Fortunately, for this example and other cases, **the gradient can be evaluated efficiently without any human intervention or numerical instabilities.**

In order to discuss the alternative methods we have to base the evaluation of the function and its gradient on sequential programs. Using an informal programming language we can evaluate $y = f(x)$ by the following code.

Evaluation of Product

$$x_{n+1} = x_1$$
$$For \ i = n + 2, n + 3 \ldots 2n$$
$$x_i = x_{i-n} \, x_{i-1}$$
$$y = x_{2n}$$

Here and throughout the paper we will allocate all scalar quantities in a single memory vector $\langle x_i \rangle_{i=1\ldots m}$, starting with the independent variables $\langle x_i \rangle_{i=1\ldots n}$ and ending with a single dependent variable x_m. The issue of the storage requirements for actual computer implementations will be discussed in Section 4.

Since the intermediate quantities $x_{n+i}, i = 1 \ldots n$ are smooth functions they possess gradients $\nabla x_{n+i}, i = 1 \ldots n$ with respect to the independent variables $x_1, x_2, \ldots, x_n$. In particular we have $\nabla x_{2n} = g \equiv \nabla f$ and $\nabla x_{n+1} = e_1$. Evaluating the intermediate gradients by the chain rule we obtain the following expanded program.

Forward Differentiation of Product

$$
\begin{aligned}
x_{n+1} &= x_1 \\
\nabla x_{n+1} &= e_1 \\
For \quad i &= n+2, n+3 \ldots 2n \\
x_i &= x_{i-n}\, x_{i-1} \\
\nabla x_i &= x_{i-1}\, e_{i-n} + x_{i-n}\, \nabla x_{i-1} \\
y \quad &= x_{2n} \\
g \quad &= \nabla x_{2n}
\end{aligned}
$$

This program evaluates both function and gradient simultaneously. It can be generated in a 'mechanical' fashion and is only about twice as long as the original program because each assignment to an intermediate quantity is simply augmented by the calculation of its gradient. This *forward* approach has been developed and advocated by several authors (See e.g. [3], [30], [21], and [25]). Various software implementations will be discussed in Section 4.

A simple count reveals that the calculation of our example gradient by the program above involves $\frac{1}{2}n^2$ nontrivial multiplications, so that $q \simeq n/2$. In general we must expect that the forward mode of automatic differentiation increases the number of arithmetic operations by the factor n, because each evaluation of an intermediate scalar quantity x_i is accompanied by the calculation of the corresponding gradient vector ∇x_i. Apparently Speelpenning was the first to notice that, instead of the gradient vector, only another scalar, say $\bar{x}_i$, needs to be associated with each quantity $x_i, i = 1 \cdots 2n$. In case of the product example one may define $\bar{x}_{n+i}$ as the product of all x_j with $i < j \leq n$ and then set

$$
\partial f / \partial x_i = \bar{x}_i \equiv x_{n+i-1}\, \bar{x}_{n+i}.
$$

This calculation is performed by the following extended program.

Reverse Differentiation of Product

$$
\begin{aligned}
x_{n+1} &= x_1 \\
For\ i &= n+2, n+3, \ldots, 2n \\
x_i &= x_{i-n}\, x_{i-1} && \{Forward\ Sweep\} \\
y \quad &= x_{2n} \\
\bar{x}_{2n} &= 1 \\
For\ i &= 2n, 2n-1, \ldots, n+2 \\
\bar{x}_{i-1} &= \bar{x}_i\, x_{i-n} && \{Reverse\ Sweep\} \\
\bar{x}_{i-n} &= \bar{x}_i\, x_{i-1} \\
\bar{x}_1 \quad &= \bar{x}_{n+1} \\
g \quad &= \langle \bar{x}_i \rangle_{i=1 \ldots n}
\end{aligned}
$$

This algorithm requires $3n - 3$ multiplications in order to compute the function and its gradient, so that now $q \simeq 3$. Note that there is no need for any branching when one of

Div. Diff.	Symbolic	Forward	Reverse 1	Reverse 2
6	2.0	1.5	1.00	6.80
11	9.8	2.1	1.66	4.66
21	22	3.8	1.94	3.46
31	-	5.2	2.04	3.95
41	-	7.6	2.67	3.65
51	-	-	2.88	3.82
61	-	-	-	3.76
71	-	-	-	3.80
81	-	-	-	3.83
FORTRAN	MAPLE	PASCAL-SC	PASCAL-SC	JAKEF

Table 1: Observed work ratios on Helmholtz energy for $n = 5, 10, 20, \ldots, 80$.

the variables is small. The amazing fact is that this apparently tricky algorithm for the gradient of a product can be obtained by a general, straight-forward transformation from the original function evaluation program.

2.3   Experimental Comparison on Helmholtz Energy

Before discussing the details of this transformation in the following sections, let us list some empirically observed values for the work ratio $q\{f\}$ on our first example. The numbers in Table 1 represent the ratio between the execution times of an extended program that evaluates $f(x)$ and $\nabla f(x)$ jointly and of the original program that evaluates only $f(x)$ at a given argument. The entries in the first column represent the work ratio for divided differences, namely $n + 1$ with n being the number of variables. The three numbers in the second column were obtained as follows. The Helmholtz energy function $f(x)$ was entered into the algebraic manipulation package MAPLE [6] and then differentiated symbolically using the *grad* command. On a Sun 3/140 with 16 megabytes real memory, the symbolic generation of the gradient always took several minutes, and when n was set to 30 the differentiation failed after 15 minutes due to a lack of memory space. The time for this process was not included in the listed work ratios, which reflect only the times needed to substitute the indeterminates x_i by real arguments in the expressions for $f(x)$ and $\nabla f(x)$. For example when $n = 20$ the substitution took 7.13 and 160 seconds CPU time respectively.

The results in the third and forth column were obtained on an IBM XT using the programming language PASCAL-SC [22]. Like other modern languages this extension of standard PASCAL allows the transformation of a program for the evaluation of $f(x)$ into one that evaluates $f(x)$ and $\nabla f(x)$ by a process called *operator overloading*. This approach was first implemented by Rall [26],[27] in the forward mode of automatic differentiation. We have implemented the same approach in the reverse mode as described in Section 4. Again the entries in the table do not include the compilation times for the original and extended programs but represent the ratios of the respective execution times. The forth column was obtained in almost the same way, except that the original program was written in FORTAN and then extended to the gradient routine by the precompiler JAKEF [14] (an update of Speelpennings original version JAKE [28]). The resulting pair of FORTRAN programs was run on the Sun 3 so that the execution times were naturally much smaller than those of the PASCAL-SC programs on the IBM XT. Nevertheless the comparison between runtime ratios provides some meaningful information.

The observed work ratios grow linear in the number of variables n, for divided differences, symbolic differentiation, and in the forward mode of automatic differentiation. However, in the last case the proportionality factor is only about .2 compared to 1.0 in case of the popular divided differences. The reverse mode of automatic differentiation in PASCAL-SC is always faster than the corresponding forward scheme, and the work ratio seems indeed uniformly bounded in n. The same is true for the FORTRAN version of reverse accumulation, though there the ratios are initially somewhat larger. Due to the limitation to 512K core memory, the forward and reverse implementation in PASCAL-SC can handle the Helmholtz energy function only up to 40 and 50 variables respectively. MAPLE exhausts the many times larger memory on the Sun much earlier. On the basis of our experience with MACSYMA and MAPLE we conclude that symbolic manipulators cannot be considered suitable tools for our purposes. Finally we note that a carefully handcoded routine for evaluating suitable representations of the first four derivative tensors requires only about 1.5 times the computing time of evaluating the Helmholtz energy by itself.

3   Automatic Differentiation of Composite Functions

3.1   Composite Functions and their Computational Graph

Throughout this section we consider a function $y = f(x) : \mathbf{R}^n$ that is defined by a given sequential program of the following form.

Original Program

$$For \ i= n+1, n+2, \ldots, m$$
$$x_i = f_i \langle x_j \rangle_{j \in \mathcal{J}_i}$$
$$y \ = x_m$$

Here the *elementary functions* f_i depend on the already computed quantities x_j with j belonging to the index sets

$$\mathcal{J}_i \subset \{1, 2, \ldots, i-1\} \quad for \quad i = n+1, n+2, \ldots, m$$

In other words f is the composition of $m - n$ elementary or *library* functions f_i, whose gradients

$$\nabla f_i = \langle \partial f_i / \partial x_j \rangle_{j \in \mathcal{J}_i}$$

are assumed to be computable at all arguments of interest.

For example, this is clearly the case when all f_i represent either elementary arithmetic operations, i.e. + , - , * and / or nonlinear system functions of a single argument, e.g.

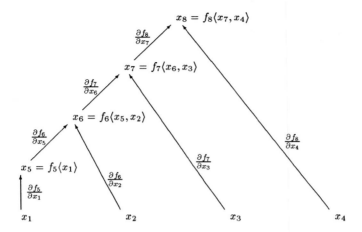

Figure 1: Graph for Product, where $f_5\langle x_1 \rangle = x_1$ and $f_i \langle x_j, x_k \rangle = x_j * x_k$ for $i = 6, 7, 8$.

logarithms, exponentials and trigonometric functions. Almost all scalar functions of practical interest can be represented in this *factorable* form, which has been used extensively by McCormick et al. [18]. Rather than restricting ourselves to unary and binary elementary functions we allow for any number of arguments $n_i \equiv |\mathcal{J}_i| < i$, where $|\cdot|$ denotes cardinality. In particular we may trivially interpret any function $f(x)$ as a composition of itself so that in the program above $f_{n+1} = f$ and $m = n + 1, n_m = n$. More importantly, this general framework allows for user defined subroutines.

Sometimes it is very helpful to visualize the original program as a *computational graph* with the vertex set $\{x_i\}_{1 \leq i \leq m}$. An arc runs from x_j to x_i exactly if j belongs to $\mathcal{J}_i$. With each arc one may associate the value of the corresponding partial derivative $\partial f_i / \partial x_j$. Because of the restriction on $\mathcal{J}_i$ one obtains an acyclic graph, whose minimal elements are the independent variables. Usually there are several linear orderings of the x_i that are compatible with the partial ordering induced by the directed graph. Whenever two elementary functions do not directly or through intermediaries depend on each others result, they can be evaluated in either order or even concurrently on a parallel machine. This aspect has been examined in [9], but will not be pursued any further here. Also, in contrast to the analysis in [15], we will not use the graph structure for our complexity bounds.

For any reasonable measure of computational work on a serial machine we may assume that

$$work\{f\} = \sum_{i=n+1}^{m} work\{f_i\} \quad .$$

In defining $work\{f\}$ one may account for the number of certain arithmetic operations as well as fetches and stores from and to memory. Now let us develop the extended programs for evaluating the gradient ∇f jointly with f.

3.2   Automatic Differentiation with Forward Accumulation

Again denoting by ∇x_i the gradient of x_i with respect to the independent variables $\langle x_j \rangle_{j=1\ldots n}$ we derive from the original program by the chain rule:

Forward Extension

$$
\begin{aligned}
&For\ i = 1,2\ldots n\\
&\qquad \nabla x_i = e_i\\
&For\ i = n+1, n+2, \ldots m\\
&\qquad x_i = f_i \langle x_j \rangle_{j \in \mathcal{J}_i}\\
&\qquad \nabla x_i = \sum_{j \in \mathcal{J}_i} \frac{\partial f_i}{\partial x_j} \nabla x_j\\
&y \quad = x_m\\
&g \quad = \nabla x_m
\end{aligned}
$$

where e_i denotes the $i - th$ Cartesian basis vector in $\mathbf{R}^n$.

Due to the assumed additivity of the work measure we find that

$$
work\{f, \nabla f\} = \sum_{i=n+1}^{m} [work\{f_i, \nabla f_i\} + n\, n_i(mults + adds)] \quad ,
$$

where the extra $n\, n_i$ arithmetic operations are needed to form ∇x_i as a linear combination of the n_i gradient vectors ∇x_j with $j \in \mathcal{J}_i$. Here we have neglected the fact that for j just above n, the gradient vectors ∇x_j will be sparse so that some arithmetic operations operations could theoretically be avoided. However, the added complexity of a suitable sparse implementation is unlikely to be justified by the savings, except in very special cases. Another possible alternative is to run through the basic loop n times, each time only evaluating the partial derivatives $\partial x_i / \partial x_j$ with respect to one particular independent variable x_j. This implementation of forward accumulation is considerably less economical in terms of computational effort but requires only about twice as much storage as the original program. We will not consider this space saver solution in the remainder of the paper.

Now suppose that the evaluation of any library function f_i requires at most $c\, n_i$ arithmetic operations, where c is a common positive constant. Then it follows from the last equation that the work ratio defined above satisfies $q\{f\} \geq 1 + n/c$. This linear growth in the number of variables was clearly observed on the Helmholtz example and is not acceptable for large problems.

3.3   Automatic Differentiation with Reverse Accumulation

In order to obtain a method with a uniformly bounded work ratio we associate with each intermediate variable x_i the scalar derivative

$$
\bar{x}_i \equiv \partial x_m / \partial x_i
$$

rather than the gradient vector ∇x_i. By definition we have $\bar{x}_m = 1$ and for $i = 1 \ldots n$

$$
\partial f(x) / \partial x_i = \bar{x}_i \quad .
$$

As a consequence of the chain rule it can be shown (see e.g. [20]) that these *adjoint* quantities satisfy the relation

$$\bar{x}_j = \sum_{i \in \mathcal{I}_j} \frac{\partial f_i}{\partial x_j} \bar{x}_i \quad ,$$

where $\mathcal{I}_j \equiv \{i \leq m : j \in \mathcal{J}_i\}$. Thus we see that $\bar{x}_j$ can be computed once all $\bar{x}_i$ with $i > j$ are known. In terms of the program structure it is slightly more convenient to increment all $\bar{x}_j$ with $j \in \mathcal{J}_i$ for a known i by the appropriate contribution $\bar{x}_i \, \partial f_i / \partial x_j$. This mathematically equivalent looping leads to the following extended program.

Reverse Extension

$$
\begin{aligned}
&For \quad i = n+1, n+2, \ldots\ldots, m \\
&\qquad\qquad x_i = f_i\langle x_j \rangle_{j \in \mathcal{J}_i} \qquad\qquad\qquad \{\text{Forward Sweep}\} \\
&\qquad\qquad \bar{x}_i = 0 \\
&\quad y \quad\ = x_m \\
&\quad \bar{x}_m \ = \gamma \\
&\quad \langle \bar{x}_i \rangle_{i=1}^n = \bar{g} \\
&For \quad i = m, m-1, \ldots, n+1 \\
&\qquad\qquad \bar{x}_j = \bar{x}_j + \frac{\partial f_i}{\partial x_j} \bar{x}_i \quad for\ all\ j \in \mathcal{J}_i \qquad \{\text{Reverse Sweep }\} \\
&\quad g \quad\ = \langle \bar{x}_i \rangle_{i=1}^n
\end{aligned}
$$

When the initial vector $\bar{g}$ is set to zero and γ equals one, then the resulting vector g is simply the gradient ∇f. Otherwise we obtain for exactly the same computational effort the more general result

$$g = \bar{g} + \gamma \, \nabla f(x) \quad .$$

In other words the above program can increment a certain multiple of the gradient ∇f to a given vector $\bar{g}$ of the same length. This is exactly the operation we have to perform for each elementary function in the reverse extension. Hence we have additivity of the computational work in that

$$work\{f, \bar{g} + \gamma \, \nabla f\} = \sum_{i=n+1}^{m} work\,\{f_i, \bar{g}_i + \gamma_i \, \nabla f_i\}$$

for arbitrary scalars γ_i and vectors $\bar{g}_i$ of length n_i. After division by the last equation of Subsection 3.1 one finds by elementary arguments that

$$Q\{f\} \equiv \frac{work\,\{f, \bar{g} + \gamma \, \nabla f\}}{work\{f\}} \leq \max_{n < i \leq m} Q\{f_i\} \quad ,$$

Note that $Q\{f\}$ is slightly larger than the work ratio $q\{f\}$ defined in Subsection 2.1. This means that the work ratio for f is bounded above by the worst ratio for any of the library functions f_i, which is clearly independent of the total number of variables n. In other words **the set of functions f for which the work ratio $Q\{f\}$ does not exceed a certain bound $\bar{Q}$ is closed with respect to composition.** This rather surprising result holds for a wide range of work functionals, provided memory space is unlimited and free. However as was mentioned above, memory access, i.e. fetches and stores, may be included as costs.

Now suppose the f_i are restricted to the elementary arithmetic operations and standard univariate functions on a modern mainframe. For sine and cosine the work ratio lies just above two, and for all other system functions it is close to 1, because their derivatives come practically free once the function itself has been evaluated. Assuming that an addition is cheaper than a multiplication and a division costs at least 50% more than a multiplication, one finds that the largest work ratio is attained for the multiplication function $f_i(x_1, x_2) \equiv x_1 * x_2$. Therefore we may use the upper bound

$$\bar{Q} \equiv Q\{x_1 * x_2\} = \frac{3\ mults + 2\ adds + 5\ fetches + 3\ stores}{1\ mult + 2\ fetches + 1\ store} \leq 5 \quad .$$

Thus we can conclude that under quite realistic assumptions **the evaluation of a gradient requires never more than five times the effort of evaluating the underlying function by itself.** Obviously the bound of 5 is somewhat pessimistic and one might expect to incur an even smaller penalty for evaluating the gradient in practice. This was found to be true in our experiments on the Helmholtz example. On the other hand the extended program may involve communications overhead, e.g. extra subroutine calls, that is not included in our work measure.

While the reverse mode is clearly superior to the forward mode in terms of computational effort, it may require a lot more storage than the latter. As coded in Subsection 3.2 the forward extension associates with each scalar variable of the original program a gradient vector of length n. Hence the storage requirement grows by the predictable factor $n+1$. This is true even is some variables are repeatedly updated during the function evaluation. In that case the associated vectors can also be overwritten by the gradient of the latest value of the variable. For example in the product program of Subsection 2.2 one would normally not allocate n extra storage locations for the partial products $x_{n+i} = x_1 \ldots x_i$ but instead store them successively in the same place. Similarly all gradients ∇x_{n+i} in the corresponding extended program could be stored in a common n-vector.

In sharp contrast the reverse accumulation in Subsection 2.2 relies on all $n-1$ partial products x_{n+i} being still available after the final function value x_{2n} has been computed. Nevertheless, for this problem both modes require essentially the same storage, and on the Helmholtz energy function reverse accumulation uses slightly less space than forward accumulation. However, the difference in the memory requirement of the two methods can be much more dramatic.

3.4 Relations to Adjoints of Initial Value Problems

Suppose the evaluation of $f(x)$ involves the numerical solution of an initial value problem

$$y'(t) = F[y(t), t, x] \quad for \quad 0 \leq t \leq 1 \quad with \quad y(0) = y_0(x) \quad ,$$

where y has r components and y_0 is a smooth function of $x \in \mathcal{R}^n$. For a scheme with fixed step size h the result $y_h(1)$ will be a differentiable function of x. Provided f depends in turn smoothly on the final values $y(1)$, the whole evaluation procedure fits (for each fixed mesh) into our framework. For simplicity let us assume that $f(x) = w^T y(1)$ with some fixed weighting vector $w \in \mathbf{R}^r$. During the numerical integration of the initial value problem with a p-stage scheme, one only has to store p vectors of length r. In the forward mode the associated gradients would increase the storage requirement for this part of the program to

$n\,p\,r$ locations. In the reverse mode we have to keep track of all r/h intermediate values, which represent a discrete approximation of the solution function $y(t)$ *for* $0 \leq t \leq 1$.

Interestingly enough this is exactly the information one needs to calculate the gradient of $\nabla f(x)$ by solving the so called adjoint differential equation [23],

$$z'(t) = -F_y^T[y(t), t, x]z(t) \quad with \quad z(1) = w,$$

where F_y denotes the Jacobian of the right hand side with respect to y. Since the boundary conditions are terminal and the sign on the right hand side is reversed, this linear system has exactly the same stability and stiffness properties as the original initial value problem. The desired gradient is given by

$$\nabla f(x)^T = z(0)^T \frac{\partial y_0}{\partial x} + \int_0^1 z(t)^T F_x[y(t), t, x]dt \quad ,$$

where F_x denotes the Jacobian of the right hand side with respect to x. Thus we see that in the limiting continuous case, the evaluation of the gradient involves a definite integration based on the solution of an additional ODE with the same dimensions as the original initial value problem. Consequently the work ratio for appropriate discretizations should be close to 2 and certainly below 5.

In fact we may interpret reverse accumulation simply as a discrete analog of the classical adjoint equations from the calculus of variations and control theory [10]. Obviously the vector y need not be finite dimensional, and one can adopt the theoretical arguments and numerical techniques to more general evolution equations in Hilbert spaces.

In terms of consistency it is probably preferable to discretize only the forward integration and then to apply reverse accumulation without explicitly referring to the adjoint differential equation at all. On the other hand separate discretizations of the original and adjoint equation allow the usage of standard software, with automatic differentiation only being used to obtain the Jacobian of the right hand side [19]. With the benefit of hindsight one could also construct an 'optimal' spline representation of $y(t)$ in order to economize on storage, especially if the integrator is adaptive and involves many tentative evaluations. Apparently nobody has studied the relative merits and computational performance of these various options.

When the differential equation is solved using an adaptive grid the actually computed function is only piecewise differentiable. As for any program that includes branching depending on values of variables, **automatic differentiation will generally yield the derivative of the smooth piece containing the current argument**. Obviously this is the best one can achieve, whereas divided differences may yield completely meaningless results if taken across a crack of the actually computed function. In transforming the original program to the extended routine with automatic differentiation, all control statements are left unaltered. In effect this means that the form of the loop in the original program may become dependent on the current argument. As pointed out by Kedem [21] errors may arise when reals are tested for equality. For example the conditional assignment

$$if \quad x \neq 0 \quad then \quad y = (1 - \cos x)/x \quad else \quad y = 0$$

would lead to the derivative $\partial y/\partial x$ at $x = 0$ being automatically evaluated as 0 rather than the correct value $1/2$. Obviously the original programming leaves something to be desired in this particular example. In our implementation of the reverse mode in PASCAL-SC tests for equality involving real variables lead to warning messages.

3.5   Estimation of the Evaluation Error

The adjoint quantities $\bar{x}_i$ can be utilized to obtain good estimates of the total error in evaluating $f(x)$. Suppose one knows that the actually computed intermediate values $\tilde{x}_i$ satisfy for each $i > n$

$$|\tilde{x}_i - f_i \langle \tilde{x}_j \rangle_{j \in \mathcal{J}_i}| \leq \delta x_i \quad .$$

Moreover, let us assume that the discrepancies between the actual inputs $\langle \tilde{x}_i \rangle_{i=1...n}$ and their ideal values $\langle x_i \rangle_{i=1...n}$ are bounded by data tolerances $\langle \delta x_i \rangle_{i=1...n}$. Then one can expect that the actually computed final value $\tilde{x}_m$ satisfies

$$|\tilde{x}_m - f(x)| \leq \sum_{i=1}^{m} |\bar{x}_i| \, \delta x_i \quad .$$

As shown by induction in [1] this inequality must hold if all functions f_i are linear and the adjoint values $\bar{x}_i$ are exact. Even though these two assumptions are rather unrealistic the right hand side above was found in [17] to provide a usually somewhat pessimistic upper bound on the total error. In that paper the local error bounds δx_i were obtained from the machine precision of the computer in question. However, other sources of local error (such as discretizations, the approximation of a transcendental function by rationals or the uncertainty of certain problem parameters) could be accounted for as well.

Since the local evaluation errors are rarely correlated and usually unbiased, it makes sense to consider them as stochastically independent random variables with zero mean and standard deviations δx_i. This assumption implies that the standard deviation of $\tilde{x}_m - f(x)$ is simply the l_2-norm of the m-vector $\langle \bar{x}_i \, \delta x_i \rangle_{i=1...m}$ rather than the l_1 norm occuring on the right hand side above. Iri et al. found that this error estimate was somewhat tighter on their test problems. Either choice is certainly far superior to the ad hoc guesses that users currently have to make in order to specify tolerances for stopping criteria in iterative methods. Therefore these error estimates could be incorporated into optimization codes, to provide optimal solution accuracy without inconveniencing the user.

3.6   Extension to Higher Derivatives

In the forward mode the Hessian $\nabla^2 x_m$ of $x_m = f(x)$ can be obtained by updating for $i = n+1...m$

$$\nabla^2 x_i = \sum_{j \in \mathcal{J}_i} \left[\frac{\partial f_i}{\partial x_j} \nabla^2 x_j + \sum_{k \in \mathcal{J}_i} \nabla x_j \frac{\partial^2 f_i}{\partial x_j \partial x_k} (\nabla x_k)^T \right]$$

starting with $\nabla x_i = e_i$ and $\nabla^2 x_i = 0$ for $i = 1...n$. Similar chain rules of differentiation apply for third and higher derivative tensors. While the inclusion of these recursive relations into the original program provides in principle little difficulty, the resulting computational effort is at least of order $(m-n)n^p$, where p is the degree of the derivative tensor. In particular the evaluation the Hessian matrix in forward mode will usually be roughly n^2 times as expensive as the function itself.

Applying the complexity bound for the reverse mode separately to each component of the gradient one finds that

$$work\{\nabla^2 f\} \leq \sum_{i=1}^{n} work\left\{\nabla\left(\frac{\partial f}{\partial x_i}\right)\right\} \leq \bar{Q} \sum_{i=1}^{n} work\left\{\frac{\partial f}{\partial x_i}\right\} \quad .$$

After division by $work\{f\}$ we obtain in agreement with the results in [17] and [11]

$$\frac{work\{\nabla^2 f\}}{work\{f\}} \leq \bar{Q} \, \frac{\sum_{i=1}^{n} work\{\partial f/\partial x_i\}}{work\{\nabla f\}} \cdot \frac{work\{\nabla f\}}{work\{f\}} \leq n\bar{Q}^2 \quad .$$

In terms of powers of n this bound is unfortunately optimal, as one can see on the simple example

$$f(x) = .5[x^T x + (a^T x)^2] \, , \, \nabla f(x) = x + (a^T x)a \, , \, \nabla^2 f(x) = I + aa^T \quad .$$

Here the function and gradient involve both $2n$ multiplication, whereas the accumulation of the Hessian requires certainly $.5n^2$ multiplications.

Fortunately, it is often sufficient to calculate derivative vectors of the form

$$\begin{aligned} \nabla^{1+p} f(x)v_1 v_2 \ldots v_p &= \nabla[\nabla^p f(x)v_1 v_2 \ldots v_p] \\ &= v_p^T (\nabla[\nabla^p f(x)v_1 v_2 \ldots v_{p-1}]) \end{aligned}$$

where the n-vectors $v_j, j = 1 \ldots p$ are given directions. For example Hessian-vector products of the form $\nabla^2 f(x) \, v_1$ can be used in the conjugate gradient method (See e.g. [8] and [20]). Second and third derivatives of the form $\nabla^2 f(x)v_1 v_2$ and $\nabla^3 f(x)v_1 v_2 v_3$ characterize the quadratic and cubic turning points [12] of bifurcation theory. Moreover, the gradients of these scalars involve terms of the form $\nabla^3 f(x)\tilde{v}_1 \tilde{v}_2$ and $\nabla^4 f(x)\tilde{v}_1 \tilde{v}_2 \tilde{v}_3$, which need be evaluated during the calculation of the turning points by Newton's method. Selected second derivatives of the Lagrangian occur in the gradient of smooth exact penalty functions [7] for constrained optimization.

According to the second equation above, the desired vector of $p + 1 - st$ derivatives is the gradient of the dot product between v_p and an analogous vector of $p - th$ derivatives. Hence it may be computed recursively using $p+1$ sweeps of reverse gradient accumulation. This shows that evaluating the left hand side above should only be about 5^{1+p} times as costly as evaluating the scalar function f itself. Thus we have exponential growth in the order of the derivative p but still no dependence on the number of variables n.

4   Computer Implementations of Automatic Differentiation

So far we have not really justified the adjective *automatic* because all program transformations were carried out *by hand*. Moreover, we can certainly not expect that the scalar function $f(x)$ is supplied by the user in form of the Original Program in Section 3.1. Also, our specification of the reverse mode via the extended program in Subsection 3.3 is not complete, because the required partial derivatives may be evaluated either during the forward or the reverse sweep. Either variant has been implemented and yields certain advantages.

4.1   Immediate versus Delayed Differentiation

The first variant might be called immediate differentiation with reverse accumulation. Provided only first derivatives are required, every elementary function is linearized at its current arguments during the forward sweep, and only the computational graph with the nodes x_i and the arc values $\partial f_i / \partial x_j$ needs to be stored in a suitable fashion. Even the nodal values x_i are no longer required after the forward sweep, and they may be overwritten by the corresponding adjoint values $\bar{x}_i$ during the reverse sweep. User defined subroutines that return their gradient together with the function value are easily incorporated.

Similarly, if there are segments of code that produce only one or two scalar values for the subsequent calculations, the corresponding gradients can be preaccumulated in a local reverse sweep. In other words, these scalars may be interpreted as $super - elementary$ functions of the variables that enter into the segment, and their gradients can be computed during the forward sweep. This applies in particular to single assignment statements with complicated right hand sides, e.g.

$$x_3 = (x_1 + 3x_2)^2 + \sin^2 x_1 \exp(.2x_2) \quad .$$

Here the the representation of x_3 as a factorable function of x_1 and x_2 involves six unary functions and three binary arithmetic operations. Thus we have originally $12 = 6 + 2 * 3$ partial derivatives as arc values. Preaccumulation of the partial derivatives $\partial x_3 / \partial x_1$ and $\partial x_3 / \partial x_2$ would cut that number to 2. Another example is the product considered in Section 2, which might occur as a super-elementary function in a larger program. Preaccumulating its gradient would essentially halve the number of arcs, whose origins, destinations and values have to be stored until the global reverse sweep.

Except in the simple cases mentioned above, the detection of suitable super-elements or $funnels$ [28] requires some combinatorial analysis of the computational graph. If the same function is evaluated over and over such a potentially very large preprocessing effort may well be justified. However, it probably will only be economical when the graph is essentially static, i.e. the control flow of the original program is largely independent of the variable values. As far as we know this kind of combinatorial optimization on the graph has not yet been implemented.

A major disadvantage of immediate differentiation is the impossibility of obtaining higher directional derivatives after the forward sweep has been completed. To this end one has to construct a complete representation of the computational graph at the current argument, rather than just its linearization. In other words one has to store the type and data dependence of each elementary function in a suitable symbol table. In a way this doubles up the structural information that is already contained in the program.

4.2   FORTRAN Precompiler

There are at least three such implementations, namely JAKEF [14], GRESS [13], and PADRE2 [17]. All three precompilers require the user to supply a source code for the evaluation of $f(x)$ in some dialect of FORTRAN. The dependent and independent variables must be nominated through explicit declarations or a naming convention. The source code is then fed to the precompiler, which analyses its arithmetic assignment statements very much like a normal compiler. As we have mentioned before the control statements

FORTRAN subroutine for evaluating product

```
      SUBROUTINE PROD(N,X,F)
      INTEGER N,I
      DOUBLE PRECISION F,GRAD
      DOUBLE PRECISION X(N)
CONSTRUCT D(F)/D(X) IN GRAD(N)                    {Nominate the dependent
      F = 1.D0                                    /independent variables }
      DO 10 I = 1,N
        F = F*X(I)
10    CONTINUE
      RETURN
      END
```

Extended FORTRAN program generated by JAKEF

```
      SUBROUTINE PRODJ(N,X,F,GRAD,YGRAD,LYGRAD,RFS,IFS,LFS)
      INTEGER LFS,IFS(LFS)                         { Lot's of extra storage }
      DOUBLE PRECISION RFS(LFS),TGRA(543)
      INTEGER N,I,LQ00,LQ01,LYGRAD,IGRAD,RGRAD,IX
      DOUBLE PRECISION X(N),F,GRAD(N),YGRAD(LYGRAD)
      IX = 544
      CALL DPINIT(IX+N,LYGRAD)                      { Initialization Routine }
      CALL DMIT0(1,RFS,IFS,LFS)                    {Storage of zero arc for
      F = 1.D0                                          constant assignment}
      LQ00 = 1
      LQ01 = N
      DO 90001 I = LQ00,LQ01                        {Loop logically unaltered}
      CALL DMIT2(1,X(I),IX+I,F,1,RFS,IFS,LFS)   {Storage of two arcs for
      F = F*X(I)                                         multiplication}
90001 CONTINUE
90000 CONTINUE
      RGRAD = 0
      CALL DPGRAD(YGRAD,LYGRAD,1,RGRAD,IGRAD,RFS,IFS,LFS)   {Accumulation
      CALL DPCOPY(GRAD,IGRAD,1,YGRAD(IX+1),N)                of gradient}
      RETURN
      END
```

remain unaltered. All calculations involving real variables are broken down into elementary arithmetic operations and univariate system functions, e.g. exponentials or trigonometric functions. For each of these elementary functions f_i the precompiler has built in expressions of the one or two partial derivatives $\partial f_i / \partial x_j$.

Using this 'knowledge' the precompiler can construct an extended FORTRAN program that evaluates the partial derivatives simultaneously with each elementary function. In the forward mode of GRESS, these local partial derivatives are used immediately to calculate the full gradient ∇x_i of the intermediate value x_i with respect to the independent variables nominated by the user. In the case of JAKEF and the reverse mode of GRESS, the local partials are stored as arc values with a suitable encoding of their origin and destination, i.e. the $j - th$ and $i - th$ node respectively. PADRE2 delays the differentiation by storing instead a symbol identifying the elementary function and the current argument, so that its first and possibly higher derivatives can be evaluated during the reverse sweep. To effect the reverse sweep the precompilers insert a call to a standard accumulation subroutine at the end of the program.

The resulting extended FORTRAN programs rely on runtime support packages containing various standard subroutines and possibly also problem specific scratch files. The user then compiles and links the whole suite to obtain an executable code for evaluating the function, its gradient, and in the case of PADRE2 also second derivatives or error estimates. As an example the next page displays the FORTRAN subroutine PROD that evaluates the product of n independent variables followed by the subroutine PRODJ obtained by precompiling PROD with JAKEF. The in-line comments on the right were added later and would naturally result in compilation errors.

Apart from the five subroutines called in the extension PRODJ there are two other subroutines in the runtime support library of JAKEF. Its total length is less than 150 lines of FORTRAN. When calling PRODJ the user has to provide the integer work arrays IFS and the real work array RFS with a sufficiently large common length LFS. The precompiler cannot provide a lower bound on LFS, because the storage requirement is usually a function of the number of variables and other problem parameters. This difficulty occurs in all reverse implementations, whereas the storage requirement in the forward mode is predictable.

Even though we have had no opportunity to test it, the recently released package GRESS, developed at Oak Ridge National Laboratory, appears to be the most versatile and user friendly precompiler for automatic differentiation that is currently available. It operates in the forward or reverse mode and allows for user defined functions as well as implicit relationships. PADRE2 is the only precompiler capable of producing second derivatives and error estimates, but as yet it is only documented in Japanese. JAKEF is quite efficient but does not allow user defined subroutines.

4.3   Operator Overloading

The use of a precompiler means in effect that the original program is compiled twice, with a rather cryptic extended source code being generated as a by product. Hence one may ask, whether it is not possible to saddle the main compiler with the task of issuing the instructions that have to be executed in order to evaluate certain derivatives. This in fact possible by a facility called *operator overloading*, which is available in most modern computer languages, including hopefully FORTRAN 8X. The key idea here is that the programmer can define new types of variables, whose occurence as arguments of an elementary function triggers the compiler to issue additional instructions. The source code itself remains essentially unchanged.

Apparently the first implementation of this kind is due to Kedem [21]. Since FORTRAN itself does not support overloading, he used the general purpose precompiler AUGMENT, which allowed the user to write the original program in a Taylor made extension of FORTAN. The resulting source code was then precompiled into standard FORTRAN by AUGMENT. Since most of its facilities are more conveniently available in modern computer languages, AUGMENT is no longer supported by its authors or anybody else. Kedem's extension of FORTRAN enabled the user to compute gradients or truncated Taylor series in the forward mode of automatic differentiation.

A few years later Rall [26] achieved a much cleaner implementation of the forward mode in the language PASCAL-SC, an extension of PASCAL for PC Compatibles distributed by Teubner and Wiley [22]. The transformation process is extremely simple. Suppose we have a standard PASCAL code for the evaluation of a function in the variables $X[1..N]$ of type REAL. Then the $X[I]$ and all real variables that depend on them are redeclared to be of the new type GRADIENT, which is completely problem independent. Each variable XJ of type GRADIENT is a record consisting of a scalar part $XJ.F$ and a vector part $XJ.D[1..N]$. At each stage of the calculation the vector part represents the gradient of the scalar part with respect to the independent variables $X[1..N]$. The vector part of the independent variable $X[I]$ is initialized as the i-th Cartesian basis vector. Whenever an argument of type GRADIENT occurs in an elementary arithmetic operation, say the assignment $Z := X * Y$, the compiler looks for an appropriate overloading of the usual elementary operation on REALs. Therefore Rall supplied small, problem independent operator declarations for every possible combination of arguments, e.g. GRADIENT*GRADIENT, REAL*GRADIENT, and GRADIENT*REAL. In the last case for example, both the scalar and vector part of the first variable are multiplies by the second variable, which is of type REAL. Unfortunately PASCAL-SC does not allow the overloading of standard functions, so that the definition of $SIN(X)$ cannot be extended to arguments X of type GRADIENT. Instead one has to introduce a new function $GSIN(X)$ that evaluates and differentiates the sine for arguments of type GRADIENT. This and some other limitations of PASCAL-SC require minor modifications of the program body. Any such changes could be avoided in a more powerful programming language such as C++.

The reverse mode of automatic differentiation can be implemented in a very similar way. Instead of GRADIENT we define a new type VAREAL that represents a record consisting of one REAL value and two pointers to other VAREALs. In contrast to the length of the vector part in GRADIENT, the size of each record of type VAREAL does not depend on the total number of independent variables. At execution time the extended program generates a doubly linked list of such records to represent the linearization of the computational graph at the current argument. Since they have to manipulate this data structure the overloaded operators for arguments of type VAREAL are logically more complicated than those for arguments of type GRADIENT in Rall's implementation. However, according to columns 2 and 3 of Table 1 in Subsection 2.3 the reverse mode is always faster than the forward mode, even when the number of variables and hence the difference in the number of arithmetic operations is small. This may partly be due to the lack of a mathematical coprocessor or floating point accelerator on the IBM PC in use. On systems with such devices the generation and manipulation of the doubly linked list might be relatively more expensive and thus shift the balance a bit in favor of the forward mode. Possibly for the same reason, it was found that recreating the list during each of several function evaluations is no more expensive than reusing the pointers from the first evaluations during subsequent calls. Overloading as such has no bearing on the execution time, because the type dependent decision which declaration of an operator applies at a particular occurence in the code is already made during the compilation.

Again using the product example, we have listed on the next page the original evaluation program in PASCAL-SC and its modification for reverse differentiation via operator overloading. The program on the left simply reads in the nine variable values and prints out their product. The program on the right does exactly the same and then prints out the nine components of the gradient at the given argument.

Reverse Automatic Differentiation by Operator Overloading in PASCAL-SC

```
PROGRAM PROD(INPUT,OUTPUT);              PROGRAM PROD(INPUT,OUTPUT);
                                         $INCLUDE VHEAD.SRC
VAR    X : ARRAY[1..9] OF REAL;          VAR    X : ARRAY[1..9] OF VAREAL;
       Y,T : REAL;                              Y : VAREAL; T : REAL;
       I : INTEGER;                             I : INTEGER;
BEGIN                                    BEGIN
                                           TAIL := NIL; SPARE := NIL;

  Y := 1;                                  Y := VARY(1);
  FOR I := 1 TO N DO                       FOR I := 1 TO N DO
  BEGIN                                    BEGIN
    READ(T);                                 READ(T);
    X[I] := T;                               X[I] := VARY(T);
    Y := Y*X[I]                              Y := Y*X[I]
  END;                                     END;

  WRITELN(Y);                              WRITELN(EVAL(Y));

                                           ACCUMULATE(Y);
                                           FOR I := 1 TO N DO
                                             WRITELN(EVAL(X[I]));

END.                                     END.
```

Program for Product Example   **Extension with Reverse Differentia-**
tion

The central sections of both codes are almost identical, except that the one on the right needs the conversion function VARY in assigning real values to variables of the new type VAREAL. Conversely the function EVAL extracts the real value from a VAREAL, which is needed in particular for output operations. The type VAREAL, the functions VARY and EVAL, the gradient accumulation procedure ACCUMULATE, the multiplication operator * between VAREALs, and the two pointer variables TAIL and NIL are all defined

in the problem independent header file VHEAD.SRC occuring in the compiler directive
$INCLUDE right at the top. The explicit initialization of TAIL and SPARE, and the two
conversion functions could be avoided in a programming language like C++, where the
assignment operator can also be overloaded. Here, any oversight in making the required
modifications will result in compile or run time errors. If the independent variables are
declared as VAREALs and program executes normally, then the gradient values should be
correct.

Compared to precompilation overloading probably requires more user sophistication
but on the other hand it clearly offers more flexibility. Provided all subprograms are compiled together, either mode of automatic differentiation in PASCAL-SC can deal with user
defined functions and even recursive procedure calls. This does not require any extension or modification of the header file. Higher derivatives and some optimization of the
computational graph can also be implemented by overloading. The forward evaluation of
general and structured Hessians in the advanced language ADA is discussed by Dixon and
Mohseninia in [8]. When the currently proposed standard for FORTRAN 8X is actually
implemented one of the major objections to operator overloading will be removed.

5   Conclusions and Discussion

Like several previous authors we conclude that in theory and practice the gradients of all
functions defined by computer programs can be evaluated cheaply and automatically. This
observation suggests the reexamination of the many arguments in the optimization literature, that rely at least implicitly on the seemingly reasonable assumption, that gradients
codes are often hard to to come by and run typically much slower than the corresponding
function routine.

Since truly derivative-free algorithms rarely have worked for more than a handful of
variables, many researchers recommend the approximation of gradients by central or one-
sided differences. Whenever this classical technique can be applied at all, we must have a
reasonably accurate evaluation algorithm, in which case automatic differentiation provides
a far superior alternative. Provided there is enough storage, reverse accumulation yields
truncation error free gradient values at less than $5/n$ times the computing time of divided
differences. This technique has been successfully implemented on problems in nuclear
engineering and oceanography with thousands or even millions of variables. Should the
function evaluation be so lengthy that the storage of all intermediate results is impossible,
then one can still employ the forward mode to achieve better accuracy at essentially the
same cost as divided differences.

Many line search procedures avoid the evaluation of the gradient at trial points before
these have been accepted as the next main iterate. This strategy could still make sense,
since we found that the gradient may well be four or five times more expensive to evaluate
then the function. Also, the cubic interpolation made possible by the value of the directional derivative at the trial point destroys the simplicity of usual quadratic interpolation.
Moreover the improved accuracy of the cubic interpolants rarely leads to a significant reduction in the overall number of evaluations or iterations. On the other hand, keeping two
evaluation routines (one without and one with the gradient) and calling them successively
at all main iterates does not seem that economical either.

Penalty functions have long been used to convert constrained optimization problems into unconstrained problems. If one wants the penalty functions to be exact, i.e. attain local minima right at the solutions of the constrained problem, then there are basically two choices. Either the penalty function nonsmooth or it depends explicitly on the gradients of the objective and constraint functions [7]. In the latter case the resulting gradient and Hessian depend on second and third derivatives of the original problem functions respectively. Since this additional level of differentiation was thought to be unacceptable, nonsmooth penalty functions have generally been preferred. However, automatic differentiation can produce the restricted second derivative terms in the gradient of smooth exact penalty functions at a reasonable cost, namely a fixed multiple of evaluating the objective and constraint functions. Therefore a suitable implementation of unconstrained BFGS could be both user friendly and efficient, especially since the troublesome Maratos effect of nonsmooth penalty functions cannot occur here.

The combination of automatic differentiation with the variable metric method BFGS recommended above may seem a strange mixture. Indeed, some researchers in automatic differentiation feel that the development of quasi-Newton methods was an emergency measure, which is outdated now that we can obtain the Hessian automatically. This seems to us a rather premature assessment. As we have seen in Subsection 3.6 the evaluation of a Hessian-vector product by either mode of automatic differentiation may be up to $5n$ times as expensive as that of the gradient. Thus we must expect that sometimes an exact or inexact Newton method based on automatic differentiation of the gradient will be less efficient than the corresponding finite difference version. In view of the trouble with negative curvature one may then prefer the simple and usually quite efficient BFGS method with line-search.

In any event automatic differentiation should allow the design of an optimization package that requires the user only to supply source code for the evaluation of the objective and constraint functions. The generation of the corresponding gradient codes, the detection of sparsity, and the determination of the maximal achievable solution accuracy, could all be done automatically. Ideally, the selection of a suitable linear equation solver for the computation of steps on large structured problems could also be left to the package.

In nonlinear least squares it is usually assumed that the calculating the gradient of the residual norm requires the evaluation of the full Jacobian. Hence, the argument goes, we might as well fully utilize this derivative information by employing a Gauss-Newton like procedure. However, as is the case for certain inverse problems [29], the Jacobian matrix may be huge and dense, whereas reverse accumulation always yields the gradient cheaply. Then nonlinear conjugate gradients or a variable metric method with limited memory is clearly the only choice. On the other hand, there are many problems, where the Jacobian is of moderate size and costs little more than the residual vector to evaluate.

Throughout this paper we have restricted our attention to a scalar valued function $f(x)$ in n variables. Naturally all results and techniques can be separately applied to the m components of a vector valued function $F(x)$. However, this approach may be far from optimal if the component functions are closely related, i.e. have many common expressions. Also, if m is significantly larger than n the forward mode of automatic differentiation is likely to be cheaper. Currently there appears to be no clearly superior strategy for the evaluation of derivative matrices (rather than vectors).

Even though the underlying mathematics are straight forward much remains to done in the field of Automatic Differentiation. With regards to general purpose differentiation software for various machine architectures, the problems are mainly of a computer science nature. However, some combinatorial analysis of the graph structure might be beneficial for the optimal evaluation of derivative matrices and the local preaccumulation of gradients, which was briefly mentioned in Subsection 4.1. Also, as in the case of evolution equations discussed in Subsection 3.4, there are probably other problem classes in which the reverse sweep has a natural interpretation and can be implemented in various ways. Finally, automatic differentiation could and should be integrated into numerical packages for special purposes, such as optimization, stiff differential equation, boundary value problems, optimal control, and path-following with bifurcation analysis. This process would be a lot simpler and more widely acceptable, if the next FORTRAN standard were to allow user-defined types with function and operator overloading.

6 Acknowledgements

In preparing this paper the author was aided by the incisive comments of Jorge Moré, Steven Wright, and Rob Womersley.

References

[1] F.L. Bauer (1974). "Computational Graphs and Rounding Errors", *SINUM*, Vol.11, No.1, pp.87-96 .

[2] W. Baur and V. Strassen (1983). "The Complexity of Partial Derivatives", *Theoretical Computer Science*, Vol. 22, pp.317-330.

[3] L.M. Beda et al (1959). "Programs for Automatic Differentiation for the Machine BESM", Inst. Precise Mechanics and Computation Techniques, Academy of Science, Moscow.

[4] D.G. Cacuci (1981). "Sensitivity Theory for Nonlinear Systems. I. Nonlinear Functional Analysis Approach", Journal of Mathematical Physics, Vol.22, No.12, pp.2794-2802.

[5] D.G. Cacuci (1981). "Sensitivity Theory for Nonlinear Systems. II. Extension to Additional Classes of Responses", Journal of Mathematical Physics, Vol.22, No.12, pp.2803-2812.

[6] B.W. Char, K.O. Geddes, G.H. Gonnet, M.B. Monegan, and S.M. Watt (1988). *MAPLE Reference Manual, Fifth Edition*, Symbolic Computation Group, Department of Computer Science, University of Waterloo, Waterloo, Ontario, Canada N2L 3G1.

[7] G. Di Pillo and L. Grippo (1986). "An Exact Penalty Method with Global Convergence Properties for Nonlinear Programming Problems", *SIAM J. Control Optim.* Vol.23, pp.72-84.

[8] L.C.W. Dixon and M. Mohseninia (1987). "The Use of the Extended Operations Set of ADA with Automatic Differentiation and the Truncated Newton Method", *Technical Report No.176*, The Hatfield Polytechnic, Hatfield, U.K.

[9] L.C.W. Dixon (1987). "Automatic Differentiation and Parallel Processing in Optimisation", *Technical Report No.180*, The Hatfield Polytechnic, Hatfield, U.K.

[10] Iu. G. Evtushenko (1982) . *Metody resheniia ekstremal'nykh zadach ikh primenenie v sistemakh optimizatsii*, Nauka Publishers, Moskow

[11] H. Fischer (1987). "Automatic Differentiation: How to compute the Hessian matrix", *Technical Report #104A*, Technische Universität München, Institut für Angewandte Mathematik und Statistik.

[12] A. Griewank and G.W. Reddien (1988). "Computation of Cusp Singularities for Operator Equations and their Discretizations", *Technical Memorandum* ANL/MCS-TM-115, Mathematics and Computer Science Division, Argonne National Laboratory, Argonne IL 60439. To appear in the special issue on *Continuation Techniques and Bifurcation Problems* of the *Journal of Computational and Applied Mathematics*.

[13] J.E. Horwedel, B.A. Worley, E.M. Oblow, and F.G. Pin (1988). *GRESS Version 0.0 Users Manual*, ORNL/TM 10835 , Oak Ridge National Laboratory, Oak Ridge, Tennessee 37830, U.S.A.

[14] K.E. Hillstrom (1985). *Users Guide for JAKEF*, Technical Memorandum ANL/MCS-TM-16, Mathematics and Computer Science Division, Argonne National Laboratory, Argonne IL 60439.

[15] M. Iri (1984). "Simultaneous Computations of Functions, Partial Derivatives and Estimates of Rounding Errors - Complexity and Practicality", *Japan Journal of Applied Mathematics*, Vol.1, No.2 pp.223-252.

[16] M. Iri, T. Tsuchiya, and M. Hoshi (1988). "Automatic Computation of Partial Derivatives and Rounding Error Estimates with Applications to Large-Scale Systems of Nonlinear Equations", To appear in *Journal of Computational and Applied Mathematics*, Vol.23

[17] M. Iri, and K. Kubota (1987). "Methods of Fast Automatic Differentiation and Applications", *Research memorandum RMI 87-0*, Department of Mathematical Engineering and Instrumentation Physics, Faculty of Engineering, University of Tokyo.

[18] R.H.F. Jackson, and G.P. McCormick (1988). "Second order Sensitivity Analysis in Factorable Programming: Theory and Applications", *Mathematical Programming*, Vol.41, No.1, pp.1-28.

[19] H. Kagiwada, R. Kalaba, N.Rosakhoo, and Karl Spingarn (1986). "Numerical Derivatives and Nonlinear Analysis", Vol. 31 of **Mathematical Concepts and Methods in Science and Engineering** Edt. A.Miele, Plenum Press, New York and London

[20] K.V. Kim, Iu.E. Nesterov, V.A. Skokov, and B.V. Cherkasskii (1984). "An efficient Algorithm for Computing Derivatives and extremal Problems" English translation of "Effektivnyi algoritm vychisleniia proizvodnykh i ekstremal'nye zaduchi", *Ekonomika i matematicheskie metody*, Vol.20, No.2, pp.309-318.

[21] G. Kedem (1980). "Automatic Differentiation of Computer Programs", *ACM TOMS*, Vol.6, No.2, pp.150-165.

[22] U. Kulisch et al (1987). *PASCAL-SC, A PASCAL Extension for Scientific Computation, Information Manual and Floppy Disk*, B.G. Teubner, Stuttgart, and John Wiley & Sons, New York.

[23] G. Leitmann (1981). *The Calculus of Variations and Optimal Control* Vol.20 of Mathematical Concepts and Methods in Science and Engineering, Edt. A.Miele, Plenum Press, New York and London

[24] D.Y. Peng and D.B. Robinson (1976). "A new two-constant Equation of State", *Ind. Eng. Chem. Fundamentals*, Vol.15, pp.59-64.

[25] L.B. Rall (1981). *Automatic Differentiation - Techniques and Applications*, Springer Lecture Notes in Computer Science, Vol.120 .

[26] L.B. Rall (1984). "Differentiation in PASCAL-SC: Type GRADIENT", *ACM TOMS* Vol.10,pp.161-184.

[27] L.B. Rall (1987). "Optimal Implementation of Differentiation Arithmetic", in *Computer Arithmetic, Scientific Computation and Programming Languages*, ed. U. Kulisch, Teubner, Stuttgart.

[28] B.Speelpenning (1980). "Compiling fast Partial Derivatives of Functions given by Algorithms", Ph.D. Dissertation, Department of Computer Science, University of Illinois at Urbana-Champaign, Urbana, IL 61801.

[29] W.T. Thacker and R.B. Long (1988). "Fitting Dynamics to Data", *Journal of Geophysical Research*, Vol.93, No.C2, pp.1227-1240.

[30] R.E. Wengert (1964). "A simple Automatic Derivative Evaluation Program". *Com. ACM*, Vol. 7,pp.463-464 .

[31] P.Wolfe (1982) ."Checking the Calculation of Gradients", *ACM TOMS*, Vol.6, No.4, pp. 337-343.

[32] B.A. Worley et al (1989). "Deterministic Sensitivity, and Uncertainty Analysis in Large Scale Computer Models", *Proceedings of 10th Annual DOE low level Waste Management Conference* in Denver, Aug.30 -Sept.1, 1988.

Recent Developments and New Directions in Linear Programming

Michael J. TODD*

Center for Applied Mathematics, School of Operations Research and Industrial Engineering, Upson Hall, Cornell University, Ithaca, NY 14853, U.S.A.

ABSTRACT. We describe recent developments and outline new directions in linear programming. We highlight research on interior algorithms in the last few years, but also discuss work related to the simplex method and the ellipsoid algorithm. Our development stresses the relationships between different method and the different geometries in which they can be viewed.

1. Introduction.

The aim of this paper is to describe some of the recent developments in linear programming and to suggest some directions in which future progress might be made. We will concentrate on the research in the last four years on interior algorithms for linear programming, stemming from the fundamental paper of Karmarkar [37], but also discuss other work related to Dantzig's simplex algorithm and the ellipsoid method of Shor, Nemirovsky and Yudin. Indeed, we wish to stress the fruitfulness of exploring the interplay between these different algorithmic approaches, between analysis, algebra and geometry, and between the many different geometries in which linear programming can be viewed.

Section 2 describes results relating to the simplex method, and the three geometrical viewpoints which have elucidated its study. In sections 3, 4 and 5 we discuss projective, affine, and path-following interior algorithms respectively. (Some connections with the ellipsoid

*
Research supported in part by NSF Grant ECS-8602534 and ONR Contract N00014-87-K-0212.

M. Iri and K. Tanabe (eds.), Mathematical Programming, 109–157.
© *1989 by KTK Scientific Publishers, Tokyo.*

method will be briefly mentioned in section 3.) These methods take a
diametrically-opposed (in fact, the concept of diameter is totally
irrelevant!) geometric view of polyhedra. They attempt to find
transformations under which the feasible region is well-approximated by
a ball centered at the current iterate. Are the polyhedra of real-life
problems close to balls, for which interior algorithms are efficient?
Or are they closer to crystals with long edges from one side to the
other, for which simplex variants work well? It appears from the
excellent behavior of both classes of algorithms that either viewpoint
is valid, and that the two geometrical perspectives--combinatorial and
metric--can peacefully coexist.

While there is a great variety of derivations of the different
interior algorithms, all generate search directions that are linear
combinations of two basic directions; a steepest descent direction for
the objective function in a transformed space, and a "centering"
direction. This observation is due to Gonzaga [32]; see also Mitchell
and Todd [55]. In general, projective methods require $O(nL)$
interations and path-following methods $O(\sqrt{n}L)$ iterations to solve a
problem with n variables and integer data with input length L.
Affine methods in their original form are not believed to be
polynomially bounded. Each iteration requires $O(n^3)$ arithmetic
operations to solve a linear system of equations in the basic methods,
but modified versions of the algorithms use a trick due to Karmarkar
using rank-one updates that needs only $O(n^{5/2})$ arithmetic operations
per iteration on average.

Hence the overall complexities of projective methods and
path-following methods are $O(n^{7/2}L)$ and $O(n^3L)$ respectively.
However, these bounds are established in very different ways.
Projective algorithms assure a fixed decrease in Karmarkar's potential
function at each iteration; by performing line searches, considerably
greater decreases can be achieved in practice with no loss in
guaranteed performance. Path-following methods stay close to a central

trajectory while driving the objective function value to its optimal
value, and, while long steps appear to be very useful in practice, they
can destroy the proximity to the central trajectory which is required
by the theoretical analysis.

Section 6 discusses ways to combine these different approaches.
In particular, we discuss a recent method of Ye [80] based on
potential functions that does not use projective transformations,

requires only $O(\sqrt{n}L)$ iterations, and allows large steps. We offer
some concluding remarks and discuss open problems in section 7.

While I have tried to highlight some new ideas that I believe will
be fruitful, this paper is not intended to be a comprehensive survey.
There are several important topics I have chosen not to cover, and some
significant contributions to the area of interior algorithms have not
been cited. Thus the material presented provides a biased but, I hope,
coherent view of some recent important developments. Two particular
omitted topics are the linear-time algorithms for linear programming
when one dimension is fixed, due to Megiddo and others, and the
strongly polynomial algorithm to solve combinatorial linear programs
developed by Tardos [67]. These topics are covered in the excellent
survey paper of Megiddo [52], which also discusses different models of
computation and the resulting notions of complexity in depth. The
reader may also wish to consult the state-of-the-art survey paper of
Goldfarb and Todd [29].

2.   Simplex-related results.

The traditional way to study the simplex method from a geometric
viewpoint is to consider the polyhedron of its feasible solutions.
Then the sequence of solutions generated by a simplex variant
corresponds to a path proceeding from vertex to vertex along edges of
the polyhedron. Many results on the worst-case or average-case number
of iterations required can be obtained and understood most clearly from
this perspective; indeed, many authors write of "the" geometry of the
simplex method when discussing this view.

It is not our intention to provide a wide overview of the vast
literature relating to the efficiency of the simplex method; a

comprehensive survey has been provided by Shamir [64]. Nor will we discuss in depth the question of the diameter of polyhedra, for which we refer to the excellent paper of Klee and Kleinschmidt [42]. We will, however, pick out some important recent themes.

Thoroughout this paper, we will denote by d the dimension of a polyhedron, by n the total number of inequalities defining it, and by m the difference $n-d$. Barring rank-deficiency or degeneracy, we can define such a polyhedron by n inequalities in d unrestricted variables, by m inequalities in d nonnegative variables, or by m equations in n nonnegative variables.

The diameter of a polyhedron is the maximum, over all pairs of vertices, of the minimum length of an edge-path joining these vertices. The celebrated Hirsch conjecture (see Dantzig [16; p.160]) claims that the diameter of a bounded polyhedron of the dimensions above does not exceed m. Note that lower bounds on maximum diameters provide bounds on the number of iterations required by the best simplex-type method applied to the worst problem of a given size. The conjecture has only been proved for very limited cases ($d \leq 3$ or $m \leq 5$) and is believed to be false in general [42]. However, there are some significant classes of polyhedra for which the conjecture holds, or at least a polynomial bound on the diameter is known. In addition, there is a surprising recent result of Lee [48]: for every simple d-polytope P with n facets, one can construct a simple (d+1)-polytope Q with $n+1$ facets, of which one is congruent to P, with diameter at most $2(n-d)$. (A d-polytope is _simple_ if every vertex lies on exactly d facets.)

Klee and Kleinschmidt [42] list classes of polyhedra for which the Hirsch conjecutre is known to hold. These include Leontief substitution polyhedra, dual transportation polyhedra, and certain transportation polytopes, for example assignment polytopes. In terms of polynomial simplex algorithms, Orlin [61] has given a dual simplex method requiring $O(v^3 \log v)$ iterations for the minimum cost network flow problem, Goldfarb and Hao [26] an $O(ve)$-iteration primal algorithm for the maximum flow problem and Roohy-Laleh [63] an $O(v^3)$-iteration primal and Balinski [9] an $O(v^2)$-iteration dual simplex algorithm for the assignment problem. (Here v denotes the number of

nodes and e the number of edges in the graph or network.)
Kleinschmidt, Lee and Schannath [44] have extended Balinski's method to
certain transportation problems. Of course, there exist many
non-simplex polynomial algorithms for such problems, but they are
beyond the scope of this paper.

   Before we turn to the other topic of this section, the expected
behavior of the simplex method, we want to stress that the study of
polytopes in general and their diameters in particular has benefitted
greatly from another geometrical view, that of the polar polytope. If

$P = \{x \in \mathbb{R}^d : a_i^T x \leq 1 \text{ for } i = 1,2,\ldots,n\}$, its polar is $P^0 :=$

$\text{conv}\{a_1, a_2, \ldots, a_n\}$, and if P is simple, P^0 is simplicial -- all

its facets are simplices. In this case, the boundary complex of P^0
is a simplicial complex, and methods of (algebraic or piecewise-linear)
topology can be applied. Again we refer to [42].

   During the late seventies and early eighties, there was a burst of
activity in the probabilistic analysis of the simplex algorithm.
Notable contributions were made by Borgwardt, Smale, Adler, Haimovich,
Karp, Megiddo, Shamir and Todd. These results are surveyed in Shamir
[64] and the 0th chapter of Borgwardt's monograph [14]. Rather little
research in the areas has been performed in the last few years, but
significant problems remain.

   Borgwardt considers problems in the form

$$\max \quad c^T x$$
$$a_i^T x \leq 1, \quad i = 1,2,\ldots,n$$
$$(x \geq 0)$$

with or without nonnegativity constraints, where c and each $a_i \in \mathbb{R}^d$.
He describes closely related algorithms for these problems, which
combine a variable-dimension phase I procedure with a parametric
objective phase II method. In [14] he proves that, if c and each a_i
are distributed independently and identically according to a
rotationally symmetric distribution, then the expected number of

iterations required is $O(d^4 n^{1/(d-1)})$.

The main drawback of the result is that not only are problems feasible with probability 1 (this is reasonable), but a feasible solution – the zero vector – is a priori known and used by the algorithms. Recently, Borgwardt [15] has extended the result to the cases of right-hand sides b_i which are uniformly distributed (and not almost surely positive); nevertheless, the probabilistic model is quite restrictive.

In contrast to Borgwardt's model, that considered by Adler, Karp and Shamir [2], Adler and Megiddo [3] and Todd [68] is rather general. This "sign invariant" model considers the problem in the form

$$\max \quad c^T x$$
$$Ax \leq b$$
$$x \geq 0$$

where A is m × d, and assumes that the data are generated according to any distrubtion that yields nondegeneracy with probability one and is invariant with respect to changes in sign in rows or columns of $\begin{bmatrix} 0 & c^T \\ b & A \end{bmatrix}$. The algorithm applied is a lexicographic variant of Dantzig's self-dual method, and the result obtained independently in all studies is that the expected number of iterations required is $O(\min\{m,d\})^2$. Adler and Megiddo, under slightly stronger probabilistic assumptions, establish a lower bound of the same order of magnitude.

The sign-invariant model has an equally serious disadvantage. For large m and d, almost all problems are infeasible (if $m/d \geq 1 + \epsilon > 1$) or unbounded (if $m/d \leq 1 - \epsilon$). For $m/d = 1$, about half are infeasible, about half are unbounded, and $\Omega(d^{-\frac{1}{2}})$ have optimal solutions; as pointed out by Adler and Megiddo, this allows an $O(d^{5/2})$ expected bound for such problems, conditioned on the existence of an optimal solution. Nevertheless, such a distribution does not appear to be realistic.

The analyses and geometric viewpoint of these papers is quite diverse. Borgwardt translates his algorithm to the polar polytope P^0

and applies his probabilistic analysis in that setting. Adler Karp and
Shamir base their analysis on a sequential application of earlier
results of Haimovich and Adler, for which the geometry of the primal
feasible polyhedron and the associated arrangement of hyperplanes is
the natural viewpoint. Todd's analysis is purely combinatorial, since
he came to the algorithm via oriented matroids. Finally, Adler and
Megiddo were motivated by Smale's earlier estimation of the volumes of
cones that arise naturally in analyzing Lemke's algorithm, and these
considerations are again reminiscent of Borgwardt's approach, although
in a primal-dual setting.

Finally, we wish to describe a model and result described to the
author by G. Debreu [17] in 1976. While in one respect it is very
limited (m = 2!), it is very general in others and it raises a very
significant open problem: to extend the result to larger m. In
addition, it views the simplex method from a third geometric
perspective, that of the columns of the coefficient matrix when the
problem is expressed in standard form. This geometry is, after all,
why the method is called the simplex method, and only the insights from
this geometrical view led Dantzig to believe that the method might be
efficient for large-scale problems (see Dantzig [16],pp.160-166).

The problem under consideration is

(P)
$$\min \sum_j \gamma_j x_j$$
$$\sum_j \alpha_j x_j = \beta$$
$$\sum_j x_j = 1$$
$$x_j \geq 0 \text{ all } j.$$

We have intentionally omitted the range of summation for j, since   j
can range over an arbitrary index set, although we assume that
$\alpha_1 = \min \alpha_j$ and $\alpha_2 = \max \alpha_j$ exist and are known and that $\{(\alpha_j, \gamma_j)\}$
is compact. The probabilistic model makes no further assumptions on
the γ_j's and α_j's, but assumes that β is uniformly distributed in
$[\alpha_1, \alpha_2]$. Of course, we cannot get a bound on the number of iterations,
so we consider only the rate of convergence.

The simplex variant used is Dantzig's most negative reduced cost rule, starting with the basic variables x_1 and x_2. The entering variable will then be x_3, where (α_3, γ_3) minimizes $\overline{\gamma}_j := \gamma_j - \alpha_j \lambda - \mu$ with λ and μ chosen so that $\overline{\gamma}_1 = \overline{\gamma}_2 = 0$. It follows that all points (α_j, γ_j) lie on or above the line $\gamma = \alpha\lambda + \mu$ through (α_3, γ_3), and (α_1, γ_1) and (α_2, γ_2) lie on a parallel line - see figure 1.

The new basic variables will be x_1 and x_3 or x_3 and x_2, and in either case, the distribution of β conditional on the new basis being feasible will be uniform in the appropriate range; $[\alpha_1, \alpha_3]$ or $[\alpha_3, \alpha_2]$. Moreover, if v_0, v_1 and z_* denote the old, new and optimal objective function values, then figure 1 shows that $(v_1 - z_*)/(v_0 - z_*) \le (\alpha_3 - \beta)/(\alpha_3 - \alpha_1)$ (or $(\beta - \alpha_3)/(\alpha_2 - \alpha_3)$), which is uniformly distributed in $[0,1]$. Hence $E(\ell n(v_1 - z_*) - \ell n(v_0 - z_*))$ $< \int_0^1 \ell n\, t\, dt = -1$. Since the situation at the end of the first iteration is exactly the same probabilistically as before, the argument can be continued to show that, after k iterations,

$$E(\ell n(v_k - z_*)) \le E(\ell n(v_0 - z_*)) - k.$$

Here 14 iterations will on average give 6 more significant digits of accuracy. This remarkable result of Debreu is independent of the distribution of the coefficient matrix or the cost vector and independent of the number of variables, subject to the availability of an oracle to choose the entering variable.

If one attempts to generalize the argument to higher m, an immediate problem arises. Suppose $\{\alpha_j\}$ are now $(m-1)$-vectors, and all α_j's lie in the simplex with vertices $\alpha_1, \alpha_2, \ldots, \alpha_m$. Suppose β is an $(m-1)$-vector distributed uniformly in this simplex. Then if x_{m+1} is the entering variable, one can check that the conditional distribution of β is again uniform in the appropriate subsimplex (like that with vertices $\alpha_2, \ldots, \alpha_{m+1}$) and that

$$E(\ell n(v_1 - z_*) - \ell n(v_0 - z_*)) \leq - \frac{1}{m-1}.$$

This appears highly satisfactory, and seems to lead to $O(m)$
iterations to achieve a specified accuracy. However, one cannot
guarantee that the <u>next</u> entering vector, say α_{m+2}, lies in the
appropriate subsimplex (when $m = 2$ the intervals are indeed nested),
and from then on, the uniform assumption fails and the argument breaks
down. A resolution of this difficulty would be very worthwhile.

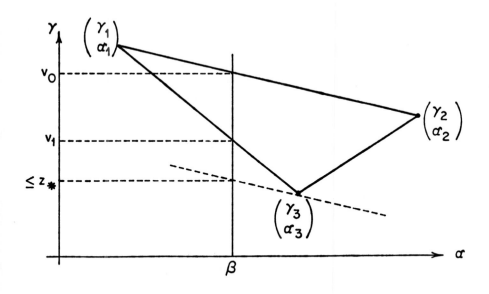

Figure 1. The objective improvement in a simplex iteration.

3. Projective interior methods.

Henceforth we will deal with problems in standard form for ease of comparison. We therefore consider

$$\min\ c^T x$$

(P) $$A x = b$$

$$x \geq 0$$

where A is m × n. We assume that (P) has a nonempty compact set of optimal solutions, that a strictly positive feasible solution x^0 is known, and that A has full row rank. The last two assumptions can be relaxed – as in the simplex method, the third is purely for notational convenience – but the first appears to be essential for convergence proofs. We also assume for simplicity that the objective function is not constant on the feasible region of (P); then if x is a strictly positive feasible solution and z_* is the optimal value of (P), $c^T x > z_*$. For ease in stating results, we assume that $x^0 = e$, the vector of ones. When we give bounds on the number of iterations or arithmetic complexity of solving (P), we assume that the data A, b and c are all integer, with input length L. In this case, it is well known that if we have a feasible solution $\overline{x}$ with $c^T\overline{x} - z_* \leq 2^{-L}$, we can obtain an optimal solution by refining $\overline{x}$ to a vertex.

The dual of (P) is the inequality-constrained problem

$$\max\ b^T y$$

(D) $$A^T y \leq c.$$

There are many ways to convert (P) to a form suitable for application of Karmarkar's method; we shall follow a convenient one that leads to the standard-form variant developed independently by Anstreicher [4], Gay [24], Gonzaga [30], Jensen and Steger [66] and Ye and Kojima [81]. A similar approach, not even requiring a feasible

solution, but which reduces to the methods above if the initial
solution is feasible, was proposed by de Ghellinck and Vial [18]. We
homogenize the problem by introducing a new variable x_{n+1} to arrive at

$$\text{min} \quad \tilde{c}^T \tilde{x}$$

(P)
$$\tilde{A}\,\tilde{x} = 0$$

$$\tilde{g}^T \tilde{x} = 1$$

$$\tilde{x} \geq 0$$

where

$$\tilde{c}^T = (c^T, 0)$$

$$\tilde{A} = (A, -b), \text{ and}$$

$$\tilde{g}^T = (0^T, 1).$$

The first m constraints in $(\tilde{P})$ are referred to as subspace or
homogeneous constraints, while the final equality is the normalizing
constraint. It can be seen that the dual $(\tilde{D})$ of $(\tilde{P})$ is also a
simple reformulation of (D).

At the nth iteration, we have available a strictly positive
feasible solution $\tilde{x}^k (= ((x^k)^T, 1)^T)$ to $(\tilde{P})$, and we form the diagonal
matrix $\tilde{X}_k := \text{diag}(\tilde{x}^k)$ with diagonal entries equal to the components
of $\tilde{x}^k$. Then, with the data $\bar{c}^T = \tilde{c}^T \tilde{X}_k$, $\bar{A} = \tilde{A}\tilde{X}_k$, and $\bar{g}^T = \tilde{g}^T \tilde{X}_k$, we
have the equivalent rescaled problem

$$\text{min} \quad \bar{c}^T \bar{x}$$

(P̄)
$$\bar{A}\,\bar{x} = 0$$

$$\bar{g}^T \bar{x} = 1$$

$$\bar{x} \geq 0$$

in terms of the variables $\bar{x} = \tilde{X}_k^{-1} \tilde{x}$; the current solution is $\bar{x} = e$, where   e   denotes a vector of ones of the appropriate dimension.

We would like to view   e   as a "central" solution, so that a step can be taken without concern for the complicating nonnegativity constraints; alternatively, so that the feasible region is well-approximated by a ball centered at e. However, this depends very much on the form of the constraints of $(\overline{P})$. Hence we now make a projective transformation to make   e   "more central". This transformation can also be called a projective scaling so the name projective scaling algorithm is also used.

Suppose we knew that the optimal value of (P) (hence of $(\tilde{P})$ and $(\overline{P})$) were zero. Then we would basically be seeking a nonzero solution to $\overline{c}^T\overline{x} = 0$ and the subspace constraints – the particular normalization does not matter too much. We could therefore replace $\overline{g}^T\overline{x} = 1$ with $e^T\overline{x} = n+1$; the feasible region is now contained in the unit simplex $S := \{\overline{x} \in \mathbb{R}^{n+1}: e^T\overline{x} = n+1, \overline{x} \geq 0\}$, for which $\overline{x} = e$ is certainly central. (This replacement of the normalizing constraint corresponds to a projective transformation of the feasible region.)

The key geometrical fact is now that the inscribed (in S) ball

$B_r := \{\overline{x} \in \mathbb{R}^{n+1}: e^T\overline{x} = n+1, \|\overline{x}-e\| \leq r\}$ and the circumscribed ball

$B_R := \{\overline{x} \in \mathbb{R}^{n+1}: e^T\overline{x} = n+1, \|\overline{x}-e\| \leq R\}$ satisfy $R/r = n$. Indeed, $R = \sqrt{n(n+1)}$ and $r = \sqrt{(n+1)/n}$. The same is true for the lower-dimensional balls $B_r \cap N(\overline{A})$ and $B_R \cap N(\overline{A})$, where $N(\overline{A})$ denotes the null space of $\overline{A}$; moreover, these balls respectively are contained in and contain the feasible region of $(\overline{P})$ with the new normalizing constraint. Since the optimal value is assumed to be zero, the minimum of $\overline{c}^T\overline{x}$ over the containing ball is nonpositive, and hence its minimum over the contained ball is at most $(1 - \frac{1}{n})$ times its value at the center   e. See Figure 2. Moreover, the minimizing point over $B_\alpha \cap N(\overline{A})$ is easily calculated for any $\alpha > 0$: it is $\overline{x}(\alpha) := e - P_{\overline{B}}\,\overline{c}/\|P_{\overline{B}}\,\overline{c}\|$, where $P_{\overline{B}}$ is the matrix projecting onto the null space of $\overline{B} := \begin{bmatrix} \overline{A} \\ e^T \end{bmatrix}$.

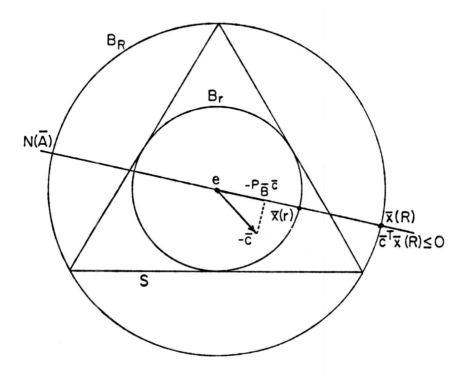

Figure 2.   An iteration in the projective interior algorithm.

We now return to the assumption that the optimal value of $(\overline{P})$ is known to be zero. Of course, if it is known to be z_*, we could replace $\overline{c}$ by $\overline{c} - z_*\overline{g}$ in $(\overline{P})$ to reduce it to zero, and then the analysis would be identical. If z_* is unknown, we use a <u>lower bound</u> z_k to z_*. (For now, we assume a suitable lower bound z_0 is known.) It is then natural to consider

$$\overline{x}(\alpha) := e - \alpha\, P_{\overline{B}}(\overline{c} - z_k\overline{g}) / \| P_{\overline{B}}(\overline{c} - z_k\overline{g}) \| \qquad (3.1)$$

for $\alpha = r$. Then we find that $(\overline{c} - z_k\overline{g})^T\overline{x}(r) = (\overline{c} - z_k\overline{g})^T e - r\| P_{\overline{B}}(\overline{c} - z_k\overline{g}) \|$, so we will make good progress toward the lower bound as long as this norm is reasonably large.

Next note that $P_{\overline{B}} = P_{e^T}\, P_{\overline{A}}$, and that

$$\overline{p} := P_{\overline{A}}(\overline{c} - z_k\overline{g}) = \overline{c} - z_k\overline{g} - \overline{A}^T y(z_k) \ \text{where} \ \ y(z) := (\overline{A}\,\overline{A}^T)^{-1}\overline{A}(c - z\overline{g}). \,(3.2)$$

Now the dual to $(\overline{P})$ is

$$\max \qquad z$$

$(\overline{D})$

$$\overline{A}^T y + \overline{g}z \leq \overline{c},$$

which is just $(\tilde{D})$ with scaled constraints. If $(y(z_k), z_k)$ is <u>not</u> feasible in $(\overline{D})$, we set $z_{k+1} = z_k$. It then follows that $\overline{p}$ has a nonpositive component, so that

$$P_{\overline{B}}(\overline{c} - z_{k+1}\overline{g}) = P_{e^T}\overline{p} = \overline{p} - e\left(\frac{\overline{p}^T e}{n+1}\right)$$

has a component at most $-\overline{p}^T e/(n+1)$, and hence

$$\|P_{\overline{B}}(\overline{c}-z_{k+1}\overline{g})\|_2 \geq \|P_{\overline{B}}(\overline{c}-z_{k+1}\overline{g})\|_\infty \geq \overline{p}^Te/(n+1), \qquad (3.3)$$

where $\overline{p}^Te = (\overline{c}-z_{k+1}\overline{g})^T P_{\overline{A}} e = (\overline{c}-z_{k+1}\overline{g})^Te \geq 0$. (Since e is feasible in $(\overline{P})$, it lies in the null space of $\overline{A}$, and its objective value $\overline{c}^Te$ is at least $z_{k+1} = z_{k+1}\overline{g}^Te$.) We have therefore shown that

$$(\overline{c}-z_{k+1}\overline{g})^T\overline{x}(r) \leq (\overline{c}-z_{k+1}\overline{g})^Te \ (1 - \frac{r}{n+1}),$$

a result very similar to that obtained before when the optimal value was known.

If $(y(z_k),z_k)$ is feasible in $(\overline{D})$, we merely increase z from z_k, adjusting $y(z)$ as in (3.2), until some constraint in $(\overline{D})$ is tight. Since we have a feasible solution to $(\overline{D})$, this value z_{k+1} is again a valid lower bound, and since a constraint is tight, the new $\overline{p}$ has a nonpositive component and the argument above can be repeated. Using duality in this way to generate and update lower bounds was first proposed in Todd and Burrell [72].

We now have two arguments explaining the reduction of the order of $(1 - \frac{1}{n})$ in the (transformed) objective function. The geometric argument used the fact that containing and contained balls had radii in the ratio n, while the algebraic argument basically loses two factors of $\sqrt{n+1}$; one in going from $(\overline{c}-z\overline{g})^Te = \overline{p}^Te \cong \|\overline{p}\|_1$ (if the lower bound is updated, $\overline{p} \geq 0$) to $\|\overline{p}\|_2 \geq \|P_{\overline{B}}(\overline{c}-z \ \overline{g})\|_2$, and one in going from the latter 2-norm to $\|P_{\overline{B}}(\overline{c}-z\overline{g})\|_\infty$.

Unfortunately, when we transform the new point $\overline{x}(r)$ back to the feasible region for (P), we may not achieve a corresponding decrease in the objective function $\tilde{c}^T\tilde{x}$. The reason is that linear objective functions, while invariant under the scaling operation from $(\tilde{P})$ to $(\overline{P})$, are not invariant under the projective transformation corresponding to the replacement of the normalizing constraint. Hence,

instead of working with the objective function, we use Karmarkar's
potential function

$$\tilde{f}(\tilde{x};z) := (n+1)\ell n(\tilde{c}^T\tilde{x} - z) - \sum_{1}^{n+1} \ell n \, \tilde{x}_j$$

$$= \sum_{1}^{n+1} \ell n \, (\frac{(\tilde{c}-z\tilde{g})^T\tilde{x}}{\tilde{x}_j}), \qquad (3.4)$$

defined on strictly positive feasible solutions $\tilde{x}$ with $z \leq z_*$. This
function combines (a nonlinear tranformation of) the objective function
with a barrier term repulsing $\tilde{x}$ from the boundary of the feasible
region. We can view $f(\cdot;z_*)$ as an exact barrier function, similar to
the exact penalty functions of nonlinear programming.

Moreover, f is seen to be homogeneous of degree zero in its
second definition, and it easily follows that, if $\overline{x}$ is the
corresponding renormalized, rescaled point

$$\overline{x} = \frac{n \, \tilde{X}_k^{-1}\tilde{x}}{e^T \, \tilde{X}_k^{-1}\tilde{x}},$$

then $f(\tilde{x};z)$ differs by a constant from

$$\overline{f}(\overline{x};z) := (n+1) \, \ell n(\overline{c}-z\overline{g})^T\overline{x} - \sum_{1}^{n} \ell n \, \overline{x}_j.$$

Hence, by decreasing $\overline{f}$ by a constant, we will decrease f in the
original space by a constant. it can be shown (see the references
cited above) that $\overline{f}$ can be so decreased by moving from e to $\overline{x}(\alpha)$,
for some $0 < \alpha < 1$. In essence, the argument above shows that the
first term decreases by a linear term in $\alpha > 0$. Since the second term

is minimized on S at $\bar{x}$ = e, its increase is only of second order in α for small $\alpha > 0$. Hence a suitable choice of α provides the required constant decrease. Indeed, several authors have shown that

$$\alpha = \frac{n-1}{2n-3} r \quad \text{decreases} \quad f \quad \text{by .3.}$$

Suppose that the feasible region of $(\tilde{P})$ is bounded, so that max $\{e^T\tilde{x}:\ \tilde{A}\tilde{x} = 0,\ \tilde{g}^T\tilde{x} = 1,\ \tilde{x} \geq 0\} = n\mu$ is finite. Then it can be shown that, after k iterations, we have

$$\tilde{c}^T \tilde{x}^k - z_k \leq \mu\ \exp(-.3k/(n+1))(\tilde{c}^T e - z_0).$$

This provides the $O(nL)$ bound on the number of iterations required stated in the introduction. In practice, one can terminate when the duality gap given on the left is sufficiently small.

For ease of comparison, we record here that moving from e towards $\bar{x}(\alpha)$ is equivalent in terms of the original problem (P) to moving from x^k in the direction

$$d_{PRO} := -X_k P_{AX_k} X_k c + \nu X_K P_{AX_k} e \qquad (3.5)$$

where $\nu := (c^T X_k^2 A^T (AX_k^2 A^T)^{-1} b - z_{k+1})/(1 + b^T(AX_k^2 A^T)^{-1} b)$ and $X_k := \text{diag}(x^k)$. We see that d_{PRO} is a combination of two directions, of which we shall see more in subsequent sections. The importance of these two directions was stressed by Gonzaga [32]. The first is a projected steepest descent direction for the objective function, which appears in the affine scaling method, while the second is a "centering" direction related to the logarithmic barrier function.

In [30], Gonzaga obtains a problem of the form $(\tilde{P})$ without increasing the dimension, and his method of updating z_k and setting ν is slightly different; his direction, however, is again of the form d_{PRO} in (3.5). Finally, if z_k is updated, then by setting $y^{k+1} := y(z_{k+1})$

we find that (y^{k+1}, z_{k+1}) is feasible in (D) with $z_{k+1} \leq b^T y^{k+1}$.
Moreover, it can be shown that

$$y^{k+1} = (AX_k^2 A^T)^{-1} AX_k^2 c - \nu (AX_k^2 A^T)^{-1} b. \qquad (3.6)$$

To conclude this section, we describe a selection of recent results relating to the projective method. Several papers are concerned with extending the applicability of the method to more convenient forms and relaxing the assumptions made. We have already described two; using duality to relax the assumption of known optimal value [72] and the standard-form variant [4,18,24,30,66,81]. In addition, Anstreicher [4] showed how to apply the method to fractional linear programming problems, how to generate an initial lower bound during the course of the algorithms and how to ensure monotonicity in the objective function. De Ghellinck and Vial [18] described an algorithm that achieved feasibility along with optimality, and Anstreicher [5] proposed another approach to this problem. In addition, de Ghellinck and Vial renormalize to hold the first part of the potential function fixed and thereby achieve a short proof of a large decrease in the potential function, without having to balance the two terms as in the standard approach outlined above. However, this analysis lacks the geometric insights afforded by other viewpoints.

The algorithm we have described requires $O(nL)$ iterations (or $O(nq)$ iterations to achieve an improvement in the duality gap of a factor of 2^q), and $O(n^3)$ arithmetic operations per iteration to perform the projections. However, the projections on successive iterations are into $N(AX_{k-1})$ and $N(AX_k)$ respectively. One might hope that by approximating X_k a reduction in complexity could be achieved. Indeed, by optimizing over inscribed and circumscribing ellipsoids of the form $\{\overline{x}: \overline{A}\overline{x} = 0, e^T\overline{x} = n, \|D(\overline{x}-e)\| \leq \delta\}$, where the diagonal entries of the diagonal matrix D are suitably close to 1, one can still achieve a constant reduction in potential. Moroever, the resulting direction is (cf. (3.5))

$$\hat{d}_{PRO} := -\hat{X}_k P_{A\hat{X}_k} \hat{X}_k c + \hat{\nu} \hat{X}_k P_{A\hat{X}_k} \hat{X}_k^{-1} X_k e$$

for a suitable scalar $\hat{\nu}$, where $\hat{X}_k = D^{-1} X_k$. By choosing D appropriately, we may be able to arrange that many diagonal entries of $\hat{X}_k$ are equal to those of $\hat{X}_{k-1}$ in which case the key matrix $A\hat{X}_k^2 A^T$ differs from $A\hat{X}_{k-1}^2 A^T$ by a matrix of low rank. This is indeed possible if we also allow a common scalar multiple, and Karmarkar [37] shows that on average only $O(\sqrt{n})$ rank one updates need be made. In this way the overall complexity of the algorithm is reduced to $O(n^{7/2}L)$ arithmetic operations, compared to $O(n^4 L)$ for the ellipsoid method (Khachian [41]). In practice, the projective method works much better than its worst-case bound, partly because line searches on α permit much larger decreases in the potential function. It is therefore worthwhile to point out that Anstreicher [7] has shown how the above complexity improvement can still be achieved with a safeguarded line search on α.

Mention of the ellipsoid method above prompts us to remark that there is indeed a rather close relationship between the two algorithms, as described in [70]. Indeed, both have as their heart the solution of a weighted least-squares problem at each iteration (y^k in (3.2) is the solution to

$$\min \|\tilde{X}_k(\tilde{A}^T y - (\tilde{c}-z_k\tilde{g}))\|$$

for example). In the ellipsoid method, only one weight changes from one iteration to the next, so that the solution can be updated in $O(n^2)$ work; but $O(n^2 L)$ iterations are required. In the projective method, only $O(nL)$ iterations are needed (and in practice far fewer) while $O(n^3)$ operations per iteration are needed in the basic algorithm. The previous paragraph shows that a lower-rank change can be made without increasing the bound on the number of iterations. However the advantage of projective methods seems far greater than a

factor of $\sqrt{n}$ in practice, since the ellipsoid algorithm typically
performs similarly to its worst-case bound.

While the elliposid method appears to be a hopelessly inefficient
practical tool in linear programming (while at the same time a very
powerful theoretical tool in combinatorial optimization--see Grötschel,
Lovasz and Schrijver [34]), it has inspired further insights into
projective methods. Both Todd [69] and Ye [79] have shown how
ellipsoids containing all optimal dual solutions to (P) can be
generated, and Ye shows that the logarithmic volume of a containing
ellipsoid in the space of dual slacks is exactly Karmarkar's potential
function. Such ellipsoids can be used for determining primal variables
that are zero at optimality.

In practical computation, the most costly step is the projection
shown in (3.1) or (3.3). It is therefore important to be able to use
inexact projections, and Goldfarb and Mehrotra [27,28] have shown how
to do this while maintaining a polynomial bound. The approach of de
Ghellinck and Vial is valuable here, too, since it does not require
feasibility; see Vial [78]. There has arisen a feeling that projective
algorithms are inferior in practice to affine methods; see [50,59] for
instance. However, this judgement appears to be based largely on
results quoted by Tomlin [74], an early study that used a reformulation
that seems less efficient than that described above.

Much recent work has been concerned with the effect of changing
the potential function (3.4), in particular increasing the weight on
the objective function logarithm to $n+\sqrt{n}$; see Todd and Ye [73],
Gonzaga [33] and Ye [80]. Apparently such a choice of potential
function was first proposed by Karmarkar [38]. Homogeneity can be
maintained by changing the coefficient of the "constant term" $\ell n \; \tilde{x}_{n+1}$
to $\sqrt{n}$. This is analogous to adding $\sqrt{n}$ homogenizing variables
instead of one, which corresponds to adding the constraint $-b^T y + z \leq 0$
$\sqrt{n}$ times in the dual to $(\tilde{P})$ and is thus similar to Renegar's
path-following method [62] discussed in Section 5. We describe Ye's
work [80] in Section 6.

The excellent practical behavior of variants of Karmarkar's
algorithm suggests that it may be possible to improve the worst-case
complexity bound of $O(nL)$ iterations. Indeed, computational experience
with projective methods indicates that simple line searches lead to
average decreases in potential function that grow, perhaps even
linearly, with dimension n. However, it is easy to construct examples
in which only a fixed decrease of the potential function is possible,
and Anstreicher [6] and McDiarmid [49] have provided the best possible
bound for the decrease, about .7. Moreover, their analysis (see in
particular [49]) shows what is required to obtain larger decreases;
many iterations in which the components of $P_{\overline{B}}(\overline{c}-z_k\overline{g})$ are roughly
equal. This is also suggested by our analysis of the reduction in
$(\overline{c}-z_k\overline{g})^T\overline{x}$ -- see for example (3.3). Recently, Asic, Kovacevic-Vujcic
and Radosavljevic-Nikolic [8] have analyzed the original Karmarkar
algorithm and shown that, at least in the nondegenerate case,

$P_{\overline{B}}\overline{c}/\|P_{\overline{B}}\overline{c}\|$ converges to a vector with just two values for its
components, a positive one for those components of x that are zero at
the optimum and a negative one for those that are positive. This
yields an asymptotic reduction of $\Omega(\sqrt{n})$ for the potential function.

Finally, since part of our aim is to illustrate the synergy of
different approaches to linear programming, we cannot resist citing
[71], where a surprising connection is drawn between a variant of
Karmarkar's projective algorithm and the decomposition principle of
Dantzig and Wolfe for the simplex method.

4. Affine Interior Methods.

Let us consider again the standard form problem

$$\max\ c^T x$$
$$(P) \qquad\qquad Ax = b$$
$$x \geq 0$$

and apply the scaling directly without first reformulating the problem as in the previous section. Suppose x^k is a known strictly positive feasible solution to (P); then e is such a solution to the rescaled problem

$$\min \quad \hat{c}^T \hat{x}$$

(P̂)

$$\hat{A} \; \hat{x} \; = \; b$$

$$\hat{x} \geq 0$$

where

$$\hat{c}^T \; = \; c^T X_k$$

$$\hat{A} \; = \; A X_k$$

and X_k is again a diagonal matrix whose diagonal entries are the components of x^k . Because an affine scaling is used, these methods are also called affine scaling algorithms.

We can now replace the nonnegativities by the more restrictive constraint $\|\hat{x} - e\| \leq \beta \leq 1$ and solve the resulting problem to get

$$\hat{x}(\beta) \; = \; e \; - \; \beta \; P_{\hat{A}} \; \hat{c}/\|P_{\hat{A}} \; \hat{c}\|; \tag{4.1}$$

see figure 3. However, while we now have a ball contained in the feasible region, it is far from clear how a containing ball can be manufactured, and this poses a considerable difficulty in analyzing the resulting algorithm. From (4.1) we see that

$$\hat{c}^T \; \hat{x}(\beta) \; = \; \hat{c}^T e \; - \; \beta \; \|P_{\hat{A}} \; \hat{c}\|,$$

but it seems impossible to obtain a lower bound for this norm as in (3.3). Nevertheless, we can scale back to get

$$x^{k+1} \; = \; X_k \; \hat{x}(\beta) \tag{4.2}$$

and continue the iterations.

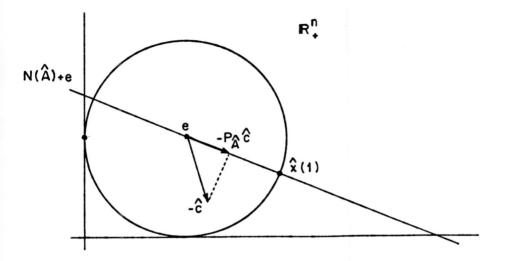

Figure 3. An iteration in the affine interior algorithm.

This is the _affine scaling algorithm_, which, as a natural simplification of Karmarkar's method, was proposed by a number of authors, including Vanderbei, Meketon and Freedman [77] and Barnes [10], soon after the appearance of Karmarkar's work. However, it turned out that the method had been proposed almost twenty years before, by a student of Kantorovich, Dikin [19,20].

From (4.1) and (4.2), we see that x^{k+1} is obtained from x^k by a move in the direction

$$d_{AFF} := -X_k P_{AX_k} X_k c,\qquad (4.3)$$

which is the first direction appearing in the linear combination (3.5) yielding d_{PRO}.

As far as the choice of β goes, it seems reasonable to choose a large value to decrease the objective function more as long as $\hat{x}(\beta)$ remains positive. In fact, Dikin [19,20] chooses $\beta = 1$ and shows that, if $\hat{x}(\beta)$ has a zero component it is in fact optimal. Most other authors allow β to be greater than 1, as long as all components of $\hat{x}(\beta)$ remain greater than $1-\gamma > 0$; this gives

$$\beta = \gamma \|P_A \hat{c}\| / \max_j (P_A \hat{c})_j$$

and corresponds to moving a proportion of the way γ to the boundary of the feasible region as long as $P_A \hat{c}$ has a positive component; otherwise, of course, (P) is unbounded.

The convergence of the method has been shown in [77] for this choice of β, and for fixed $\beta < 1$ in [10], both under the assumption of primal and dual degeneracy. However, Dikin's convergence proof [20] in 1974, for the case $\beta = 1$, only assumes primal nondegeneracy and yields the stronger result that the iterates converge to a point in the relative interior of the optimal set. Dikin's result has been elucidated by Vanderbei and Lagarias [76].

Adler, Karmarkar, Resende and Veiga [1] suggested applying the affine algorithm to the dual problem

$$\max \quad b^T y$$

(D) $$A^T y + s = c$$

$$s \geq 0$$

since, if numerical errors are made in computing the updated y, a simple change in s will maintain feasibility. (We remark that Gay [24] proposed a similar dual projective algorithm, and that, if one is willing to compute (a representation of) a basis inverse, one can allow mild inaccuracies in primal algorithms while regaining feasibility by adjusting the values of basic variables.) By eliminating the unrestricted variables y, applying the primal affine algorithm to the slack variables s, and then expressing the result in terms of the variables y again, we are led to an algorithm proceeding as follows.

Suppose we have a feasible solution (y^k, s^k) to (D) with $s^k > 0$. Let S_k be the diagonal matrix with diagonal entries the components of s^k. Then compute

$$d_y = (AS_k^{-2}A^T)^{-1}b;$$

$$d_s = -A^T d_y; \qquad\qquad (4.4)$$

and

$$y^{k+1} = y^k + \alpha d_y, \quad s^{k+1} = s^k + \alpha d_s,$$

where

$$\alpha = \gamma \min_{j:(d_s)_j<0} (s^k_j/(-d_s)_j)$$

and $0 < \gamma < 1$. This choice of α amounts to moving a proportion γ of the way to the boundary of the feasible region; if d_s is nonnegative, then it is easily seen that (D) is unbounded (and hence (P) is infeasible).

We have not yet discussed how to terminate the primal or dual affine algorithm. Of course, it would be desirable to have bounds as in the projective method so that we could stop with a guarantee of near-optimality. However, such bounds appear as elusive as containing balls, and in practice the iterations are truncated when

$$\frac{|c^T x^{k+1} - c^T x^k|}{\max\{1, |c^T x^k|\}} \quad \text{or} \quad \frac{|b^T y^{k+1} - b^T y^k|}{\max\{1, |b^T y^k|\}}$$

is below some tolerance. This appears to work well in most cases, although there is no assurance of quality of the resulting solution.

The most theoretically satisfactory way to obtain bounds is via
duality, and one might hope that feasible solutions to (D) (in the
primal algorithm) or (D) (in the dual algorithm) would be generated.
In particular, if one really wants to solve (P), then applying the dual
affine algorithm seems useless without some way to recover an optimal
solution to (P). In fact, in either algorithm, dual estimates can be
obtained which, under nondegeneracy assumptions, converge to the
optimal solutions to the respective problems. For the primal
algorithm, we set

$$y_{AFF} = (AX_k^2A^T)^{-1}AX_k^2c \qquad (4.5)$$

so that $d_{AFF} = -X_k^2(c - A^Ty_{AFF})$. In the dual algorithm, we set

$$x_{AFF} = -S_k^{-2}d_s = S_k^{-2}A^T(AS_k^{-2}A^T)^{-1}b. \qquad (4.6)$$

Note how y_{AFF} is the first part of the feasible dual solution
y^{k+1} in (3.6); the second part is clearly related to the direction d_y
in (4.4), with X_k^2 instead of S_k^{-2}. Similarly, if we take a step of
size 1 from x^k in the direction $-X_kP_{AX_k}e$ of the second part of
d_{PRO} in (3.5), we arrive at $X_k^2A^T(AX_k^2A^T)^{-1}b$, which is again closely
related to x_{AFF} above.

Because of results of Megiddo and Shub [53] concerning the
associated continuous trajectories, it is not believed that the primal
and dual affine algorithms above are polynomially bounded. However,
Monteiro, Adler and Resende [59] have recently described a variant of
these algorithms that operates in primal-dual space and which has an
$O(nL^2)$ iteration bound. Barnes, Chopra and Jensen [11] add "centering"
steps based on a potential function to the affine algorithm to achieve
a bound of $O(nL)$ iterations.

As noted above, the algorithms appear to work well in practice. Vanderbei, Meketon and Freedman [77] give experimental results for the primal affine method, while Adler, Karmarkar, Resende and Veiga [1] and Monma and Morton [56] give extensive results for the dual affine method, which are generally favorable in comparison to the MINOS code [60] which implements the simplex method for linear programming problems.

5.   Path-following Algorithms

In section 3 and 4 we described directions d_{PRO} and d_{AFF}, defined at each strictly positive feasible solution $x = x^k$ to (P) (for definiteness, take z_{k+1} to be a constant, say $z_{k+1} = z_*$, in (3.5)). Many authors realized that insight into the algorithms could be obtained by studying the corresponding vector field associating d_{PRO} or d_{AFF} with x and the associated trajectories of $\dot{x} = d_{PRO}(x)$ or $\dot{x} = d_{AFF}(x)$, $x(0) = x^0$. Such trajectories were investigated by Karmarkar, by Bayer and Lagarias [12,13,47], by Megiddo [51], and by Megiddo and Shub [53].

Bayer and Lagarias conduct their analysis for a general linear programming problem. For simplicity, we confine ourselves to the forms (P) and (D), with the assumptions of Section 3 holding. Let $F(P)$ and $F(D)$ denote the feasible region of (P) and (D) respectively, and let $F_+(P)$ and $F_+(D)$ denote their relative interiors. Under our assumptions, $F_+(P) = \{x: Ax = b, x > 0\}$ and $F_+(D) = \{y: s = c - A^T y > 0\}$. For the primal problem (P), we introduce the pure barrier function

$$\beta_P(x) := -\Sigma_j \log x_j \qquad (5.1)$$

defined on $F_+(P)$. This function is minimized where its gradient, projected into the null space of A, vanishes. Bayer and Lagarias define the Legendre transform of $x \in F_+(P)$ to be this projected gradient:

$$\xi = \xi(x) := P_A \nabla \beta_p(x) \qquad (5.2)$$

They show that this is a one-to-one transformation, and, if $F(P)$ is bounded, onto the null space of A. Hence in this case there is a unique point $x \in F_+(P)$ with $\xi(x) = 0$; this point is called the center (or analytic center) of the constraints defining $F(P)$, and has also been studied by Huard [36] and by Sonnevend [65]. If e^T is in the row space of A, then $x = e$ is the center of the primal constraints.

For the dual problem (D), we let

$$\beta_D(y) := -\sum_j \log s_j, \qquad (5.3)$$

where $s = s(y) = c - A^T y$, and define the Legendre transform of $y \in F_+(D)$ to be

$$\eta = \eta(y) := \nabla \beta_D(y). \qquad (5.4)$$

Again, this is one-to-one, and onto $\mathbb{R}^m$ if $F(D)$ is bounded, in which case the unique point y with $\eta(y) = 0$ is called the center for the constraints defining $F(D)$.

Bayer and Lagarias prove some remarkable properties of the trajectories when viewed in Legendre transform coordinates. Specifically, the affine trajectories are linearized, mapping into parallel lines $\{\xi^0 - \lambda P_A c\}$. The affine trajectory through the center of the constaints defining $F(P)$, if the center exists, is called the central (affine) trajectory and is mapped into $\{-\lambda P_A c\}$. Each projective trajectory, when viewed in Legendre transform coordinates, also has a radial component pulling it towards the origin (corresponding to the center), and lies in the 2-dimensional plane through ξ^0 and the line $\{-\lambda P_A c\}$. It is asymptotic to the latter line as $\lambda \to \infty$ (if the optimal value is zero). The central trajectory is also a projective trajectory and all projective trajectories

converge to it. Hence Legendre transform coordinates provide a natural
nonlinear geometry in which to view interior algorithms for linear
programming.

The Legendre transform is an algebraic mapping, and hence affine
and projective trajectories are algebraic curves [12,13,47]. It is
therefore attractive to consider higher-order methods to follow these
curves, and power expansions have been obtained by Bayer and Lagarias
[13], Megiddo [51], Adler, Karmarkar, Resende and Veiga [1] and
Monteiro, Adler and Resende [59].

Points on the central trajectory can be characterized in several
ways. They are centers of constraints defining slices $\{x: Ax = b, c^T x$
$= z, x \geq 0\}$ or cuts $\{x: Ax = b, c^T x \leq z, x \geq 0\}$, or points that
minimize the primal barrier function (for $\mu > 0$)

$$\phi_P(x;\mu) := c^T x + \mu\beta_P(x) \qquad (5.5)$$

over $F_+(P)$; in the latter two cases we may not obtain all the points
of the central trajectory [13]. Here ϕ_P is the classical logarithmic
barrier function attributed to Frisch [23], and central trajectories
have been studied in this context also - see Fiacco and McCormick [21].
Note that $\nabla_x \phi_P(x;\mu) = c - \mu X^{-1}e$ and $\nabla_{xx}^2 \phi_P(x;\mu) = \mu X^{-2}$ so that the
projected Newton barrier direction, the solution to

$$\min \nabla_x \phi_P(x;\mu)^T d + \tfrac{1}{2}d^T \nabla_{xx}^2 \phi_P(x;\mu d)$$
$$Ad = 0$$

is

$$d = -XP_{AX}(Xc - \mu e); \qquad (5.6)$$

the similarity between (5.6) and (3.5) was observed by Gill, Murray,
Saunders, Tomlin and Wright [25]--but note that ν in (3.5) may not be
positive. The first part, $-XP_{AX}Xc$, is again the direction of the
affine algorithm, while the second part, $\mu XP_{AX}e$, is seen to be the
projected Newton direction for the pure barrier function $\beta_P(x)$.

The optimality conditions for x to minimize $\Phi_P(x;\mu)$ are that $Ax = b$, $x > 0$ and $P_A\nabla_x\Phi_P(x;\mu) = 0$, which can be written as $c-\mu X^{-1}e = A^Ty$ for some y, or, letting $s := \mu X^{-1}e$,

$$Ax = b$$
$$A^Ty+s = c \qquad\qquad (5.7)$$
$$XSe = \mu e$$
$$x > 0, \ s > 0$$

where $S := \text{diag}(s)$. This system has a remarkably symmetric form; note that $y \in F_+(D)$. Indeed, it is easily seen that (5.7) gives the optimality conditions for y to solve the dual barrier function problem:

$$\max \Phi_D(y;\mu) := b^Ty - \mu\beta_D(y) \qquad\qquad (5.8)$$

for $y \in F_+(D)$. Megiddo [51] first described these primal-dual central pathways, and his work has inspired several algorithms.

Megiddo and Shub [53] study the boundary behavior of affine and projective trajectories. In particular, they show that there is an affine trajectory that approaches arbitrarily close to every vertex of the Klee-Minty cube [43], while projective trajectories avoid this behavior. This result suggests that the affine algorithm may not be polynomial, but since it takes discrete steps in the direction given by the affine trajectory at the current point rather than following the trajectory closely, this remains an open question.

Now we turn to the algorithms that try to follow the central trajectory, in primal, dual or primal-dual space. The first to analyze such a method was Renegar [62], whose algorithm assumes an inequality-form linear program like (D). Suppose z_0 is a strict lower bound to the optimal value of (D), and suppose that the center y^0 of the system of inequalities

$$A^T y \leq c$$

$(D(z))$ $-b^T y \leq -z$

$$\vdots \qquad \vdots$$

$$-b^T y \leq -z$$

for $z = z_0$ is known. (Here the inequality $-b^T y \leq -z$ is repeated n times; note that the center depends on the defining system of inequalities, not just the feasible region.) Then $b^T y^0 > z_0$, so that z_0 can be updated. Renegar shows that, if

$$z_1 = \delta \, b^T y^0 + (1 - \delta) z_0 \quad \text{for} \quad \delta = 1/13\sqrt{n} \qquad (5.9)$$

then a single Newton step from y^0 to approximate the center of the system $(D(z_1))$ yields a point y^1 that is sufficiently close to the center that the procedure can be iterated.

Specifically, let $\bar{y}^1$ be the center of the system $(D(z_1))$, and let $\bar{s}^1$ be the corresponding slack vector (in this system, so that $\bar{s}^1 \in \mathbb{R}^{2n}$). Let s^1 similarly correspond to y^1. Then, with $\bar{S}_1 := \text{diag } \bar{s}^1$,

$$\| \bar{S}_1^{-1} s^1 - e \| \leq 1/46. \qquad (5.10)$$

Renegar proves that this sense of proximity to the center can be maintained throughout the algorithm, as the lower bounds are increased and Newton steps are taken to approximate the resulting centers. Hence the method, while taking discrete steps, remains close to the central trajectory. Of course, (5.10) is only a theoretical criterion, since $\bar{y}^1$ and $\bar{s}^1$ cannot be computed exactly.

If z_* denotes the optimal value of (P) and (D), Renegar shows
that all y^k's lie in $F_+(D)$ and

$$z_* - b^T y^k \leq (1 - \frac{1}{28\sqrt{n}})^k (z_* - z_0).$$

From this, it follows that $O(\sqrt{n}L)$ iterations suffice to obtain
$(z_*-b^T y^k) \leq 2^{-L}(z_*-z_0)$. This is an improvement of a factor of $\sqrt{n}$
over the bound for projective methods. Renegar computed the Newton
step exactly, which requires $O(n^3)$ arithmetic operations per
iteration and an overall complexity of $O(n^{7/2}L)$. Subsequently, Vaidya
[75] showed how to compute the Newton steps inaccurately, as in
Karmarkar's modified algorithm discussed in section 3, to reduce the
complexity to $O(n^3 L)$.

Independently, Gonzaga [31] developed an algorithmm of the same
complexity that stayed close to the central primal trajectory by
considering the barrier function ϕ_P in (5.5). Let $\bar{x}^k$ solve the
problem

$$\begin{array}{c} \min \phi_P(x;\mu) \\ (BP(\mu)) \qquad Ax = b \\ x > 0 \end{array}$$

for $\mu = \mu_k$. Gonzaga's algorithm generates a sequence $\{\mu_k\} \to 0$ of
barrier parameters and $\{x^k\}$ of approximate solutions to $(BP(\mu_k))$
satisfying

$$\|\bar{X}_k^{-1} x^k - \text{ell}\| \leq .015 \qquad (5.11)$$

for all k. Given such a solution x^k, Gonzaga updates μ_k by

$$\mu_{k+1} = (1 - \frac{.005}{\sqrt{n}})\mu_k \qquad (5.12)$$

and then lets x^{k+1} be the result of a single Newton step from x^k for the problem $(BP(\mu_{k+1}))$. Hence in $O(\sqrt{n}L)$ steps, μ_k is reduced by a factor of 2^{-L}, and $c^T x^k - z_*$ is closely related to $n\mu_k$. Finally, Gonzaga uses the idea of inexactly computing the Newton step to achieve an overall complexity of $O(n^3 L)$.

Several authors have devised algorithms to follow the primal-dual central pathways of Megiddo [51] defined by (5.7), which we denote by $(PD(\mu))$. Suppose we have an approximate solution to (x^k, y^k, s^k) to $(PD(\mu_k))$ for some $\mu_k > 0$, where the only inaccuracy is in the nonlinear centering condition $XSe = \mu e$. Moreover, suppose

$$\|X_k S_k e - \mu_k e\| \le \alpha\mu_k \qquad (5.13)$$

for suitably small $\alpha > 0$. The idea is then to set

$$\mu_{k+1} = (1 - \sigma)\mu_k \qquad (5.14)$$

and then take a single Newton step from (x^k, y^k, s^k) to get an approximate solution to $(PD(\mu_{k+1}))$, say $(x^{k+1}, y^{k+1}, s^{k+1})$. As long as nonnegativity is preserved for x^k and s^k, we will maintain feasibility in the primal and dual systems since the corresponding equations in $(PD(\mu))$ are linear. Hence $c^T x^k \le z_* \le b^T y^k$, and it is easy to check that

$$b^T y^k - c^T x^k = (x^k)^T s^k.$$

By (5.13), therefore, the duality gap is bounded by $\mu_k(n + \alpha\sqrt{n})$. Hence controlling μ_k as in (5.14) will control the duality gap also.

   An algorithm of this form was first proposed by Kojima, Mizuno and
Yoshise [45].   They took σ in (5.14) to be a constant, and found that
they had to use a modified Newton step with a step size proportional to
$1/n$ to preserve (5.13).   This led to a method requiring $O(nL)$
iterations.   Monteiro and Adler [57] showed that by taking σ
proportional to $1/\sqrt{n}$ a full step could be taken, and this led to an
$O(\sqrt{n}\ L)$-iteration method.   Monteiro and Adler also extended their
method to solve convex quadratic programming problems [58].
Independently of Monteiro and Adler's work, Kojima, Mizuno and Yoshise
[46] improved their method so that it only required $O(\sqrt{n}L)$ iterations
and extended it to solve positive semi-definite linear complementarity
problem, which include linear and convex quadratic programming
problems.   All these methods use inexact solutions of systems to
achieve an overall complexity of $O(n^3 L)$ arithmetic operations.
   The path-following algorithms discussed in this section all
require an initial center or approximate center of the primal or dual
system of inequalities, or of both.   By suitably transforming the
original problem, such centers can be obtained, by methods discussed in
the cited references.
   Suppose that the appropriate linear systems are solved exactly.
Then Renegar's method requires the solution of systems involving the
matrix

$$AS_k^{-2} A^T,$$

Gonzaga's method uses the coefficient matrix

$$AX_k^2 A^T,$$

and the primal-dual methods of [45,46,57,58] solve systems with the
matrix

$$AX_k S_k^{-1} A^T .$$

The scaling matrix $X_k S_k^{-1}$ is the "geometric mean" of the primal
scaling matrix X_k^2 and the dual scaling matrix S_k^{-2}. Also note the
similarity to the matrices arising in the projective (see (3.6)) and
affine (see (4.4) and (4.5)) methods.

Finally, let us discuss briefly the $O(\sqrt{n} L)$ iteration bound of
the path-following methods. First note that this bound is in a sense a
conscious decision of the algorithms, in that it follows from Renegar's
choice of update for z_k (5.9) or the rule for updating μ_k in (5.12)
or (5.14). In contrast, Karmarkar's bound of $O(nL)$ is less
transparent, although it is explained to some extent by the geometric
and algebraic reasoning given in section 3. However, this naturally
raises the question whether a better improvement can be achieved, and
where the $\sqrt{n}$ "comes from". Unfortunately, I can see no simple
geometric justification. However, Gonzaga [31] shows that, if $x(\mu)$

denotes the solution to $(BP(\mu))$, then $\dfrac{d\ \overline{X}_k^{-1} x(\mu)}{d\mu} = P_{A\overline{X}_k} e/\mu$, so that to

bound $\| \overline{X}_k^{-1} x(\mu) - e \|_2$ by a constant, the proportional change in μ
should be bounded by $1/\| P_{A\overline{X}_k} e \|_2 \geq 1/\| e \|_2 = 1/\sqrt{n}$, so that again the
effects of different norms seems to be important. Further discussion
of this can be found in Kojima, Mizuno and Yoshise [45] and Todd and Ye
[73].

6. Combining the Approaches.

Although they all operate in the interior of the feasible region,
involve similar computation at each iteration, and are based on very
closely related concepts, the algorithms of the previous three sections
have somewhat diverse motivations. The projective methods employ
projective transformations and rely on the decrease of a potential
function, the affine methods use only diagonal scaling and projected

gradient ideas, while the path-following methods take small steps to
stay close to a central trajectory and achieve their (currently best)
complexity by measuring the optimality gap or duality gap directly. Is
there a way to combine the most favorable and simple features of these
three classes of algorithm? We shall see that there is.

First we remark that, for practically useful algorithms with
theoretical guarantees, projective methods have a considerable
advantage over path-following methods. Indeed, line searches in
practice allow large decreases in the potential function to be
realized, and constant decreases are guaranteed from any interior
feasible point. On the other hand, path-following methods would
certainly benefit from much bolder increases in the lower bound z_k
or decreases in the barrier parameter μ_k, but then the measure of
closeness given by (5.10) or (5.11) or (5.13) is likely to be violated
and the next iteration may not even be well-defined. (A large-step
variant of Monteiro and Adler's algorithm has been implemented by
McShane, Monma and Shanno [50] with very encouraging practical results,
but the theoretical guarantee has gone.) The key here is the use of
the potential function; projective transformations seem to be a less
fundamental concept.

To reinforce this argument, we note that there are two polynomial
variants of the affine scaling method. Barnes, Chopra and Jensen [11]
use centering steps as well as objective function steps, and measure
progress in centering with a potential function. Their method requires
$O(nL)$ steps. Monteiro, Adler and Resende [59] work in primal-dual
space using a combination of the primal and dual affine directions.
Their analysis forces them to take very short steps, leading to $O(nL^2)$
iterations.

Potential functions seem to provide an extremely useful guarantee
of progress. Gonzaga [33] describes an algorithm that is almost as
simple as the primal affine variant, taking steps in the projected
gradient direction

$$d^k = -X_k P_{AX_k} X_k \nabla f_\rho(x^k) \qquad (6.1)$$

for the potential function

$$f_\rho(x) := \rho \ell n \ c^T x - \Sigma_j \ \ell n \ x_j,$$ (6.2)

where the optimal value is assumed to be zero. If $\rho = n + \sqrt{n}$, a constant decrease in the potential function can be achieved, leading to an $O(nL)$ iteration bound. Note that no projective transformation need be used. Also, ρ above is greater than the value n+1 that is used in projective algorithms, thus putting a greater emphasis on decreasing the objective function.

Earlier, Ye and Todd [73,82] introduced a primal-dual potential function. In discussing containing ellipsoids in the path-following methods, they show [82] that the primal-dual algorithm of [46,57,58] decreases the primal-dual potential function

$$g_\rho(x,s) := \rho \ \ell n \ x^T s - \Sigma_j \ \ell n \ x_j - \Sigma_j \ \ell n \ s_j$$ (6.3)

by $\cdot 1\sqrt{n}$ when $\rho = 2n$. In [73], they describe a "centered" projective algorithm that operates in primal-dual space using the "average" scaling $X_k(S_k)^{-1}$, uses projective transformations and strives to decrease a potential function, and yet with fixed step sizes stays close to the central trajectories and requires only $O(\sqrt{n}L)$ iterations. It uses g_ρ with $\rho = n + \sqrt{n}$, and indeed this value of ρ is critical to the analysis. Note that this is less than the number (2n) of barrier terms, so this method puts less weight on the duality gap $x^T s$ and more on the centering. Also, g_ρ can be written

$$g_\rho(x,s) = (\rho - n)\ell n \ x^T s - \Sigma_j \ \ell n \ (\frac{x_j s_j}{x^T s})$$ (6.4)

where the n terms $x_j s_j / x^T s$ sum to 1; the weight $\rho - n = \sqrt{n}$ on the duality gap term is less than the number n of the primal-dual barrier terms, and these terms try to keep all $x_j s_j$'s approximately equal, c.f. (5.7). It might be hoped that this algorithm would allow line searches on g_ρ thus maintaining its attractive iteration bound while allowing better progress in practice. Unfortunately, however, a constant decrease in g_ρ can only be guaranteed when (x,s) is close to the central trajectory in primal-dual space. Hence, while this method is based on a potential function, it seems to require short steps.

   This difficulty has been resolved in recent exciting work of Ye [80]. Ye's analysis depends on the relationship between primal, dual, and primal-dual potential functions. Let us fix $\rho = n + \sqrt{n}$ and write these functions in terms of x and y. We have the primal function

$$f_P(x,\underline{z}) := \rho \, \ell n(c^T x - \underline{z}) - \sum_j \ell n \, x_j \qquad (6.5)$$

for $x \in F_+(P)$ and $\underline{z} \leq z_*$; the dual function

$$f_D(y,\overline{z}) := \rho \, \ell n(\overline{z} - b^T y) - \sum_j \ell n \, s_j \qquad (6.6)$$

where $s = s(y) = c - A^T y$, $y \in F_+(D)$ and $\overline{z} \geq z_*$; and the primal-dual function

$$
\begin{aligned}
f_{PD}(x,y) &:= \rho \, \ell n(c^T x - b^T y) - \sum_j \ell n \, x_j - \sum_j \ell n \, s_j \\
&= \rho \, \ell n \, (x^T s) - \sum_j \ell n \, x_j - \sum_j \ell n \, s_j \qquad (6.7) \\
&= \sqrt{n} \, \ell n \, (x^T s) - \sum_j \ell n \, (\frac{x_j s_j}{x^T s})
\end{aligned}
$$

where s is as above, for $x \in F_+(P)$ and $y \in F_+(D)$. Note that, if $\underline{z} = b^T y$ for $y \in F_+(D)$, then

$$f_P(x;z) = f_{PD}(x,y) + \sum_j \ell n \, s_j,$$

while if $\overline{z} = c^T x$ for $x \in F_+(P)$, then

$$f_D(y;\overline{z}) = f_{PD}(x,y) + \sum_j \ell n \, x_j.$$

Suppose that an initial x^0, y^0 can be found with $f_{PD}(x^0,y^0) = O(\sqrt{n}L)$. (This can be achieved as in the primal-dual path-following methods.) Since $-\sum_j \ell n(x_j s_j / x^T s)$ is nicely bounded below, if we can decrease f_{PD} by a constant at each iteration we will have $(x^k)^T s^k \leq 2^{-L}$ in $O(\sqrt{n}L)$ iterations, by (6.7). Ye achieves this as follows.

Given (x^k, y^k) with associated dual slack vector s^k, first compute

$$p = P_{AX_k} \, X_k \, \nabla_x \, f_P(x^k, \underline{z}_k)$$

where $\underline{z}_k = b^T y^k$. If $\|p\|$ is sufficiently large (at least a certain constant), then just updating x^k to

$$x^{k+1} = x^k - \beta \, X_k p / \|p\|$$

(compare with (6.1)) leads to a constant decrease in $f_P(\cdot, \underline{z}_k)$; hence, with $y^{k+1} = y^k$, f_{PD} is decreased by a constant in moving from (x^k, y^k) to (x^{k+1}, y^{k+1}).

Let

$$y(z_k) = (AX_k^2 A^T)^{-1} \left(AX_k^2 c - \frac{c^T x^k - z_k}{\rho} b\right).$$

Ye shows that $\|p\|$ will be large enough if $y(z_k)$ is not feasible in (D); if $y(z_k)$ is feasible but $\|X_k s(z_k) - \Delta e\| \geq \alpha\Delta$ where

$$\Delta := \frac{(x^k)^T s(z_k)}{n} = \frac{c^T x^k - b^T y(z_k)}{n}$$ and $\alpha \in (0,1)$ is some constant; or if $\Delta \geq (1 - \frac{\alpha}{2\sqrt{n}})\Delta_k$, where $\Delta_k = \frac{c^T x^k - b^T y^k}{n}$. Here $s(z_k)$ is the slack vector corresponding to $y(z_k)$. Hence the only case in which a constant decrease in the potential function cannot be achieved by a primal move is when $y(z_k)$ is feasible, $(x^k, y(z_k), s(z_k))$ is approximately centered (compare the condition above with (5.13)), and the duality gap of x^k and $y(z_k)$ is appreciably smaller than that of x^k and y^k. In this case, Ye proves that replacing y^k by $y(z_k)$ achieves a constant decrease in the primal-dual potential function. Hence, either a primal or a dual step achieves the desired goal.

Ye's analysis is very short and simple. It achieves an $0(\sqrt{n}L)$ bound on the number of iterations while allowing great freedom in line searches to decrease the potential function, and it does not require the algorithm to stay close to any central trajectories. Since the analysis takes place in primal-dual space, it would be nice to develop a symmetric version that made moves in both primal and dual spaces. However, it may be impossible to do this efficiently (i.e., using only a single projection matrix) unless the primal-dual pair is approximately centered, which appears to require taking short steps. Certainly understanding Ye's algorithm better, and perhaps finding an appropriate geometry in which to view it, is an important task for future research. Freund [22] has analyzed the effect of different choices for ρ and argues that $\rho = n + \sqrt{n}$ provides the best bound $0(\sqrt{n} L)$ on the number of iterations.

7. Conclusions and open questions.

In the preceding sections we have attempted to give an overview of
several recent developments in linear programming. We have tried to
stress the insights to be obtained from an appropriate geometry in
which to view a certain problem or technique. For example, the simplex
method 'looks good' in the space of the columns of the coefficient
matrix A (the simplex interpretation of the method). Karmarkar's
projective algorithm appears natural in the transformed space where the
current iterate is the center of the simplex, and path-following
methods seem attractive when viewed in Legendre transform coordinates.
Questions concerning polyhedra in general and simplex method variants
in particular can profitably be considered from several different
viewpoints.

At the same time, duality, the cornerstone of the simplex method,
is also a key ingredient in other algorithms. It provides bounds for
termination criteria and so that potential functions can be used when
optimal values are unknown; "centered" solutions naturally provide
feasible dual solutions; and primal-dual potential functions appear to
provide better bounds than either primal or dual ones. (Duality also
appears in disguise in the ellipsoid method.)

Several questions remain open, and some can even be attempted with
some hope of success. The Hirsch conjecture remains a significant
unsolved problem in the study of polytopes; perhaps more practically
important is the question of the existence of a simplex pivot rule
requiring only polynomially many iterations, possibly for new special
classes of linear programming problems. Can we obtain good expected
bounds on the number of iterations for more practically reasonable
probability distributions for the data? Is it possible to extend
Debreu's argument described in section 2 to general m?

One key question for Karmarkar's projective algorithm is whether
the complexity bound can be further improved. This is closely related
to the problem of finding lower bounds for the complexity. A similar

question for path-following methods may require the analysis of
higher-order methods as in [59]. Is there an exponential-step example
for the discrete-step affine method? Can we obtain geometrical insight
that "explains" the $O(\sqrt{n}L)$ iteration bound of several recent interior
methods? A more symmetric primal-dual version of Ye's recent algorithm
would also be of considerable interest.

One very fruitful application of linear programming in recent
years has been to combinatorial optimization problems. Using
sophisticated techniques of polyhedral combinatorics to identify and
generate classes of strong cutting planes for particular problems
within a linear programming system has proved remarkably successful in
solving much larger instances than were thought possible a few years
ago; see Hoffman and Padberg [35]. Can interior algorithms be used to
solve the linear programming problems that result, and in particular,
to reoptimize efficiently after the addition of constraints and/or
variables? Some work along these lines for perfect matching and linear
ordering problems has recently been done by Mitchell [54] using
variants of Karmarkar's algorithm. In addition, Karmarkar has given
lectures on a different approach to such problems using nonconvex
quadratic programming - see, e.g., [39].

References

1.   Adler, I., Karmarkar, N., Resende, M.G.C., and Veiga, G. (1986),
     "An implementation of Karmarkar's algorithm for linear
     programming," Working Paper, Operations Research Center,
     University of California, Berkeley, California.

2.   Adler, I., Karp, R.M. and Shamir, R. (1983), "A simplex variant
     solving an $m \times d$ linear program in $O(\min(m^2,d^2))$ expected number
     of pivot steps," Report UCB/CSD 83/158, Computer Science Division,
     University of California, Berkeley, California.

3.   Adler, I. and Megiddo, N. (1984), "A simplex algorithm whose
     average number of steps is bounded between two quadratic functions
     of the smaller dimension," Journal of the Association for
     Computing Machinery 32, 871-895.

4.   Anstreicher, K.M. (1986), "A monotonic projective algorithm for
     fractional linear programming," Algorithmica 1, 483-498.

5. Anstreicher, K.M. (1986), "A combined 'phase I-phase II' projective algorithm for linear programming," manuscript, Yale School of Organization and Management, New Haven, CT, to appear in Mathematical Programming.

6. Anstreicher, K.M. (1986), "The worse-case step in Karmarkar's algorithm," manuscript, Yale School of Organization and Management, New Haven, CT, to appear in Mathematics of Operations Research.

7. Anstreicher, K.M. (1987), "A standard form variant, and safeguarded linesearch, for the modified Karmarkar algorithm," manuscript, Yale School of Organization and Management, New Haven, CT.

8. Asic, M.D., Kovacevic-Vujcic, V.V., and Radosavljevic-Nikolic, M.D. (1987), "Asymptotic behaviour of Karmarkar's method for linear programming," manuscript, Faculty of Organizational Science, Belgrade University, Yugoslavia, to appear in Mathematical Programming.

9. Balinski, M.L. (1985), "Signature methods for the assignment problem," Operations Research 33, 527-536.

10. Barnes, E.R. (1986), "A variation on Karmarkar's algorithm for solving linear programming problems," Mathematical Programming 36, 174-182.

11. Barnes, E.R., Chopra, S., and Jensen, D.L. (1988), "A Polynomial Time Version of the Affine Scaling Algorithm," IBM T.J. Watson Research Center, Yorktown Heights, NY.

12. Bayer, D. and Lagarias, J.C. (1987), "The nonlinear geometry of linear programming, I. Affine and projective scaling trajectories," to appear in Transactions of the American Mathematical Society.

13. Bayer, D. and Lagarias, J.C. (1987), "The nonlinear geometry of linear programming, II. Legendre transform coordinates and central trajectories," to appear in Transactions of the American Mathematical Society.

14. Borgwardt, K.H. (1987), The Simplex Method: A Probabilistic Analysis. Springer-Verlag, Berlin.

15. Borgwardt, K.H. (1987), "Probabilistic Analysis of the Simplex Method," in Operations Research Proceedings 1987, 16th DGOR Meeting, pp. 564-576.

16. Dantzig, G.B. (1963), Linear Programming and Extensions, Princeton University Press, Princeton, New Jersey.

17. Debreu, G. (1976), private communication.

18. de Ghellinck, G. and Vial, J.-Ph. (1986), "A polynomial Newton method for linear programming," _Algorithmica_ 1, 425-453.

19. Dikin, I.I. (1967), "Iterative solution of problems of linear and quadratic programming," _Doklady Akademiia Nauk SSSR_ 174, 747-748 [English translation: _Soviet Mathematics Doklady_ 8, 674-675].

20. Dikin, I.I. (1974), "On the convergence of an iterative process," _Upravlyaemye Sistemi_ 12, 54-60 (in Russian).

21. Fiacco, A.V. and McCormick, G.P. (1968), _Nonlinear Programming: Sequential Unconstrained Minimization Techniques_, Wiley, New York.

22. Freund, R. (1988), "Polynomial-time algorithms for linear programming based only on primal scaling and projected gradients of a potential function," manuscript, Sloan School of Management, M.I.T., Cambridge, Massachusetts.

23. Frisch, K.R. (1955), "The logarithmic potential method of convex programming," Memorandum, University Institute of Economics, Oslo, Norway.

24. Gay, D. (1987), "A variant of Karmarkar's linear programming algorithm for problems in standard form," _Mathematical Programming_ 37, 81-90.

25. Gill, P.E., Murray, W., Saunders, M.A., Tomlin, J.A. and Wright, M.H. (1986), "On projected Newton barrier methods for linear programming and an equivalence to Karmarkar's projective method," _Mathematical Programming_ 36, 183-209.

26. Goldfarb, D. and Hao, J. (1988), "A primal simplex algorithm that solves the maximum flow problem in at most mn pivots and $O(n^2m)$ time," manuscript, Department of Industrial Engineering and Operations Research, Columbia University, New York.

27. Goldfarb, D. and Mehrotra, S. (1988), "A relaxed version of Karmarkar's method," _Mathematical Programming_ 40, 289-315.

28. Goldfarb, D. and Mehrotra, S. (1988), "Relaxed variants of Karmarkar's algorithm for linear programs with unknown optimal objective value," _Mathematical Programming_ 40, 183-195.

29. Goldfarb, D. and Todd, M.J. (1989), "Linear Programming," to appear in _Handbooks in Operations Research and Management Science_, Vol. 1, Optimization (G.L. Nemhauser, A.H.G. Rinnooy Kan, and M.J. Todd, eds.), North-Holland, Amsterdam.

30. Gonzaga, C. (1985), "A conical projection algorithm for linear
     programming," manuscript, Department of Electrical Engineering and
     Computer Science, University of California, Berkeley,
     California, to appear in <u>Mathematical Programming</u>.

31. Gonzaga, C. (1988), "An algorithm for solving linear programming
     in $O(n^3L)$ operations," in: <u>Progress in Mathematical Programming</u>
     (N. Megiddo, ed.), Springer-Verlag, Berlin, 1-28.

32. Gonzaga, C. (1987), "Search directions for interior linear
     programming methods," Memorandum UCB/ERL M87/44, Electronics
     Laboratory, College of Engineering, University of
     California, Berkeley, California.

33. Gonzaga, C. (1988), "Polynomial affine algorithms for linear
     programming," Report ES-139/88, Universidade Federal do Rio de
     Janeiro, Brazil.

34. Grötschel, M., Lovasz, L., and Schrijver, A. (1988), <u>The Ellipsoid
     Method and Combinatorial Optimization</u>, Springer Verlag,
     Heidelberg.

35. Hoffman, K. and Padberg, M. (1985), "LP-based combinatorial
     problem-solving," <u>Annals of Operations Research</u> 4, 145-194.

36. Huard, P. (1967), "Resolution of mathematical programming with
     nonlinear constraints by the method of centers," in: <u>Nonlinear
     Programming</u> (J. Abadie, ed.), North-Holland, Amsterdam.

37. Karmarkar, N. (1984), "A new polynomial time algorithm for linear
     programming," <u>Combinatorica</u> 4, 373-395.

38. Karmarkar, N. (1986), Lecture at Cornell University, Ithaca, New
     York.

39. Karmarkar, N. (1988), "An interior-point approach to NP-complete
     problems," extended abstract, AT&T Bell Laboratories, Murray Hill,
     NJ.

40. Karmarkar, N. and Ramakrishnan, K.G. (1988), "Implementation and
     computational results of the Karmarkar algorithm for linear
     programming, using an iterative method for computing projections,"
     extended abstract, AT&T Bell Laboratories, Murray Hill, NJ,
     presented at the XIII International Symposium on Mathematical
     Programming, Tokyo.

41. Khachian, L.G. (1980), "Polynomial algorithms in linear
     programming," (in Russian), <u>Zhurnal Vychisditel-noi Matematiki i
     Matematischeskoi Fiziki</u> 20, 51-68 [English translation:
     <u>USSR Computational Mathematics and Mathematical Physics</u> 20,
     53-72].

42. Klee, V. and Kleinschmidt, P. (1987), "The d-step conjecture and its relatives," Mathematics of Operations Research 12, 718-755.

43. Klee, V. and Minty, G.J. (1972), "How good is the simplex algorithm?", in: Inequalities III (O. Shisha, ed), Academic Press, New York, 159-175.

44. Kleinschmidt, P., Lee, C., and Schannath, H. (1987), "Transportation problems which can be solved by the use of Hirsch-paths for the dual problems," Mathematical Programming 37, 153-168.

45. Kojima, M., Mizuno, S., and Yoshise, A. (1988), "A primal-dual interior point method for linear programming," in: Progress in Mathematical Programming (N. Megiddo, ed.), Springer Verlag, Berlin, 29-47.

46. Kojima, M., Mizuno, S. and Yoshise, A. (1987), "A polynomial-time algorithm for a class of linear complementarity problems," Research Report No. B-193, Department of Information Sciences, Tokyo Institute of Technology, Tokyo, Japan, to appear in Mathematical Programming.

47. Lagarias, J.C. (1987), "The nonlinear geometry of linear programming III: projective Legendre transform coordinates and Hilbert geometry," to appear in Transactions of the American Mathematical Society.

48. Lee, C. (1988), "Recent developments in the structure of polytopes," presented at the AMS-IMS-SIAM Summer Conference on Mathematical Developments Arising from Linear Programming, Bowdoin College, Maine.

49. McDiarmid, C.J.H. (1986), "On the improvement per iteration in Karmarkar's algorithm for linear programming," manuscript, Institute of Economics and Statistics, Oxford University, Oxford, to appear in Mathematical Programming.

50. McShane, K.A., Monma, C.L., and Shanno, D. (1988), "An implementation of a primal-dual interior point method for linear programming" Rutcor Research Report, Rutgers University, New Brunswick, New Jersey.

51. Megiddo, N. (1988), "Pathways to the optimal set in linear programming," in: Progress in Mathematical Programming (N. Megiddo, ed.), Springer-Verlag, Berlin, 131-158.

52. Megiddo, N. (1987), "On the complexity of linear programming," in: Advances in Economic Theory (T. Bewley, ed.), Cambridge University Press, Cambridge, 225-268.

53. Megiddo, N. and Shub, M. (1986), "Boundary behavior of interior
     point algorithms for linear programming," IBM Research Report RJ
     5319, T.J. Watson Research Center, Yorktown Heights, New York.

54. Mitchell, J.E. (1988), "Karmarkar's algorithm and combinatorial
     optimization problems," Ph.D. thesis, School of Operations
     Research and Industrial Engineering, Cornell University, Ithaca,
     NY.

55. Mitchell, J.E. and Todd, M.J. (1986), "On the relationship between
     the search directions in the affine and projective variants of
     Karmarkar's linear programming algorithm," Technical Report 725,
     School of Operations Research and Industrial Engineering, Cornell
     University, Ithaca, NY.

56. Monma, C.L. and Morton, A.J. (1987), "Computational experience
     with a dual variant of Karmarkar's method for linear programming,"
     Operations Research Letters, 6, 261-267.

57. Monteiro, R.C. and Adler, I. (1987), "An $O(n^3 L)$ primal-dual
     interior point algorithm for linear programming," manuscript,
     Department of Industrial Engineering and Operations Resarch,
     University of California, Berkeley, California, to appear in
     Mathematical Programming.

58. Monteiro, R.C. and Adler, I. (1987), "An $O(n^3 L)$ interior point
     algorithm for convex quadratic programming," manuscript,
     Department of Industrial Engineering and Operations Resarch,
     University of California, Berkeley, California, to appear in
     Mathematical Programming.

59. Monteiro, R.D.C., Adler, I., and Resende, M.G.C. (1988), "A
     polynomial-time primal-dual affine scaling algorithm for linear
     and convex quadratic programming and its power series extension,"
     manuscript, Department of Industrial Engineering and Operations
     Research, University of California, Berkeley, California.

60. Murtagh, B.A. and Saunders, M.A. (1978), "Large-scale linearly
     constrained optimization," Mathematical Programming 14, 41-72.

61. Orlin, J.B. (1984), "Genuinely polynomial simplex and non-simplex
     algorithms for the minimum cost flow problem," Technical Report
     1615-84, Sloan School of Management, MIT, Cambridge, MA.

62. Renegar, J. (1988), "A polynomial-time algorithm based on Newton's
     method for linear programming," Mathematical Programming 40,
     59-93.

63. Roohy-Laleh, E. (1981), "Improvements to the theoretical
     efficiency of the network simplex method," Ph.D. thesis, Carleton
     Univrsity, Ottawa, Ontario.

64. Shamir, R. (1987) "The efficiency of the simplex method: a
     survey," Management Science 33, 301-334.

65. Sonnevend, G. (1985), "An analytical centre for polyhedrons and
     new classes of global algorithms for linear (smooth, convex)
     programming," Proceedings of the 12th IFIP Conference on System
     Modeling and Optimization, Budapest, to appear in Lecture Notes in
     Control and Information Sciences, Springer-Verlag.

66. Steger, A. (1985), "An extension of Karmarkar's algorithm for
     bounded linear programming problems," M.S. Thesis, SUNY at
     Stonybrook, New York.

67. Tardos, E. (1986), "A strongly polynomial algorithm to solve
     combinatorial linear programs," Operations Research 34, 250-256.

68. Todd, M.J. (1986), "Polynomial expected behavior of a pivoting
     algorithm for linear complementarity and linear programming
     problems," Mathematical Programming 35, 173-192.

69. Todd, M.J. (1986), "Improved bounds and containing ellipsoids in
     Karmarkar's linear programming algorithm," Technical Report 721,
     School of Operations Research and Industrial Engineering, Cornell
     University, Ithaca, NY, to appear in Mathematics of Operations
     Research.

70. Todd, M.J. (1988), "Polynomial algorithms for linear programming,"
     in: Advances in Optimization and Control (H.A. Eiselt and G.
     Pederzoli, eds.), Springer-Verlag, Berlin, 49-66.

71. Todd, M.J. (1988), "Karmarkar as Dantzig-Wolfe," Technical Report
     782, School of Operations Research and Industrial Engineering,
     Cornell University, Ithaca, NY.

72. Todd, M.J. and Burrell, B.P. (1986), "An extension of Karmarkar's
     algorithm for linear programming using dual variables,"
     Algorithmica 1, 409-424.

73. Todd, M.J. and Ye, Y. (1987), "A centered projective algorithm for
     linear programming," Technical Report 763, School of Operations
     Research and Industrial Engineering, Cornell University, Ithaca,
     NY.

74. Tomlin, J.A. (1987), "An experimental approach to Karmarkar's
     projective method for linear programming," Mathematical
     Programming Study 31, 175-191.

75. Vaidya, P.M. (1987), "An algorithm for linear programming which
     requires $O(((m+n)n^2 + (m+n)^{1.5}n)L)$ arithmetic operations,"
     preprint, AT&T Bell Laboratories, Murray Hill, New Jersey.

76. Vanderbei, R.J. and Lagarias, J.C. (1988), "I.I. Dikin's convergence result for the affine-scaling algorithm," manuscript, AT&T Bell Laboratories, Murray Hill, NJ.

77. Vanderbei, R.J., Meketon, M.S., and Freedman, B.A. (1986), "A modification of Karmarkar's linear programming algorithm," _Algorithmica_ 1, 395-407.

78. Vial, J.-Ph. (1987), "A unified approach to projective algorithms for linear programming," manuscript, University of Geneva, Geneva.

79. Ye, Y. (1987), "Karmarkar's algorithm and the ellipsoid method," _Operations Research Letters_ 4, 177-182.

80. Ye, Y. (1988), "A class of potential functions for linear programming," manuscript, Department of Management Sciences, The University of Iowa, Iowa City, Iowa.

81. Ye, Y. and Kojima, M. (1987), "Recovering optimal dual solutions in Karmarkar's polynomial algorithm for linear programming," _Mathematical Programming_ 39, 305-317.

82. Ye, Y. and Todd, M.J. (1987), "Containing and shrinking ellipsoids in the path-following algorithm," manuscript, School of Operations Research and Industrial Engineering, Cornell University, Ithaca, NY, to appear in _Mathematical Programming_.

The Influences of Algorithmic and Hardware Developments on Computational Mathematical Programming*

J. A. TOMLIN

IBM Almaden Research Center, San Jose, CA 95120, U.S.A.

Abstract

The computational power of mathematical programming has increased enormously over the last two or three decades. Obviously, this is partly due to hardware improvements, but in large part it is also due to algorithmic advances. Even in heavily researched fields, these advances continue. This talk will discuss the influences of these sources of improvement.

1. Introduction

The theme of this memorial lecture is inspired by a chance remark made to me by Martin Beale in, I believe, 1969. In words that have stuck in my mind ever since, he said that he would "much rather work with today's algorithms on yesterday's computers than with yesterday's algorithms on today's computers."

Something like these words must have remained in Martin's mind also, for he was to express similar sentiments several times in print. For example, in his Blackett Memorial Lecture (Beale (1980)) he stated that "improvements in practical LP capability in each of the last two decades have owed more to mathematical developments than to improved computer hardware." In slightly less ringing tones, he

---

* Presented as the Martin Beale memorial lecture

M. Iri and K. Tanabe (eds.), Mathematical Programming, 159–175.
© *1989 by KTK Scientific Publishers, Tokyo.*

was to reaffirm in a late paper (Beale (1985a)) that "the steady increase in the size of (linear programming) problem that can be solved has been due as much to a better understanding of how to exploit sparseness as to larger and faster computers."

There would be little point in trying to "prove" such statements, though presumably one could attempt a statistical study to do so. What is much more interesting is the attitude that such statements represent, the context in which they were made, and their relevance to Martin Beale's work, the work of his colleagues and the development of computational mathematical programming. Martin Beale was of course the premier practical algorithm developer of mathematical programming, and as such was bound to be more interested in this aspect of problem solving, but his career also spanned the development of computers from the primitive to the micro/supercomputer range of almost the present day. As such he was in an almost unique position to evaluate the relative importance of these influences on the state of the art, though I doubt if he ever did so in any formal way. However, his informal evaluation was based on a simultaneous embrace of the three fundamental areas of our profession:

- The formulation and solution of real-world problems.

- The development of mathematical programming theory and algorithms.

- The development of advanced, reliable and versatile mathematical programming *systems* (MPSs).

Without the third of these the first is almost impossible and the second almost pointless. It is in this context, then, that we must consider the contributions made in the realms of algorithms and hardware.

When we consider the algorithms and computers of 1969 in the MPS context, it should be remembered that the computer of "yesterday" in our case was the IBM 7094 and the algorithms were those implemented in the CEIR LP/90/94 system. The computer of "today" was the Univac 1108 and the developing system was UMPIRE. While the 1108 was considerably faster than the 7094 (a fixed point add time of 0.75μs versus 4μs for the 7094 I and 2.8μs for the 7094 II, for example) the CEIR LP/90/94 system was highly developed, with facilities for some extensions of LP not found in many systems today (see Beale (1968)) for a general description). Fortunately, in the late 60's and early 70's Beale was able to lead a development effort at Scicon (also in association with mathematical programmers at British Petroleum) which lead to many significant algorithmic developments in UMPIRE and subsequently influenced the development of any other MPS which hoped to remain competitive. The interaction between the algorithmic and computer-specific aspects of these MPSs is the major topic of the rest of this paper.

2. The Simplex Method

Those interested in the very early history of LP will be intrigued by the ingenious use that Dantzig and Orchard-Hays (1954) were able to make of the Card Programmed Calculator when developing the product form inverse (PFI) version of the revised simplex method. In this first factorization scheme for LP, the basis inverse is stored as a product of elementary transformations:

$$B^{-1} = E_l E_{l-1} \dots E_2 E_1$$

where the nontrivial columns of the E_k are historically referred to as "etas".

Nearly a decade later, a major limiting factor in the LP/90 system was the use of magnetic tape as out-of-core store for these etas, as periodically produced by Gauss-Jordan elimination in INVERT and added to at each iteration. Markowitz (1957) had already shown that the number of eta nonzeros could be substantially reduced by using Gaussian elimination and sparse matrix techniques which minimized a worst case estimate of local fill-in. Unfortunately, the memory requirements of this scheme were not feasible in out-of-core systems and it was considered unrealistic for almost twenty years. However, it was discovered that even such a simple expedient as ordering the columns of the basis by density could lead to enormous improvements over the lax practice of pivoting the basis columns in matrix order. This could obviously be done with minimal effort and prompted a search for other heuristics which would reduce fill-in during INVERT. A well known outcome of this approach (see Orchard-Hays (1968)) is permutation of the basis to find forward and backward triangular segments and a "bump":

$$PBQ = \begin{pmatrix} L_1 & & \\ E & F & \\ G & H & L_2 \end{pmatrix}$$

which can be put in product form with fill-in only in the F and H segments. Martin observed that if this permuted form were factorized as:

$$\begin{pmatrix} L_1 & & \\ E & F & \\ G & H & L_2 \end{pmatrix} = \begin{pmatrix} L_1 & & \\ E & I & \\ G & & I \end{pmatrix} \begin{pmatrix} I & & \\ & F & \\ & & I \end{pmatrix} \begin{pmatrix} I & & \\ & I & \\ & H & I \end{pmatrix} \begin{pmatrix} I & & \\ & I & \\ & & L_2 \end{pmatrix}$$

etas could be created without fill-in in either L_2 or H and stored sequentially on a "backward eta file" and that fill-in on the "forward eta file" would occur only in the bump F. The two files may then be merged after INVERT. This scheme was implemented in UMPIRE and led to considerable improvement over older schemes. However, it seemed that further improvement should be possible. After some needless reinvention, the key turned out to be LU decomposition of F within the context of a slightly rearranged ordering:

$$\begin{pmatrix} L_1 & & \\ G & U_2 & H \\ E & & F \end{pmatrix}$$

The ordering of F was still heuristic and once-and-for-all, using a "merit" scheme, because of the need to use a sequential file-oriented out-of-core implementation.

At about the same time Hellerman and Rarick were developing a quite independent ordering scheme with their P^3 and later P^4 techniques (see Hellerman and Rarick (1972)) which proved very efficient on "friendly" problems but require some modification to preserve numerical stability on others. This "bump and spike" approach was also strongly motivated by the need to deal with an out-of-core limited storage system.

Another almost simultaneous, and very influential, event was the publication by Bartels and Golub (1969) of their LU decomposition approach, emphasizing updating of factorizations in a stable way. This really marked the arrival of the numerical analysts in the field, though it took some time for the influence to be fully felt. To begin with, it was not clear how the new ideas, based on dense matrix algorithms, could be implemented in MPSs designed for out-of-core solution of (then) large and sparse problems. The Forrest-Tomlin (1972)) method was able to reconcile sparse triangular updating with out-of-core requirements, though it did not guarantee a *priori* numerical stability, producing a modified factorization:

$$B^{-1} = \bar{U}^{-1} R_l \dots R_2 R_1 L^{-1}.$$

In this representation of B^{-1}, the updates R_k are elementary *row* transformations added to the PFI representation of L^{-1} and the PFI representation of $\bar{U}^{-1}$ is modified from that of the original U without insertion of new nonzeros in existing etas. The latter may occur in the Bartels–Golub form, making it unsuitable for the MPSs of the time.

Apart from basis manipulation, the other major effort in the simplex method comes with column choice (PRICE) and pivot selection (CHUZR). The traditional out-of-core technique had become multiple and partial pricing—selecting several columns while reading only part of the A-matrix file. While efficient on many problems, on others it could perform very badly. In particular the LP models developed by British Petroleum (BP) became something of a thorn in our sides, not to mention an embarrassment, since BP owned Scicon.

The now well-known BP 822 row problem (25FV47) was taking many thousands of simplex iterations and over 1 hour 20 minutes to solve on UMPIRE. However, we knew that Paula Harris was able to solve this problem in fewer iterations on her own LP code. With some difficulty we managed to persuade her to describe her algorithm in enough detail (and eventually to publish it (Harris(1973))) so that we could implement it in UMPIRE. This eventually enabled us to dramatically reduce the number of iterations and the solution time to 40 minutes. This was remarkable—because the cost of reading the entire matrix file at each iteration, in order to update and apply the Harris dynamic scaling factors, was thought to be prohibitive—and remains a spectacular, purely algorithmic achievement, obtained in the face of increased memory requirements and I/O, at a time when these were at a premium. Later, John Forrest was able to reduce this time to 22 minutes on his early SCICONIC code for the 1108 (not to be confused with the current FORTRAN-based SCICONIC/VM). Though extremely problem dependent, these achievements remain some of the most spectacular evidence for Martin Beale's advocacy of strong algorithm development in MPSs.

So far, I have belabored LP algorithmic developments in the out-of-core MPS framework. With the development of early virtual memory systems, particularly the late IBM 360/ early 370 models, this concern began to dissipate, though the memory partitions typically available were pathetic by today's standards (even on many PCs). A watershed product was the WHIZARD in-core optimizer developed by the late Dennis Rarick at Management Science Systems, with support from

Milton Gutterman at AMOCO. Among the novel features of this system were its use of the "supersparsity" concept, due to Kalan (1971)), and its ability to be called from two competing "host" systems—MPSIII and MPSX—which handled all the more dreary data management functions. What is interesting from the point of view of this discussion, however, is that algorithmically it was (and is) built around clever implementations of algorithms originally constrained by out-of-core considerations—the Hellerman-Rarick INVERT, the Forrest-Tomlin update and a more sophisticated multiple/partial PRICE strategy. Even now, after much rewriting by J.S. Welch, these remain central to this very fast and successful code. Again, this must be viewed as some vindication of the central importance of algorithms, even in changing hardware environments.

Only in about 1976 did John Reid grasp the nettle and write his LA05 FORTRAN routines implementing a full Markowitz INVERT and sparse Bartels-Golub update (see Reid (1976)). These are now incorporated in the widely distributed XMP FORTRAN code and have influenced the later versions of MINOS (see Murtagh and Saunders (1983)).

3. Special Structure LP Models

When considering special structure LP models, one is almost by definition discussing algorithmic developments. Of course, it can be argued that almost all LPs are specially structured and that this structure should be exploited. Leading exponents of this school have been G.W. Graves and G.G. Brown (see, e.g. Graves and McBride (1976) and Brown et al. (1987)) and they have achieved some outstanding successes in many types of hardware environment. It is, however, difficult and perhaps self-defeating to try and "black box" this approach in general purpose MPSs which must be distributed to users of many levels of sophistication.

Martin had his own criteria for incorporating special-purpose algorithms in a general-purpose MPS. These were (see Beale and Tomlin (1970)):

(a) They can be implemented reasonably easily

(b) They are reasonably easy to use

(c) They enable a significant class of problems to be solved more easily than they could by other means

After mixed success with Dantzig-Wolfe decomposition in LP/90/94 (Beale et al. (1965)), Martin became convinced that Generalized Upper Bounds (GUB) satisfied these conditions and they were incorporated into the standard structure of UMPIRE (see Beale (1970)). Some very large problems could indeed be solved in the out-of-core environment using GUB, but for some reason the GUB algorithm seems to have largely fallen out of favor as more powerful in-core systems have developed. This is perhaps an example of backward progress from the algorists point of view.

There can be no question of backward progress when it comes to pure and generalized network problems. The original work of Glover and Klingman and of Bradley et al. (1977) and Brown and McBride (1984) in developing network simplex methods has led to orders-of-magnitude improvement even over earlier special purpose algorithms. One would truly be better off running these codes on the 7094 than older algorithms even on today's mainframes. Gratifyingly, it has proved possible to integrate much of this network technology into a general purpose MPS rather neatly (Tomlin and Welch (1985)).

4. Nonlinear Programming

I shall pass fairly quickly over nonlinear programming, partly because the scope and importance of algorithm development are so obviously and overwhelmingly important and partly because Powell (1987) has already given us a definitive summary of Martin Beale's technical contributions in this area.

A quite large proportion of Martin's work in nonlinear programming was devoted to refinements of separable programming—an activity essential to treatment of nonlinearities in early MPSs. Except for use in nonconvex optimization (see below) separable programming is largely superseded now by more sophisticated approaches such as those in MINOS, or by various developments in what used to be known as the method of approximation programming (MAP) but is now usually called sequential linear programming (SLP).

I believe that many people still do not appreciate fully one of Martin's practical contributions to SLP—his method of formulating and defining nonlinearities. Where Griffith and Stewart (1961) had written their constraints in the form:

$$\sum_j a_{ij} x_j + g_i(z_1, \ldots, z_q) = b_i,$$

entirely segregating the nonlinear variables z_k in every row (whether they appear nonlinearly in it or not), the new formulation (Beale (1974)) called for writing what appeared to be a linear program, with standard looking constraints:

$$\sum_j a_{ij} x_j = b_i,$$

but allowing the the a_{ij} and b_i to be functions of a set of nonlinear variables $y_1, \ldots, y_r$. If all a_{ij} are constant and only the b_i are functions of the y_k, then nothing has really changed. But if some of the nonlinearities can be thrown into the $a_{ij} x_j$ terms the number of nonlinear variables may be dramatically reduced.

A nice example of this reduction occurs in financial models which keep track of accounts over time. If we denote by x_t the balance at the end of period t, by u_t and v_t the withdrawals and deposits, and by y the (variable) interest rate, then the material balance constraints:

$$x_t = (1 + y)x_{t-1} - u_t + v_t$$

would define T nonlinear variables for T periods (y and $x_1, \ldots, x_{T-1}$) with the old approach but only one (y) with the new.

The net effect of Martin's reformulation is not only a mathematical advance but a great improvement in ease of use of MPS model management capabilities for nonlinear models.

5. Integer and Nonconvex programming

It takes considerable nerve to contemplate solving discrete programming problems, essentially by enumerative methods, especially on primitive computers. As Johnson (1978) has remarked, the difficulty is not necessarily one of degree, but whether anything useful can be accomplished at all.

If there is any area in mathematical programming in which raw computer speed might be thought irrelevant it is (mixed) integer programming (MIP). After all, with n zero-one variables potentially leading to 2^n branches, a hundred-fold speed-up could be wiped out by adding only 7 new variables to the problem. Nevertheless, problem solving power has increased substantially from the relative handful of variables usually handled by the CEIR LP/90/94 implementation of the sixties

(Beale and Small (1965)), to dozens, then hundreds in the seventies, to thousands today.

The great majority of this improvement is due to algorithm development and new mathematical results. Many multiple-choice and separable nonconvex programming problems, which could in principle be handled by integer variables, can be handled in a far superior way by means of "special ordered sets (SOS)" (see Beale and Tomlin (1970) and Beale and Forrest (1976)). By explicitly treating *sets* of variables, users are not only able to apply more information to the branching process, but the worst case behavior when n binary variables (say) are treated as n/m sets of m variables is $m^{\frac{n}{m}}$ branches, and the SOS use of a "divide and conquer" principle reduces this to more like $(\log_2 m)^{\frac{n}{m}}$, though in practice the behaviour is usually much better. SOS have proved remarkably amenable to extension. In particular "linked ordered sets" and "chains of linked ordered sets" are able to handle more complex nonconvex nonlinearities (see Beale (1979,1985b)).

Some of the most influential work in the last decade has been in the development of strong formulations and cutting planes to reduce the "integer gap" in branch and bound methods. Of particular consequence to MPSs has been the development by Crowder et al (1983) of the PIPX extension of MPSX for solving large and difficult pure zero-one models. While it is true that pure zero-one models make up a fairly small proportion of the discrete models we see in practice, this shows what can be done by applying mathematical ingenuity in this framework. More recently Van Roy and Wolsey (1987) have extended many of these ideas to the mixed integer case—a development which potentially has even greater impact. It is perhaps fitting that this latter work was implemented in conjunction with Scicon's MPS (SCICONIC/VM) and that Hattersley has recalled (in Dantzig and Tomlin (1987)) that this approach was among Martin's interests in his last year.

The preponderance of algorithmic and mathematical work should not be allowed to give the impression that hardware developments have been completely irrelevant to MIP. The early branch and bound systems were almost constrained by sequential file (usually tape) technology to use a last in–first out (LIFO) search strategy. It later became apparent (see e.g. Benichou et al (1977), Forrest et al (1974)) that more flexible search strategies (in addition to better heuristics) could lead to tremendous improvements. In addition, branch and bound schemes inevitably benefit from improvements in LP capability, whether algorithmic or at the hardware level. To anticipate a later section, use of vector processing for pricing in the dual algorithm for reoptimization can have significant effects on performance and strategy choice.

6. Microcomputing

A future widespread use of minicomputers in MP was forecast by both Crowder (1976) and Beale (1980), but the really impressive development at the smaller-scale end has been the rapid and chaotic growth in the use of microcomputers (PCs). PCs have made MP a feasible, easier to use, desk-top tool for almost anybody who wants it, or can be persuaded to want it. For a few hundred dollars, software is available which not only provides a respectable optimizer but model generation capabilities as well.

Perhaps the best reference in this fast-moving field is to Bausch and Brown (1988) who, with the right combination of new (80386 based) hardware and software tools have succeeded in bringing quite large scale MP (including some integer programming) to the PC. Any attempt to give a broader survey is doomed to almost instant obsolescence.

A particularly dramatic example of the impact of hardware developments in this area was noted by my colleague John Forrest. He noted in UMPIRE performance figures (*circa* 1971) that an LP model from BP with 2164 rows (including 590 simple bound rows) was shown as solving in 4 hours 52 minutes on the 1108 at a cost of 2920 pounds sterling (!). For a much lower present cost you can *buy* a 80386-based computer which will solve a very similar problem in 2 hours 30 minutes (using the same software tools as Bausch and Brown). The same problem solves in less than 3 minutes on an IBM 3090VF.

7. Supercomputing

At the other extreme from the microcomputer we have the "supercomputer", usually now in the form of a vector multiprocessor with a small number of processors. The impact of such machines can only be substantial (Duff (1986)), but it took some time to see how they could be applied effectively to MP. The sparsity pattern of most LPs is such that they cannot use existing vector processors efficiently (because of start-up times) without some restructuring of both data and algorithms. This was accomplished by Forrest (1987) in the MPSX framework, essentially only vectorizing the PRICE routine. However, this was enough to double the speed on some models, and more importantly, to employ strategies such as the Harris Devex scheme which would otherwise have been too expensive. Other parts of the simplex method have since been vectorized in our research codes (see Forrest and Tomlin (1988)).

The other important impact of supercomputing is in conjunction with interior point methods. When Karmarkar's projective scaling method (Karmarkar (1984)) began to receive wide publicity, claims were made that this new approach is particularly suited for parallel processing. AT&T's current activities certainly suggest that they still hold this view. Let us consider why this might be so.

We now understand that the essential computational feature of interior methods is the use of a projected Newton barrier method (see Gill et al (1986)), which effectively means the solution of large sparse least squares problems. This has placed LP back even more firmly in the main stream of sparse matrix research, particularly as it applies to supercomputers. Most of those working in the interior point field use the Cholesky decomposition of a normal matrix AD^2A^T, either directly or as a preconditioner in an iterative method, and find that this decomposition consumes most of the work (often 80% and up) except for special problems. Work on sparse Cholesky decomposition has been proceeding determinedly in non-MP fields with some success (see Duff et al (1986) and Ashcraft et al (1987)).

Some of the experiments we have carried out illustrate the importance of hardware characteristics in algorithm development and implementation (Forrest and Tomlin (1988)). Two critical aspect of sparse matrix calculations on vector processors (even those having hardware indirect addressing, such as the IBM 3090VF) are the vector length (number of nonzeros) and number and locality of references to memory. One way to work with longer vectors is to use the operations list (or map) approach to Cholesky factorization; an interpretive implementation of the GNSO approach of Gustavson et al. (1970). In a recent computational study of such schemes, Gay (1988) concluded that they confer very little advantage in a scalar environment, despite the success claimed in some quarters. However, in a vector environment this is not necessarily the case.

Let us suppose our Cholesky factors of the permuted normal matrix:

$$LL^T = PAD^2A^TP^T$$

are stored in packed column order. Now consider any updated column $l_{.k}$ with p subdiagonal nonzeros, and number them by their offset from the first subdiagonal nonzero as $0, 1, \ldots, (p-1)$. Alternatively, we may denote the nonzeros as occurring on rows $i_0, i_1, \ldots, i_{p-1}$. When we pivot on the diagonal for column k, the $p(p-1)/2$ off-diagonal elements in positions $(i_1, i_0), \ldots, (i_{p-1}, i_{p-2})$ will be updated in a conventional "push ahead" factorization. Now the elements updated by this column could be considered in row major order, so that the single element $l_{i_1 i_0}$ in row i_1 is updated, then the two elements $l_{i_2 i_0}, l_{i_2 i_1}$ in row i_2, and so on up to the elements $l_{i_{p-1} i_0}, \ldots, l_{i_{p-1} i_{p-2}}$ in row $(p-1)$. Let us suppose we have a list of the locations of these elements in L (the operations list or map).

Vector processing may be used by referring to two index vectors of offset numbers:

$$\begin{pmatrix} 1 & 2 & 2 & 3 & 3 & 3 & 4 & 4 & 4 & 4 & \ldots\ldots & p-1 & p-1 & \ldots & p-1 \\ 0 & 0 & 1 & 0 & 1 & 2 & 0 & 1 & 2 & 3 & \ldots\ldots & 0 & 1 & \ldots & p-2 \end{pmatrix}$$

All subdiagonal nonzeros affected by $l_{.k}$ may then be updated by loading two $p(p-1)/2$-length vectors of the nonzeros in this column corresponding to these offsets, multiplying them together and subtracting the result from the locations pointed to by the map.

This procedure should be expected to make much better use of the vector facility than one which works with vectors of length p, if a large memory for the map array can be accommodated. Very large memories are indeed available, but they are usually hierarchical. When we applied the scheme, some significant improvement was seen on problems where p was relatively small. However for larger, denser problems the solution times were *worse* than a straightforward sparse Cholesky scheme.

The cause of this behavior had to be in memory access to L via the cache. A partitioning scheme, combined with a column-wise updating approach, leads to some relief from this problem, but this is clearly a case of apparently contradictory algorithm requirements (for vector speed) and architecture (degradation when faced with widely scattered data).

In contrast to the above situation, considerable success was achieved by exploiting dense matrix substructure (see Dongarra (1988) and Ashcraft et al (1987)). We have been able to achieve significant speed-up over straightforward victimization of the sparse code by using a block-type approach to the densest segment of the

Cholesky factors. It seems ironic that after decades in the pursuit of sparsity we now look for dense processing to take advantage of some of the more powerful new algorithms and architectures.

8. Conclusions

In ending this survey I must say that Martin's belief in the supremacy of algorithms has been right more often than not, but that one cannot guarantee that such a situation will continue in every field of MP. The development of new classes of hardware and new techniques of software engineering are bound to change the cost, sophistication, availability and usefulness of MPSs and other approaches to large-scale optimization. However, without the contributions of Martin Beale our field would have been poorer in every sense.

Acknowledgements

I am indebted to my colleague John Forrest for his encouragement, his recollections, and for supplying some unpublished computational results.

Bibliography

Ashcraft, C.C., Grimes, R.G., Lewis, J.G., Peyton, B.W. and Simon, H.D. (1987). Progress in sparse matrix methods for large linear systems on vector supercomputers. *International J. of Supercomputer Applics.* **1**, pp. 10–30.

Bartels, R.H. and Golub, G.H. (1969). The simplex method of linear programming using the *LU* decomposition, *Comm. ACM* **12**, 266-268.

Bausch, D.O. and Brown, G.G. (1988). A PC environment for large-scale programming, *OR/MS Today* **15**, no. 3, 20-25.

Beale, E.M.L. (1968). *Mathematical Programming in Practice*, Pitmans, London and Wiley, New York.

Beale, E.M.L. (1970). Advanced algorithmic features for general mathematical programming systems, in *Integer and Nonlinear Programming*, (J. Abadie, ed.), North Holland Publishing Company, Amsterdam, 119-137.

Beale, E. M. L. (1974). A conjugate-gradient method of approximation programming, in *Optimization Methods for Resource Allocation* (R.W. Cottle and J. Krarup, eds.) 261-277, English Universities Press.

Beale, E.M.L. (1979). Branch and bound methods for mathematical programming systems, *Annals of Discrete Mathematics* **5**, 201-219.

Beale, E.M.L. (1980). The Blackett Memorial Lecture 1980. Operational research and computers: a personal view, *J. Oper. Res. Soc.* **31**, 761-767.

Beale, E.M.L. (1985a). The evolution of mathematical programming systems, *J. Oper. Res. Society* **36**, 357-366.

Beale, E.M.L. (1985b). Integer programming, in *Computational Mathematical Programming*, (K. Schittkowski, ed.), NATO ASI Series F: Computer and System Sciences, Vol. 15, Springer-Verlag, Berlin, 1-24.

Beale, E.M.L. and Forrest, J.J.H. (1976). Global optimization using special ordered sets, *Math. Prog.* **10**, 52-69.

Beale, E.M.L., Hughes, P.A.B. and Small, R.E. (1965). Experience in using a decomposition program, *Comput. J.* **8**, 13-18.

Beale, E.M.L. and Small, R.E. (1965). Mixed integer programming by a branch and bound technique. In *Proc. 1965 IFIP Congr.* (W.A. Kalenich, ed.), vol. 2, Spartan Books, Washington, D.C., 450-451.

Beale, E.M.L. and Tomlin, J.A. (1970). Special facilities in a general mathematical programming system for non-convex problems using ordered sets of variables, in *Proceedings of the Fifth International Conference on Operational Research*, (J. Lawrence, ed.), Tavistock Publications, London, 447-454.

Benichou, M., Gauthier, J. M., Hentges, G. and Ribière, G. (1977). The efficient solution of large-scale linear programming problems — some algorithmic techniques and computational results, *Math. Prog.* **13**, pp. 280-322.

Bradley, G.H., Brown, G.G. and Graves, G.W. (1977). Design and implementation of large-scale primal transshipment algorithms, *Management Sci.* **24**, 1-34.

Brown, G.G., Graves, G.W. and Honczarenko, M.D. (1987). Design and operation of a multicommodity production/ distribution system using primal goal decomposition, *Management Sci.* **33**, 1469–1480.

Brown, G.G. and McBride, R.D. (1984). Solving generalized networks, *Management Sci.* **30**, no. 12.

Crowder, H. (1976). Impact of future computer technology on mathematical programming, *Proc. SHARE* **47**, Montreal, 1513-1519.

Crowder, H., Johnson, E.L. and Padberg, M.W. (1983). Solving large-scale zero-one linear programming problems, *Operations Res.* **31**, 803-834.

Dantzig, G.B. (1963). *Linear Programming and Extensions.* Princeton University Press, Princeton, NJ.

Dantzig, G.B. and Orchard-Hays, W. (1954). The product form of the inverse in the simplex method, *Math. Comp.* **8**, 64-67.

Dantzig, G.B. and Tomlin, J.A. (1987). E.M.L. Beale, FRS; friend and colleague, *Math. Prog.* **38**, 117–131.

Dongarra, J. J. (1988). "Designing algorithms for dense linear algebra problems on high performance computers." Paper presented at the workshop on *Supercomputers and large scale optimization*, University of Minnesota, Minneapolis, MN.

Duff, I.S. (1986). "The use of vector and parallel computers in the solution of large sparse linear equations." AERE-R.12393, Harwell Lab., Oxfordshire, England.

Duff, I. S., Erisman, A. M., and Reid, J. K. (1986). *Direct Methods for Sparse Matrices.* Oxford University Press, London.

Forrest, J.J.H. (1987). "Linear programming using the IBM 3090 vector facility". Paper presented at the Beale Memorial Symposium, The Royal Society, London.

Forrest, J.J.H., Hirst, J.P.H. and Tomlin, J.A. (1974). Practical solution of large mixed integer programming problems with UMPIRE, *Management Sci.* **20**, 736–773.

Forrest, J. J. H. and Tomlin, J. A. (1972). Updating triangular factors of the basis to maintain sparsity in the product form simplex method. *Math. Prog.* **2**, 263–278.

Forrest, J. J. H. and Tomlin, J. A. (1988). "Vector processing in simplex and interior methods for linear programming." Paper presented at the workshop on *Supercomputers and large scale optimization*, University of Minnesota, Minneapolis, MN.

Gay, D. M. (1988) "Massive memory buys little speed for complete, in-core sparse Cholesky factorization", AT&T Bell Laboratories, Numerical Analysis Manuscript 88-04, Murray Hill, NJ. Presented at the 25th ORSA/TIMS joint meeting, Washington, D.C., April, 1988.

Gill, P. E., Murray, W., Saunders, M. A., Tomlin, J. A. and Wright, M. H. (1986). On Projected Newton Barrier Methods for Linear Programming and an Equivalence to Karmarkar's Projective Method, *Math. Prog.* **36**, 183–209.

Graves, G.W. and R.D. McBride (1976). The factorization approach to large-scale linear programming, *Math. Prog.* **10**, 91-111.

Griffith, R.E. and Stewart, R.A. (1961). A nonlinear programming technique for the optimization of continuous processing systems, *Management Sci.* **7**, 379-392.

Gustavson, F. G., Liniger, W. M., and Willoughby, R. A. (1970). Symbolic generation of an optimal Crout algorithm for sparse systems of linear equations. *J. ACM* **17**, 87–109.

Harris. P.M.J. (1973). Pivot selection methods of the Devex LP code, *Math. Prog.* **5**, 1–28.

Hellerman, E. and Rarick, D. C. (1972). Reinversion with the Partitioned Preassigned Pivot Procedure, in *Sparse Matrices and Their Applications* (D.J. Rose and R.A. Willoughby, eds.), Plenum Press, New York, 67–76.

Johnson, E.L. (1978). Some considerations in using branch and bound codes. In *Design and implementation of optimization software* (H.J. Greenberg, ed.), Sijtoff and Noordhoff, The Netherlands, 241–247.

Kalan, J. E. (1971). Aspects of large-scale in-core linear programming. in *Proceedings of the ACM Annual Conference, Chicago, IL*, ACM, New York, 304–313.

Karmarkar, N. (1984). A new polynomial-time algorithm for linear programming, *Combinatorica* **4**, 373–395.

Markowitz, H. M. (1957). The elimination form of the inverse and its application to linear programming, *Management Sci.* **3**, 255- 269.

Murtagh, B. A. and Saunders, M. A. (1983). *MINOS 5.0 User's Guide*, Report SOL 83-20, Department of Operations Research, Stanford University, California.

Orchard-Hays, Wm. (1968). *Advanced linear programming computing techniques.* McGraw-Hill, New York.

Powell, M.J.D. (1987). A biographical memoir of Evelyn Martin Lansdowne Beale, FRS, *Biographical Memoirs of Fellows of the Royal Society* **33**, 23–45

Reid, J. K. (1976). "FORTRAN subroutines for handling sparse linear programming bases", Report AERE-R.8269, Harwell Lab., Oxfordshire , England.

Tomlin, J. A. and Welch, J. S. (1985). Integration of a primal simplex network algorithm with a large scale mathematical programming system, *ACM Trans. on Math. Softw.* **11**, 1–11.

Van Roy, T.J. and Wolsey, L.A. (1987). Solving mixed integer programming problems using automatic reformulation, *Operations Res.* **35**, 45–57.

Geometry of Numbers and Integer Programming

László LOVÁSZ*

Department of Computer Science, Eötvös Loránd University, Budapest, Hungary and Princeton University, Princeton, New Jersey, U.S.A.

Abstract. The geometry of numbers arose around the end of the last century as a tool in number theory for solving problems in diophantine approximation and diophantine equations. One of its fundamental problems is to find a lattice point in a convex body. This problem is also central in integer programming and combinatorial optimization. Until very recently, little was known about the connections of these two applications of the same general problem. Recently the methods of the geometry of numbers have penetrated integer programming, and new insight provided by integer programming and complexity theory has prompted substantial improvements over classical results in the geometry of numbers.

0. Introduction.

One of the basic problems in integer programming can be phrased like this: given a convex body in the euclidean r-space, decide whether it contains a lattice point (and find one if it does). In this generality, however, this problem is also important in another branch of mathematics, namely number theory. In fact, as a branch of number theory called the geometry of numbers, problems like this have been studied by great mathematicians like Gauss, Hermite and Dirichlet, and in particular by Minkowski. The results were used in diophantine approximation, i.e., the approximation of numbers by rational numbers with small denominators, and in the theory of diophantine equations. For a detailed treatment of the Geometry of Numbers, see Cassels (1959) and Gruber and Lekkerkerker (1987).

---

* Research supported in part by the Hungarian National Research Fund No. 1820.

M. Iri and K. Tanabe (eds.), Mathematical Programming, 177–201.

Integer programming is of course a much younger discipline, and it is rather surprising how little use it has made of results from the geometry of numbers. One reason for this is that integer programming is interested in a simple lattice (the lattice of integer points) and a convex body defined by a complicated, and in most applications highly structured, convex body (like the travelling salesman polytope), while in number theory, one usually has very simple convex bodies (balls or cubes) and complicated lattices (spanned by irrational vectors). Of course, every lattice can be transformed into the lattice of integers; then the "simple" bodies in number-theoretic applications become combinatorially still simple, but geometrically ill-behaved — long and thin. This perhaps explains why the methods used in the geometry of numbers are rather geometric in nature (volume, width, tricky affine transformations etc.). By now, integer programming also began to develop a general theory, concentrating mainly on how to make use of the special (often combinatorial) properties of the convex body (which in almost all cases of interest was a polyhedron). To study the integer solvability of linear inequalities, notions like total unimodularity, total primal and dual integrality, Hilbert bases, cutting planes were introduced.

The breakthrough in making a connection between these fields was the result of H.W.Lenstra, jr. (1983), who showed that if we fix the number of variables, then integer programs can be solved in polynomial time. He applied methods from the geometry of numbers, in particular, basis reduction in lattices. His work (and some independent development in combinatorial optimization) raised interest in the algorithmic aspects of diophantine approximation. Lenstra, Lenstra and Lovász (1982) showed that a crucial step in Lenstra's algorithm, the basis reduction, can be carried out in polynomial time even if the dimension is part of the input. This yielded polynomial-time algorithms for simultaneous diophantine approximation, polynomial factorization and a number of further problems. (There is, unfortunately, an error factor in this algorithm which grows exponentially with the dimension, and thereby blocks several other possible applications.) These algorithmic results have lead to renewed interest in lattice geometry, and to improvements of several classical results.

In spite of these developments, the geometric and arithmetic (combinatorial) methods in integer programming have remained rather separated. The polynomial time solvability of an integer program in bounded dimension and (say) of a flow problem appear to be independent facts.

The main approach in the geometric direction is to show that lattice-point-free convex bodies must be "small" in one sense or the other. E.g. for the "homogeneous" version (when we have a 0-symmetric convex body and we are looking for a *non-zero* lattice point in it), Minkowski's classical theorem gives a bound on the volume of such a body. On the other hand, the arithmetic direction looks for conditions on the matrix describing the body that guarantee that if the body is non-empty then it also contains a lattice point. Total unimodularity of the matrix is the most important example.

In this paper we first survey some of the recent developments in geometric methods for our basic problem. Section 1, which is of an introductory nature, introduces the necessary background material from the Geometry of Numbers, and mentions some of the recent developments. In Section 2 we turn to our main question, the study of lattice-point-free bodies, and discuss what appears to be the main approach to this problem, at least along the geometric lines.

This should be followed by some sections discussing the results of the arithmetic and combinatorial direction, but there are several survey articles on this subject (e.g. Pulleyblank 1983, Schrijver 1983), and in particular an excellent monograph (Schrijver 1986). So we skip these.

The last two sections describe two approaches which seem to relate to both the geometric and arithmetic study of lattice points in convex bodies, and thereby hold out promises of some unification. In Section 3, we examine *maximal* lattice-point-free convex sets. They turn out to be related to the so-called Frobenius Problem, a notorius unsolved problem in algorithmic number theory.

In Section 4, we study the *neighbor* relation of lattice points with respect to a given matrix. This was introduced by Scarf (1981, 1986); one motivation for this was to restrict the range of lattice points that need to be searched if we want to make a local augmentation in an integer program. The neighbors of a lattice point turn out to generalize the notion of the Hilbert basis of a cone, but they are also closely related to the successive minima of Minkowski. The proof of some of the recent results concerning neighbors mixes geometric and arithmetic considerations.

1. Minkowski's Theorem and short lattice vectors

This section gives the necessary definitions and a handful of results from the geometry of numbers.

By a *convex body*, we mean a closed, convex, bounded, full-dimensional set in $\mathbb{R}^n$. vol(K) denotes the volume of K. If K is a convex body centrally symmetric about the origin (for short, 0-symmetric), then it may be viewed as the unit ball of a norm $\| \cdot \|_K$ on the space.

Given n linearly independent vectors $a_1, \ldots, a_n$, the *lattice* generated by them is the set of all of their linear combinations with integral coefficients:

$$L(a_1, \ldots, a_n) = \{x_1 a_1 + \ldots + x_n a_n : x_1, \ldots, x_n \in \mathbb{Z}\}.$$

The set $\{a_1, \ldots, a_n\}$ is called a *basis* of the lattice. A lattice can have many bases, but the following quantity, called the *determinant* of the lattice, depends only on the lattice:

$$\det(L) = |\det(a_1, \ldots, a_n)|.$$

Geometrically, the determinant of L is the area of the parallelpiped spanned by the vectors of any basis of L, or, more generally, the area of any region whose translates by lattice vectors tile the plane. Informally, det L is the area per lattice point.

The *standard lattice* in $\mathbb{R}^n$ consists of all integral vectors. This is the most important lattice for the applications, and also every other lattice could be easily transformed into it. So in most cases, it would suffice to consider the standard lattice. Sometimes, however, it will be more convenient to allow arbitrary lattices.

The set of all vectors $v \in \mathbb{R}^n$ such that $v \cdot z$ is an integer for every $z \in L$, is again a lattice, called the *dual* (or *polar*) of L, and denoted by L^*. The dual of the standard lattice is itself. It is not difficult to see that

$$(L^*)^* = L, \qquad \det(L)\det(L^*) = 1.$$

Let K be a 0-symmetric convex body. Define its *polar* by

$$K^* = \{x \in \mathbb{R} : x \cdot y \leq 1 \text{ for all } y \in K\}.$$

This is again a 0-symmetric convex body. It is true that $(K^*)^* = K$. The relationship between the volumina of K and K^* is more complicated. The product $\mathrm{vol}(K)\mathrm{vol}(K^*)$ is not constant, but it is bounded by functions of the dimension. It is quite natural to conjecture that this product is maximized when K is a ball and minimized when K (or K^*) is a cube; in other words, that

$$\frac{\Pi^n}{\Gamma\left(\frac{n}{2}+1\right)^2} \geq \mathrm{vol}(K)\mathrm{vol}(K^*) \geq \frac{4^n}{n!}.$$

The first of these assertions was proved by Santaló (1949), but the second is still unsettled. A recent important result of Bourgain and Milman (1985) gives a lower bound of the form $(c/n)^n$, which has the same form as the conjectured lower bound (and the upper bound), and will be good enough for our purposes.

The fundamental result in the classical "geometry of numbers" is Minkowski's Theorem:

1.1 Theorem. *Let L be a lattice and K, a 0-symmetric convex body in $\mathbb{R}^n$. Assume that K contains no interior lattice point other than its center. Then*

$$\mathrm{vol}(K) \leq 2^n \det L.$$

Another way to put this fundamental result is: for every lattice L and 0-symmetric convex body K, there is a non-zero lattice vector not longer than $2\left(\det L/\mathrm{vol}(K)\right)^{1/n}$ in the norm defined by K. Let $\lambda_1(K, L)$ denote the length of the shortest lattice vector with respect to the norm defined by K; equivalently, $\lambda_1(K, L)$ is the least number t for which the body tK contains a non-zero lattice point. So Minkowski's Theorem says:

$$\lambda_1(K, L) \leq 2\left(\det L/\mathrm{vol}(K)\right)^{1/n}. \tag{1}$$

It is worth to note the following characteriztion of the number $\lambda_1(K, L)$: it is the largest non-negative real number t for which the bodies $\frac{t}{2}K + v$, $(v \in L)$ have no interior point in common. In this form, it seems natural to extend the definition to general (not necessarily centrally symmetric) convex bodies. However, the following simple fact shows that this would not lead to an essentially new notion:

1.2 Proposition. *Let K be a convex body and L, a lattice in $\mathbb{R}^n$. Then the largest non-negative real number t for which the bodies $tK + v$ $(v \in L)$ have no interior point in common is $\lambda_1(K - K, L)$.*
(Here $K - K = \{x - x' : x, x' \in K\}$.)

Using these remarks one can sketch the proof of Minkowski's Theorem: We know that the bodies $(\lambda_1/2)K + v$ $(v \in L)$ have no interior point in common, and that $\det(L)$ is the "area per lattice point". Hence

$$\mathrm{vol}\left((\lambda_1/2)K\right) \leq \det(L).$$

Hence Minkowski's Theorem follows immediately. (Note that this proof is non-algorithmic: it does not give any help to gind the lattice point in K.)

The value λ_1 is just the first of a sequence of important parameters of centrally symmetric convex bodies called *successive minima*. The i-th successive minimum, $\lambda_i(K, L)$ of the 0-symmetric convex body K with respect to the lattice L is the least t for which tK contains at least i linearly independent lattice vectors. It is easy to see that the successive minima can be obtained by the following greedy procedure: for $i = 1, \ldots, n$, we let v_i be the shortest non-zero vector in the lattice (with respect to the norm $\|.\|_K$), not contained in the linear span of $v_1, \ldots, v_{i-1}$, and let $\lambda_i(K, L)$ be the length of v_i.

Successive minima enabled Minkowski to formulate and prove an extension of Theorem 1.1 that is substantially deeper:

$$\lambda_1(K, L)\lambda_2(K, L) \ldots \lambda_n(K, L) \leq 2^n \det(L)/\text{vol}(K).$$

While we will not use this inequality, successive minima will be important for some of the results on lattice-point-free bodies.

An interesting corollary of Minkowski's Theorem, to be used later on, can be derived by writing up inequality (1) also for the polar body and the dual lattice:

$$\lambda_1(K^*, L^*) \leq 2\big(\det(L^*)/\text{vol}(K^*)\big)^{1/n}.$$

Multiplying these inequalities, we obtain:

$$\lambda(K, L)\lambda_1(K^*, L^*) \leq 4 \left(\frac{\det(L)\det(L^*)}{\text{vol}(K)\text{vol}(K^*)} \right)^{1/n}.$$

Here $\det(L)\det(L^*) = 1$ and $\text{vol}(K)\text{vol}(K^*) \geq (\text{const}/n)^n$ by the theorem of Bourgain and Milman (1986), and hence it follows that

1.3 Corollary. *For every lattice L and 0-symmetric convex body K,*

$$\lambda_1(K, L)\lambda_1(K^*, L^*) \leq \text{const} \cdot n.$$

Lagarias, Lenstra and Schnorr (1988) prove a number of powerful inequalities of this type relating various successive minima of a lattice and its dual. Let us mention one that will be used in the sequel:

1.4 Theorem. *For every lattice L and 0-symmetric convex body K,*

$$\lambda_1(K, L)\lambda_n(K^*, L^*) \leq \text{const} \cdot n^2.$$

We conclude this section with a brief discussion of the algorithmic aspects of these classical results. First, we have to say a few words about how lattices and convex bodies are given when they serve as inputs to our algorithms. It is natural to assume that a lattice is given by a basis (which, in algorithmic considerations, is assumed to have rational entries). Accordingly, the contribution of the lattice to the size of the input is the number of bits needed to write down the numerators and denominators of entries of the basis vectors. With convex sets, the situation is more difficult. We want to allow solution sets of linear inequlities, convex hulls of given sets of vectors, balls, ellipsoids etc. It turns out that *it is enough to have a subroutine (oracle) to decide whether or not a point belongs ot K*, together with some minor technical information. This seemingly very weak oracle allows us then to solve other algorithmic questions, like finding the maximum of any linear objective function over K in polynomial time. We do not go into the details of this, but refer instead to Lovász (1986) and Grötschel, Lovász and Schrijver (1988).

We shall be interested in designing algorithms whose running time is polynomially bounded in the input size and in the running times of the oracles used to specify convex bodies.

It is quite easy to pose an algorithmic version of Minkowski's first theorem:

1.5 Problem. *Given a lattice L and a convex body K, centrally symmetric with respect to the origin, such that $\mathrm{vol}(K) > 2^n \det L$, find a non-zero lattice point in K.*

Recalling the formulation of Minkowski's Theorem in terms of λ_1, we can formulate the following basic algorithmic problem.

1.6 Problem. *Given a lattice and a norm, find the shortest non-zero vector of the lattice.*

A particularly simple related question is to find an algorithmic version of Corollary 1.3:

1.7 Problem. *Given a lattice L and an 0-symmetric convex body K, find a non-zero vector $u \in L$ and a non-zero vector $v \in L^*$ such that*

$$\|u\|_K \cdot \|v\|_{K^*} \le \mathrm{const} \cdot n.$$

Clearly Problem 1.6 is stronger than Problems 1.5 and 1.7. (They are, however, equivalent if we allow a relative error polynomial in n; see Lovász 1986). Unfortunely, no polynomial-time algorithm is known to solve these fundamental problems, even if the norm is very simple (euclidean or l_∞ etc.). In fact problem 1.6 is known to be NP-complete for the maximum norm (van Emde-Boas 1981). We don't know if problems 1.5 and 1.7 are in P or NP-hard for any norm.

It is important, however, that the successive minima λ_i can be determined in polynomial time *if the dimension is fixed*. More exactly, for every $\epsilon > 0$ we can determine them with relative error at most ϵ, in time polynomial in the input size of L and K (including the running time for the oracle describing K) and in $\log \epsilon$. In the case when K is given by a system of linear inequalities with rational coefficients, this yields a polynomial time algorithm to calculate the λ_i exactly. The algorithm is a combination of Lenstra's method (see in the next section) with an easy binary search.

For the case when n varies, only approximation algorithms are knwon that work in polynomial time. We do not go into the details of these; they are discussed in detail in Lovász (1986), Schrijver (1986), Grötschel, Lovász and Schrijver (1988) and Kannan (1987). We shall quote two results and introduce as much from the machinery as needed in the sequel.

First we mention that Problem 1.7 is polynomial time solvable if we allow a relative error which depends only on the dimension (Lenstra, Lenstra and Lovász 1982):

1.8 Theorem. *Given a lattice and a norm, we can find in polynomial time a vector that is at most 2^n times as long as the shortest non-zero vector of the lattice.*

The following result can be viewed as a (weak) algorithmic version of Minkowski's theorem (Grötschel, Lovász and Schrijver 1988):

1.9 Theorem. *Given a lattice L and a convex body K, centrally symmetric with respect to the origin, such that* $\text{vol}(K) > 2^{n^2} \det L$, *one can find in polynomial time a non-zero lattice point in K.*

The problem of finding a short vector in a lattice, or — somewhat stronger — of finding a basis consisting of short vectors, is a classical problem in number theory, and also central in modern algorithmic number theory. Since currently every efficient procedure works with exponentially large relative error, and every non-euclidean norm can be approximated by a euclidean norm with relative error $O(\sqrt{n})$, we may from now on restrict our attention to the euclidean norm, i.e., the case when K is a ball.

Let $b_1, \ldots, b_n$ be any basis of the lattice L. The *Gram-Schmidt orthogonalization* of this basis is the (unique) orthogonal basis $b_1^*, \ldots, b_n^*$ with the property that each b_i has a representation

$$b_i = \mu_{i,1} b_1^* + \mu_{i,2} b_2^* + \ldots + \mu_{i,i-1} b_{i-1}^* + b_i^*.$$

(in particular $b_1^* = b_1$.)

It is clear that $|b_i|^2 = \sum_j \mu_{i,j}^2 |b_i^*|^2$ (where $\mu_{i,i} = 1$), and so if we want a "short" basis we should try to choose it so that the $\mu_{i,j}$ be small. Motivated by this, we call the basis *weakly reduced* if $|\mu_{i,j}| \leq 1/2$ for every $1 \leq j < i \leq n$. If we have any basis, we can easily transform it into a weakly reduced basis by subtracting appropriate integer multiples of basis vectors from basis vectors with larger index. Note that this procedure does not change the Gram-Schmidt orthogonalization of the basis.

Unfortunately, this easy reduction procedure does not give a basis that would be good enough (e.g. the smallest vector in the resulting basis may be much longer than the shortest vector in the lattice). Therefore, we also want to achieve that the lengths of the orthogonalized vectors do not drop too fast. There is a number of additional ways to reduce the basis, and we shall not go into the details of this here. We mention that the reduction in Lenstra, Lenstra and Lovász (1982), which can be carried out in polynomial time, gives a basis whose Gram-Schmidt orthogonalization satisfies

$$|b_i^*| \geq 2^{(j-i)/2} |b_j^*|$$

for every $1 \leq j < i \leq n$. Hence it is not difficult to deduce that for a basis reduced in this sense, the i-th basis vector is at most $2^{n/2}$ times as long as the i-th successive minimum. A reduction with very good properties, but unfortunately not polynomial time, is the *Korkine-Zolotarev reduction*. In this, we choose each b_i so that b_i^* is a shortest vector in the lattice obtained by projecting L onto the orthogonal complement of $b_1, \ldots, b_{i-1}$ (see Lagarias, Lenstra and Schnorr 1988). See Cassels (1959), Gruber and Lekkerkerker (1987) and also Schrijver (1986) for a discussion (also from an algorithmic as well as a historical point of view) of lattice reduction techniques.

2. Lattice points in convex bodies and covering minima

We address the following general question: given a lattice L and a convex body K, find a lattice point in K (or conclude that none exists). Note that even the question of deciding whether K contains a lattice point is difficult; in fact, any algorithm to decide this can be used in combination with an easy bisection method to *find* the lattice point in K (if it exists).

Often it is more convenient to consider bodies that contain no *interior* lattice points. We shall call such a body *lattice-free*. Of course, by shrinking such a body by arbitrarily little, we get one that is entirely disjoint from the lattice.

We can also formulate our main question in a more classical, structure-oriented fashion: *how does a lattice-free convex body look like?* We cannot expect a final answer to this question, but we can expect to obtain results asserting that if a convex body contains no lattice point then it must be "small" in some sense.

Such an answer is potentially useful in designing an algorithm for deciding if a given convex body K contains a lattice point. We know K either contains a lattice point or it is "small". In the letter case we can try to apply some brute force search for a lattice point in K. We shall see that — in a substantially more involved manner — H.W.Lenstra's celebrated integer programming algorithm has this flavor.

Let us introduce the following parameter of a convex body: we denote by $\mu(K, L)$ the least non-negative number t for which the bodies $tK + v$ $(v \in L)$ cover the whole space. This number is called the *covering radius* of K with respect to L. (Note the analogy with the geometric interpretation of the first minimum. The name comes from the special case when K is the unit ball about 0. In this case, the covering radius is the maximum distance of any point of the space from the lattice.)

There are some simple (and rather weak) inequalities relating the covering radius to the successive minima defined in the previous section. Note that the covering radius is meaningful for all convex bodies, while the successive minima were defined only for centrally symmetric bodies. Proposition 1.2 gives a geometric meaning of the first successive minimum of the difference body $K - K$ in terms of packing of K. In this section too we usually consider the successive minima of the difference body of K.

2.1 Proposition. *Let K be a convex body, μ its covering radius, and let $\lambda_1, \ldots, \lambda_n$ be the successive minima of the difference body $K - K$. Then*

$$\lambda_n \le \mu \le \lambda_1 + \ldots + \lambda_n.$$

The connection between the covering radius and lattice-free bodies is given by the following simple fact:

2.2 Proposition. *$\mu(K, L)$ is the largest non-negative number t for which tK has a lattice-free translation.*

The problem of computing the value $\mu(K, L)$ is unsolved. This is of course NP-hard in general, but even in fixed dimension and in the case when K is a polytope, it does not seem to be reducible to integer programming (unlike the shortest lattice vector problem). Kannan (1988) gave a strongly polynomial approximation scheme for the covering radius (i.e., and algorithm to compute it with relative error ϵ, in time polynomial in the input size of L and K and in $1/\epsilon$).

Returning to the study of lattice-free bodies, we first remark that there is no immediate analogue of Minkowski's Theorem to lattice-free convex bodies: such a body can have an arbitrarily large volume (e.g. consider a flat but very long rectangle between two consecutive lattice lines in the plane).

If we interpret "small" differently then we do get a result of the type we were looking for. Let K be a convex body in $\mathbb{R}^n$ and $v \in \mathbb{R}^n$. We call the difference $\max\{v \cdot x : x \in K\} - \min\{v \cdot x : x \in K\}$ the *width* of K along the vector v, and its minimum over all vectors v in the dual lattice L^*, the *lattice width* of K. (The minimum width along unit vectors v is the usual geometric width of K.) It is easy to see that the lattice width of K is just $\lambda_1((K - K)^*, L^*)$. We shall refer to the following result as the *Flatness Theorem*:

2.3 Theorem. *Let L be a lattice and K, a convex body in $\mathbb{R}^n$, containing no point from L. Then the lattice width of K is bounded by a constant $f(n)$ dependending only on the dimension.*

Geometrically, the Flatness Theorem says that is K is a lattice-free convex body, then there is a lattice hyperplane H such that K intersects at most $f(n) + 1$ of the hyperplanes parallel to H. We can also formulate the Flatness Theorem as an inequality for the covering radius:

$$\mu(K, L)\lambda_1((K - K)^*, L^*) \leq f(n).$$

The Flatness Theorem goes back to Khinchine (1948), who proved it (as a quantitative extension of Kronecker's theorem on inhomogeneous diohantine approximation) when K is a ball with $f(n) \approx (n!)^2$. H.W.Lenstra (1983) used an algorithmic version of this result in his integer programming algorithm: he showed that for fixed n, and $f(n) \approx 2^{n^2}$, one can either find in polynomial time either a lattice vector in K or a dual lattice vector v with the property in the theorem. Grötschel, Lovász and Schrijver (1982) showed that this can in fact be achieved in polynomial time even for variable n. Babai (1986) improved this algorithmic result to $f(n) \approx 3^n$.

If we only want existence results then better bounds on $f(n)$ are known. Hastad (1988), using some new powerful results of Lagarias, Lenstra and Schnorr (1988) proved the above theorem with $f(n) \approx n^{2.5}$. Currently the strongest result appears to be that of Kannan and Lovász (1986, 1988), that asserts that the theorem is valid with $f(n) = O(n^2)$. (We shall sketch the proof of this result below.)

The best possible value of $f(n)$ is perhaps linear in n. The convex body K defined by the inequalities $x \geq 0$, $1 \cdot x \leq n$ contains no integral interior point. On the other hand, its lattice width is n. For, let v be any non-zero integral vector and assume that (say) its first entry $v_1 \neq 0$. Then the linear objective funtion $v \cdot x$ assumes the value 0 as well as the value nv_1 on K, and so $f(n)$ certainly cannot be replaced by anything less than n. (One may think that this is the worst possible construction, but already for the plane it can be worse; we shall return to this later.)

Unfortunately, these polynomial bounds cannot be achieved at the moment by polynomial time algorithms.

Now Theorem 2.3 is used in Lenstra's algorithm and its subsequent refinements in the following way. Suppose that we want to decide whether or not K contains a lattice point. Let us use an algorithmic version of Theorem 2.3 to find either a lattice point in K (in which case we are done), or a dual lattice vector v as in the theorem. In the latter case, the value of vx for any possible vector $x \in K \cap L$ is an integer between $\min\{v \cdot x : x \in K\}$ and $\max\{v \cdot x : x \in K\}$. Hence we can "branch" and reduce our problem to at most $f(n) + 1$ $(n - 1)$-dimensional subproblems: find a lattice point in $K \cap \{x : v \cdot x = t\}$ ($t \in \mathbb{Z}$, $\min\{v \cdot x : x \in K\} \leq t \leq \max\{v \cdot x : x \in K\}$).

This recurrence leads to an algorithm with a tree-structure with $f(n) \cdot f(n-1) \cdot \ldots \cdot f(1)$ nodes. If n is fixed then this is polynomial time; but for variable n, it becomes badly exponential. In number-theoretic problems (inhomogeneous diophantine approximation), n is often not large and the difficulty lies in the structure of the lattice. In these cases, such an approach may be practical. But in most combinatorial optimization problems, the number of variables is large (typically the interesting range starts with hundreds), and so the above algorithm is impractical.

This application also shows the importance of obtaining an algorithmic result with a polynomial $f(n)$. This would still give an exponentially large tree (as it is to be expected, since integer programming is NP-complete), but at least one could hope for some ways of pruning the tree and thereby improving the running time. As it stands, already one branching is forbidding in most cases.

To sketch the proof of the fact that $f(n) = O(n^2)$, let us introduce the following numbers, called the *covering minima*, which generalize the covering radius, and are in many respects analogous to the successive minima. For $1 \leq k \leq n$, let $\mu_k = \mu_k(K, L)$ denote the least positive number t for which the set $tK + L$ intersects every $(n - k)$-dimensional affine subspace.

It is evident that these numbers are invariant under the translations of K.

The last of the covering minima, μ_n is just the ordinary covering radius $\mu(K, L)$. On the other hand, the first of these numbers relates to the lattice width:

2.4 Proposition.

$$\mu_1(K, L) = \frac{1}{\lambda_1\big((K - K)^*, L^*\big)}.$$

In fact, let $t < \mu_1(K, L)$. Then there exists a hyperplane $w \cdot x = \omega$ such that $tK + L$ does not intersect this hyperplane. It is easy to argue that we may normalize the equation of this hyperplane so that w is in L^*, and in fact a primitive vector of this lattice. Then the hyperplanes $w \cdot x = \omega + k$, $k \in \mathbb{Z}$, are also avoided by the set $tK + L$. Hence the width of tK along w is at most 1, and so the width of K along w is at most $1/t$. So the lattice width of tK is at most $1/t$. Since this holds for every $t < \mu_1(K, L)$, it follows that $\mu_1(K, L) \leq 1 / \lambda_1((K - K)^*, L^*)$. The reverse inequality follows by a similar argument.

In terms of the covering minima, the Flatness Theorem (with $f(n) = O(n^2)$) can be formulated as follows:

$$\mu_n \leq \text{const} \cdot n^2 \cdot \mu_1.$$

More generally, one can show that

$$\mu_k \leq \text{const} \cdot k^2 \cdot \mu_1 \tag{1}$$

for every $1 \leq k \leq n$. This form of the theorem suggests an induction proof based on the inequality

$$\mu_{k+1} \leq \mu_k + \text{const} \cdot k \cdot \mu_1. \tag{2}$$

To prove (2), one first establishes

2.5 Lemma.

$$\mu_n \leq \mu_{n-1} + \lambda_1(K - K, L).$$

To prove this lemma, denote $\lambda_1 = \lambda_1(K - K, L)$ and let $v \in L \cap \lambda_1 \cdot (K - K)$, $v \neq 0$ (such a lattice vector exists by the definition of λ_1). We can write v as the difference of two vectors in $\lambda_1 \cdot K$; since the inequality we want to prove is invariant under translation, we may assume that these two vectors are v and 0, i.e.,

$$0 \in K \quad \text{and} \quad v \in \lambda_1 \cdot K.$$

What we want to prove is that $(\mu_{n-1} + \lambda_1)K + L$ covers the space. Let $p \in \mathbb{R}^n$. Draw a line through p parallel to v. By the definition of μ_{n-1}, this line intersects one of the bodies $\mu_{n-1} \cdot K + u$, $u \in L$; withour loss of generality we may assume that it intersects $\mu_{n-1} \cdot K$. Let w be a point of intersection. Then $p - w$ is parallel to v and so it can be written as $p - w = \delta v$ with some real number δ. Now

$$p = w + \delta v = w + (\delta - \lfloor \delta \rfloor)v + \lfloor \delta \rfloor v \in \mu_{n-1} \cdot K + \lambda_1 \cdot K + L = (\mu_{n-1} + \lambda_1) \cdot K + L.$$

(In the last step we have used that K is convex.) This proves Lemma 2.5.

Now to prove (2), we use Lemma 2.5 in combination with Corollary 1.3:

$$\lambda_1(K - K, L) \leq \text{const} \cdot n \, / \, \lambda_1((K - K)^*, L^*).$$

Hence by Lemma 2.5 and Proposition 2.4,

$$\mu_n \leq \mu_{n-1} + \text{const} \cdot n \cdot \mu_1.$$

This proves (2) for $k = n - 1$. The general case can be reduced to this by considering an appropriate projection. As remarked, inequality (1) and hence the Flatness Theorem follow by induction.

The only non-algorithmic step in this proof is the choice of the vector v, i.e., the shortest non-zero lattice vector with respect to the norm defined by $K - K$. Note that we have not really used here that v was shortest; we only needed that it was not longer than $O(n)\lambda_1((K - K)^*, L^*)$. In other words, if we could find an algorithmic proof of Corollary 1.3 (i.e., a polynomial time algorithm to construct the two "short" vectors), then this would also yield an algorithmic proof of the Flatness Theorem with $f(n) = n^2$, i.e., it would yield a polynomial time algorithm to find either a lattice point in a given convex body K, or a dual lattice vector along which K has width $O(n^2)$. Of course, we do have an algorithmic proof of Corollary 1.3 if we replace n by an exponential function of n; accordingly, we have a polynomial time algorithm in the Flatness Theorem if we replace n^2 by an exponential function.

The key to the proof above was the inequality (2), and it is natural to ask if one could tighten it; a natural inequality would be

$$\mu_{k+1} \leq \mu_k + \mu_1. \tag{3}$$

Unfortunately, this is false even for $k = 1$; Hurkens (1988) constructed a lattice-free triangle in the plane with lattice width $1 + \frac{2}{\sqrt{3}} > 2$. He also proved that this is the worst possible case, i.e.,

$$\mu_2 \leq (1 + \frac{2}{\sqrt{3}})\mu_1$$

for every convex body in the plane. It is not known what is the smallest value c_k for which

$$\mu_{k+1} \leq \mu_k + c_k \mu_1$$

always holds (Hurkens' result says that $c_1 = 2/\sqrt{3}$). It may be the case, however, that the attractive inequality (3) does hold for *centrally symmetric* bodies K. This was proved for $k = 1$ by Kannan and Lovász (1988).

The covering minima have applications other than the proof of the Flatness Theorem. Using them, Kannan (1983) and Kannan and Lovász (1986) obtained the following improvement of the Flatness Theorem. By looking at the example showing that the lattice width of a lattice-free body may be as large as the dimension, we notice that this width is attained along n linearly independent dual lattice vectors. One may expect that if the lattice width of a lattice-free body is close to the upper bound then there will be many vectors in the dual lattice along which the width is not too large. More exactly, the following is true.

2.6 Theorem. *Let L be a lattice and K, a convex body in $\mathbb{R}^n$, containing no point from L. Then there exists an integer k, $1 \leq k \leq n$, and k linearly independent vectors $v_1, \ldots, v_k$ in the dual lattice such that for each $1 \leq i \leq k$,*

$$\max\{v_i \cdot x : x \in K\} - \min\{v_i \cdot x : x \in K\} = O(k^3 \log(k+1)).$$

The interesting point here is that the right hand side does not depend on n. For example, if k is bounded then the width is bounded by an absolute constant.

Note, however, that the theorem gives us no controll over the value of k here. Consider the standard lattice and the convex set in $\mathbb{R}^n$ defined by $x \geq 0$, $x_1 + \ldots + x_m \leq m$, $x_{m+1} \leq N, \ldots, x_n \leq N$. Then only m linearly independent dual lattice vectors realize small lattice width; on the other hand, the lattice width of this body is m. So the choice of k in the theorem must be between $m^{1/3}$ and m.

This theorem too has an algorithmic version: if we replace the right hand side by 2^k then we can find, in polynomial time, either a lattice point in K or k linearly independent lattice vectors as in the theorem.

Using covering minima, we can formulate a version valid for all k:

2.7 Theorem. *Let L be a lattice and K, a convex body in $\mathbb{R}^n$, containing no point from L. Then for every $1 \leq k \leq n$ there exist k linearly independent vectors $v_1, \ldots, v_k$ in the dual lattice such that for each $1 \leq i \leq k$,*

$$\max\{v_i \cdot x : x \in K\} - \min\{v_i \cdot x : x \in K\} \leq \frac{\operatorname{const} k^2}{\mu_{n-k} - \mu_{n-k-1}}.$$

Theorem 2.6 follows on noticing the elementary fact that for at least one k, $\mu_{n-k} - \mu_{n-k+1} \geq \operatorname{const}/(k \log^2 k)$. Theorem 2.7 can be derived from Lemma 2.5, using Theorem 1.1.

We can use this result to obtain an improved Lenstra type algorithm to find a lattice point in a convex body as follows: we either find a lattice point in K and stop, or find, for some k, k linearly independent dual lattice vectors in whose direction K has width at most 2^k. This means that we can branch into $(2^k)^k$ subproblems, each in dimension $n - k$. If k

is bounded, this is a bounded number of subproblems. If k is large, we gain a lot because the dimension drops substantially. See Kannan and Lovász (1986) for more details.

Even though the volume of lattice-free convex bodies is not bounded by any function of the dimension and the lattice, the following recent result of Kannan and Lovász (1988) gives a property involving the volume, a bit in the style of Minkowski's Theorem. (This is again proved using the covering minima.) The result is best explained in terms of the covering radius. Trivially, a body whose translates by lattice vectors cover the space must have volume at least $\det(L)$. Hence the covering radius satisfies the inequality

$$\mu(K, L) \geq \left(\frac{\det(L)}{\mathrm{vol}(K)} \right)^{1/n}.$$

This may be a very poor lower bound since, as remarked, $\mathrm{vol}(K)$ may be arbitrarily large while L is fixed and $\mu(K, L) > 1$. To obtain a stronger lower bound, consider any integer k, $1 \leq k \leq n$, and a projection Π of the whole space on a k-dimensional subspace. Then it is also easy to see that

$$\mu(K, L) \geq \mu(\Pi K, \Pi L) \geq \left(\frac{\det(\Pi L)}{\mathrm{vol}(\Pi K)} \right)^{1/k}.$$

Let $\varphi(K, L)$ denote the maximum of the right hand side for all choices of k and Π. Then the following holds:

2.8 Theorem. *If L is a lattice and K, a convex body in $\mathbb{R}^n$, then*

$$1 \leq \frac{\mu(K, L)}{\varphi(K, L)} \leq n.$$

We do not know an example where the ratio μ/φ would exceed $O(\log n)$. The standard lattice, together with the body defined by $x \geq 0$, $\sum_i x_i/i \leq 1$ gives a ratio of approximately $\log n$.

If we use covering minima, we can formulate a result valid for every k:

2.9 Theorem. *Let L be a lattice and K, a convex body in $\mathbb{R}^n$. Then for every k, there exists a projection Π of R^n on an appropriate k-dimensional subspace such that*

$$\mathrm{vol}(\Pi K) \leq (\mu_{n-k} - \mu_{n-k-1})^k \det(\Pi L).$$

The previous theorem follows if we notice that, trivially, there is a k for which $\mu_{n-k} - \mu_{n-k-1} \geq \mu_n/n$. By considering other ways to partition μ_n, we can obtain other versions of this result. For example, we can bound the volume by a function of k only:

2.10 Corollary. *If L is a lattice and K, a lattice-free convex body in $\mathbb{R}^n$, then there exists a k, $1 \leq k \leq n$ and a projection Π of the whole space on some subspace of dimension k such that*

$$\mathrm{vol}(\Pi K) \leq \mathrm{const} \cdot k^k (\log k)^2 k \det(\Pi L).$$

Assume now that K is centrally symmetric with respect to a point a. We have already remarked that in this case the covering minima seem to behave nicer; but also there is a way to formulate a stronger version of the Flatness Theorem. Let us consider K as the unit ball of a norm $\|.\|_K$. Then K is lattice-free if and only if the distance $d(a, L)$ of a (measured in this norm) from the lattice L is at least one. The Flatness Theorem says that in this case there is a lattice hyperplane H such that the distance between consecutive lattice hyperplanes parallel to H is at least $1/f(n)$. In this formulation it is natural to sharpen this property of H by requiring that a should be at a distance at least $f(n)$ from the two lattice hyperplanes parallel to H next to a. In other words this means that there is a projection Π of $\mathbb{R}^n$ onto a 1-dimensional subspace so that the distance $d(\Pi a, \Pi L)$ of the projection of a from the projection of L (measured in the metric determined by the projection of K) is at least $1/f(n)$.

Such a sharpening of the Flatness Theorem is indeed true. Let $\|.\|$ be an arbitrary norm on $\mathbb{R}^n$ and let $d(.,.)$ denote the corresponding distance function.

2.11 Theorem. *Let L be a lattice in $\mathbb{R}^n$ and $a \in \mathbb{R}^n$. Then there exists a projection Π of the space onto a 1-dimensional subspace such that*

$$d(\Pi a, \Pi L) \geq \frac{1}{g(n)} d(a, L),$$

where $g(n)$ is a constant depending only on the dimension.

Let $\delta(a, L)$ denote the maximum of $d(\Pi a, \Pi L)$ over all projections Π onto a one-dimensional subspace. Note that trivially $\delta(a, L) \leq d(a, L)$, so we see that Theorem 2.11 characterizes the distance of a point from a lattice up to a factor of $g(n)$.

This result — for the case when K is a ball, with $g(n) = (n!)^2$ — was proved by Khinchine (1948). The better value $g(n) = (\text{const})^n$ follows from Babai's nearest lattice point algorithm (1986; see below). This also provides a polynomial time algorithm to find the projection. Hastad (1988) proved the theorem for the case when K is a ball (i.e., when the norm is euclidean) with $g(n) = n^2$. Using general results on the ellipsoid method (see Grötschel, Lovász and Schrijver 1988), the theorem also follows for a general centrally symmetric convex body with $g(n) = n^{2.5}$.

Let us sketch the idea of Hastad's proof, because it involves an interesting analysis of heuristics to find a "nearby" lattice point. Fix a basis $b_1, \ldots, b_n$ of the lattice (we shall have to choose this basis reduced in an appropriate sense later on); let $b_1^*, \ldots, b_n^*$ be its Gram-Schmidt orthogonalization. Recall the relation

$$b_i = \mu_{i,1} b_1^* + \ldots + \mu_{i,i-1} b_{i-1}^* + b_i^*. \tag{4}$$

Given a point a in the space, we would like to find a lattice point closest to it in the euclidean norm. This problem is NP-complete, so it makes sense to look for simple heuristics. Babai (1986) formulated two of these.

ROUNDING OFF. Write $a = \sum_i \beta_i b_i$. Let α_i be the integer nearest to β_i. Take $z = \sum_i \alpha_i b_i$ as the lattice point approximating a.

NEAREST PLANE (rephrased). Write

$$a = \gamma_1 b_1^* + \gamma_2 b_2^* + \ldots + \gamma_n b_n^*. \tag{6}$$

Subtract integer multiples of (5) (for various values of i) from (6) until we obtain an equation

$$a - \sum \phi_i b_i = \gamma_1' b_1^* + \gamma_2' b_2^* + \ldots + \gamma_n' b_n^*,$$

where $|\gamma_i'| \leq 1/2$. (This can be achieved by the same procedure as weak reduction; I leave the details to the reader.) Now take $z = \sum_i \phi_i b_i$ as the lattice point approximating a.

(This heuristics may be viewed as repeated projection on the nearest lattice plane parallel to the subspace spanned by $b_1, \ldots, b_{n-1}$; hence the name.)

The error in the result of either of these procedures depends on the basis we choose. Babai proved that if the basis is reduced in the sense of Lenstra, Lenstra and Lovász (1982) then both heuristics have a relative error of the form (const)^n (i.e., $|a - z| \leq c^n d(a, L)$). This follows from the estimate

$$|a - z| \leq c^n \delta(a, L).$$

He used the following easy fact: for the lattice point z obtained by the Nearest Plane heuristics we have

$$|a - z| \leq \sqrt{|b_1^*|^2 + \ldots + |b_n^*|^2}.$$

If the vectors in the orthogonalized basis are very short (shorter than $O(\delta(a, L)n^{3/2})$) then of course we get the better bound $|a - z| \leq O(n^2 \delta(a, L))$.

For the case when the orthogonalized basis vectors are "long", Hastad proves a substantially deeper fact about the other heuristic.

2.12 Lemma. *Assume that the dual of the basis $b_1, \ldots, b_n$ is weakly reduced and that for each i, $|b_i^*| \geq 12\delta(a, L)\sqrt{n}$. Let z be obtained from a by the Rounding Off heuristic. Then*

$$|a - z| \leq 12\delta(a, Z).$$

Now Hastad's proof is concluded by splitting the space into two parts. Given L and a, we select a maximal lattice subspace U such that the Gram-Schmidt orthogonalization of the Korkine-Zolotarev basis of $L \cap U$ consists of vectors shorter than $12\delta(a, L)n^{3/2}$. Project the lattice L onto the orthogonal complement of U, to get the lattice L/U, and construct the dual Korkine-Zolotarev basis in L/U. It is not difficult to show (using Corollary 1.3), that the vectors in the Gram-Schmidt orthogonalization of this basis are longer than $12\delta(a, L)n^{1/2}$.

We can apply the Rounding Off heuristics in the projection L/U and then the Nearest Plane heuristics in the sublattice $L \cap U$. For details, refer to Hastad (1988).

Unfortunately, Hastad's proof does not give a polynomial time algorithm to find the hyperplane in question, mainly because no polynomial time procedure is known to find the Korkine-Zolotarev basis of a lattice.

3. Maximal lattice-free bodies

If we want to study lattice-free convex bodies, it is natural to restrict our attention to *maximal* closed convex sets containing no interior lattice point. (We have to allow unbounded sets in order to guarantee that every lattice-free convex body be included in one of these sets.) Such a set will be called for short *maximal lattice-free*. We shall restrict our attention to the standard lattice $\mathbb{Z}^n$.

Example. Let us describe all maximal lattice-free convex sets in the plane. It is a little tedious but not too hard to see that they are of four possible forms:

(a) A strip $c \le ax + by \le c + 1$ where a and b are coprime integers and c is an integer.

(b) A line $ax + by = c$ where a/b is irrational and $c \notin a\mathbb{Z}^n + b\mathbb{Z}^n$.

(c) Take three lattice points u_1, u_2 and u_3 forming a triangle containing no further lattice point (this is equivalent to saying that $\det \binom{u_1 - u_2}{u_1 - u_3} = \pm 1$). Draw a line through u_i separating u_{i+1} and $u_i + u_{i+1} - u_{i-1}$, for $i = 1, 2, 3$ (we define $u_4 = u_1$). Then these three lines form a maximal lattice-free triangle.

(d) Take four lattice points u_1, u_2, u_3 and $u_4 = u_1 + u_3 - u_2$ forming a parallelogram containing no further lattice point (this is equivalent to saying that $\det \binom{u_1 - u_2}{u_1 - u_3} = \pm 1$). Draw a line through each u_i supporting the parallelogram. Then these lines form a maximal lattice-free quadrilateral.

Let us observe some elementary but important properties of maximal lattice-free convex sets. We see that they may be bounded (examples (c,d)) or unbounded (examples (a,b)). However, the unbounded case can be reduced to the bounded case by the following. First, let us give a general construction for unbounded maximal lattice-free sets. Let $b_1, \ldots, b_n$ be any basis of the lattice; let U_1 [L_1] be the linear subspace [lattice] spanned by $b_1, \ldots, b_k$ and U_2 [L_2], the linear subspace [lattice] spanned by $b_{k+1}, \ldots, b_n$. Let K_1 be a maximal convex set in U_1 free of the lattice L_1. Then the set $K_1 + U_2$ is a maximal lattice-free convex set for the whole space. We call this set a *cylinder above* K_1.

3.1 Proposition. *Every unbounded maximal lattice-free convex set is a cylinder above a bounded maximal lattice-free convex set in some lattice subspace.*

The following is a simple but very inportant fact about maximal lattice-free bodies.

3.2 Proposition. *Every maximal lattice-free convex set is a polyhedron.*

To see this, let K be maximal lattice-free. By Proposition 3.1, we may assume that K is bounded. For each lattice point v, there is a closed half-space H_v that includes K and contains v on its boundary. Then $K \subseteq \cap_v H_v$. Since the righ hand side is clearly a lattice-free closed convex set, we must have equality by the maximality of K. So $K = \cap_v H_v$, and we want to show that a finite number of halfspaces gives the same intersection.

Let a be any interior point of K. Every semiline starting from a meets one of the halfspaces H_v, since K is bounded. So by a compactness argument, we can select a finite number of halfspaces $H_{v_1}, \ldots, H_{v_p}$ such that every semiline starting from a meets one of them, i.e., $K' = \cap_i H_{v_i}$ is bounded. Now K' contains only a finite number of lattice points $u_1, \ldots, u_q$, and hence, $K'' = (\cap_i H_{v_i}) \cap (\cap_j H_{u_j})$ is a lattice-free polytope. Since trivially $K \subseteq K''$, we must have $K = K''$ by the maximality of K.

From the previous considerations the following useful proposition follows easily.

3.3 Proposition. *A convex set is maximal lattice-free if and only if it is a polyhedron that contains no interior lattice point but every facet of it contains an interior lattice point.*

The following, much more interesting result of Bell (1977) and Scarf (1977) can be viewed as a sharpening of Proposition 3.2.

3.4 Theorem. *Every maximal lattice-free convex set in $\mathbb{R}^n$ is a polyhedron with at most 2^n facets.*

The "cross-polytope" in $\mathbb{R}^n$ defined by $\sum_i |x_i| \leq n$ is a maximal lattice-free polytope with exactly 2^n facets.

Another way to put this result is the following Helly-type property of linear inequalities:

3.5 Corollary. *Let*

$$a_1 x_1 \leq \alpha_1$$
$$a_2 x_2 \leq \alpha_2$$
$$\vdots$$
$$a_m x_m \leq \alpha_m$$

be a system of linear inequalities in n variables. If every subsystem consisting of at most 2^n rows has an integral solution, then the original system has an integral solution.

So every maximal lattice-free convex set in $\mathbb{R}^n$ can be described by a system $Ax \leq b$ of at most 2^n linear inequalities. There is, hoever, not much that can be said about the matrix A here. Suppose that we have any system of strict linear inequalities $Ax < b$ that has no integral solution. Consider the inequalities in the system one by one: if an inequality can be dropped so that the remaining system still has no integral solution, drop it; if not, then relax (increase) the right hand side as much as possible without allowing an integral solution. We end up with a system $A'x < b'$ where A' consists of some rows of A and the system has no integral solution but allowing equality in any of the inequalities it will have an integral solution. It follows from Proposition 3.4 that $A'x \leq b'$ defines a maximal lattice-free polyhedron.

The previous argument suggests that it makes sense to try to derive properties of maximal lattice-free polyhedra under various hypotheses on the matrix describing it, since every system of linear inequalities with no integral solution can be relaxed to a system describing a maximal lattice-free polyhedron. So from now on we consider a system $Ax \leq b$ describing a maximal lattice-free polyhedron. We assume that A is an integral matrix. It follows easily that b is then an integral vector.

The following result of Kannan, Lovász and Scarf (1988) uses subdeterminants of A to estimate the size of maximal lattice-free polytopes.

3.6 Theorem. *Let $Ax \leq b$ describe a bounded maximal lattice-free polyhedron P. Then the diameter of P, measured in the maximum norm, is at most $n^2 \Delta_{n-1}(A)$.*

A shortcoming of this theorem is that it is not invariant under unimodular transformations, i.e. under linear transformations by an integral matrix T with determinant ± 1. Note that right-multiplying A by such a matrix does not change $\Delta_n(A)$, but may change $\Delta_k(A)$ for other indices k. The following is an invariant (and in fact slightly stronger) form of the result.

3.7 Theorem. *Let $Ax \le b$ be a minimal description of a bounded maximal lattice-free polyhedron P. Then for every vector $v \in P - P$, we have*

$$\|Av\|_\infty \le n\Delta_n(A).$$

Using a notion from section 1, we may rephrase this theorem as follows: the width of a maximal lattice-free polytope $Ax \le b$ along any row vector of A is at most $n\Delta_n(A)$.

Let us sketch the proof, which uses a method due to Cook, Gerards, Schrijver and Tardos (1986). Let $a_0 x \le \beta_0$ be any row of the given system and $v \in P - P$, we show that $|a_0 v| \le n\Delta_n(A)$. Write $v = u - w$, where $u, w \in P$. We may assume that $a_0 u > a_0 w$, so that $a_0 v > 0$. We may also assume that u lies on the facet defined by $a_0 x = \beta_0$, since otherwise, replacing u by any point on this facet increases $a_0 v$. We may further assume that u is a lattice point in the interior of this facet, since such a lattice point exists by Proposition 3.4, and we may replace u by it for free. Finally, we may assume that $u = 0$, since the assertion is invariant under translations by lattice vectors. So we have $\beta_0 = 0$ but $\beta > 0$ for every other row $ax \le \beta$.

Let us split the rows of $Ax \le b$ into two parts: let $A'x \le b'$ consist of those rows $ax \le \beta$ for which $aw < 0$ and $A''x \le b''$, of those rows $ax \le \beta$ for which $aw \ge 0$. (In particular, the row $a_0 x \le \beta_0$ is in the first group.) Consider the convex cone

$$C = \{x \in \mathbb{R}^n : \begin{pmatrix} A' \\ -A'' \end{pmatrix} x \le 0\}.$$

(Geometrically, we draw all hyperplanes through 0 parallel to a facet of P; these divide the space into a number of convex cones, and we consider one of these cones containing w.) It is easy to see that this cone is pointed. We can construct integral vectors $p_1, \ldots, p_N$ spanning the extreme rays of C as follows. Each extreme ray is the intersection of $n - 1$ hyperplanes defined by appropriate rows of A; we make up a vector from the $(n-1) \times (n-1)$ subdeterminants, with alternating signs, of the submatrix formed by these rows.

Clearly $w \in C$, and so we can write $w = \sum_i \lambda_i p_i$, where $\lambda_i \ge 0$.

Claim. If $a_0 p_i \ne 0$ for some $1 \le i \le N$, then $\lambda_i \le 1$.

Assume that $\lambda_i > 1$. Consider the lattice point p_i. We show that p_i is an interior point of P and thereby arrive at a contradiction.

First, if $ax \le \beta$ is a row in $A'x \le b'$ then by the definition of the cone,

$$ap_i \le 0 \le \beta.$$

Suppose that equality holds here. Then in particular $\beta = 0$, which only holds if $a = a_0$. But then $ap_i < 0$ by the hypothesis that $ap_i \ne 0$.

Second, if $ax \le \beta$ is a row in $A''x \le b''$ then, setting

$$\lambda'_j = \begin{cases} \lambda_i - 1, & \text{if } i = j, \\ \lambda_j, & \text{otherwise,} \end{cases}$$

we have

$$ap_i = aw - \sum_j \lambda'_j ap_j \le aw \le \beta.$$

Suppose that equality holds here. Then we must have $ap_j = 0$ for every j for which $\lambda'_j \neq 0$. Since $\lambda'_j \neq 0$ if and only if $\lambda_j \neq 0$, this implies that

$$aw = \sum_j \lambda_j a p_j = 0 < \beta,$$

so equality does not hold in $ap_i \leq \beta$. This proves the Claim.

Using this Claim, we can estimate $a_0 v$ as follows:

$$a_0 v = -a_0 w \leq \sum_j \lambda_j |a_0 p_j| \leq \sum_j |a_0 p_j| \leq n\Delta_n(A),$$

since the inner product $a_0 p_j$ is just the $n \times n$ subdeterminant of A formed by the row a_0 and the rows defining p_j. This proves Theorem 3.8.

To derive Theorem 3.6 from Theorem 3.7, consider an $n \times n$ submatrix B of A with determinant $\pm\Delta_n(A)$; assume e.g. it has determinant $\Delta = \Delta_n(A)$. Now we can write, for every vector $v \in P - P$,

$$v = B^{-1}(Bv) = \frac{1}{\Delta}(\text{Adj } B)(Bv).$$

Here Adj B is a matrix whose entries are at most $\Delta_{n-1}(A)$ in absolute value by the definition of the adjoint, while Bv is a vector whose entries at most $n\Delta$ in absolute value by Theorem 3.7. This implies Theorem 3.7.

We conclude this section with an interesting conjecture raised by H. W. Lenstra (1987). Let us fix an integral matrix A and consider all maximal lattice-free bodies of the form $Ax \leq b$. We translate them so that 0 lies on their boundary; in other words, we assume that $b \geq 0$. Obviously, b is then an integral vector. What is the structure of the set $B(A)$ of all such vectors b? In the plane one can work it out (e.g. using the results of Scarf 1981) that there is a family of $O(\log \Delta_2(A))$ segments such that $B(A)$ consists of the lattice points on these segments; moreover, these segments can all be found in polynomial time. This motivates the following conjecture:

3.8 Conjecture. Fix the dimension n. For every matrix A with n columns, there exists a family of polyhedra $P_1, \ldots, P_N$ in $\mathbb{R}^n$ and lattices $L_1, \ldots, L_N$ such that N is polynomial in the number of bits in A (perhaps even in $\Delta_n(A)$), and $B(A) = \cup_i(L_i \cap P_i)$. Hopefully, the lattices and the linear systems describing these polyhedra can be constructed in polynomial time.

Note that if $B(A)$ does indeed have this structure then — for fixed n — we can optimize any linear objective function over $B(A)$ in polynomial time (we optimize it over the lattice points in each of the polyhedra by Lenstra's algorithm, and take the best).

Lenstra points out that an affirmative answer to this question would imply a polynomial time algorithm to solve Frobenius' problem for a fixed number of variables. This classical problem is the following:

3.9 Problem. Given positive integers $a_1, \ldots, a_n$ with $\gcd(a_1, \ldots a_n) = 1$, find the largest positive integer b that cannot be written in the form

$$b = a_1 x_1 + \ldots + a_n x_n$$

with non-negative integer $x_1, \ldots, x_n$.

This problem is solved (in the form of a polynomial time algorithm) only for $n \leq 3$.

Kannan (1988) proves that the Frobenius problem can also be reduced to determining a covering radius: the largest non-expressible b is $\mu(K,L) - a_1 - \ldots - a_n$, where L is the lattice of all vectors $(x_1, \ldots, x_{n-1}$ such that $a_1 x_1 + \ldots + a_{n-1} x_{n-1}$ is divisible by a_n, and K is the convex body in $\mathbb{R}^n$ defined by the inequalities $x \geq 0$ and $a_1 x_1 + \ldots + a_{n-1} x_{n-1} \leq 1$. So the result of Kannan mentioned in the previous section yields a *fully polynomial approximation scheme* to determine the largest non-expressible b.

One can combine Lenstra's and Kannan's ideas to show that *the problem of computing the covering radius for a polytope can be solved in polynomial time if the answer to Conjecture 3.8 is in the affirmative.* In fact, the covering radius of the polytope defined by $Ax \leq b$ (with respect to the standard lattice) is the least positive τ for which the body $Ax \leq \tau b$ can be translated so that it is contained in one of the maximal lattice-free relaxations of A. If this translation is given by a vector y, then we want to solve

$$\text{maximize } \tau$$
$$\text{subject to } \tau b + Ay \leq w, \quad w \in B.$$

For each $w \in B$, this is a linear program. Now if Conjecture 3.8 is right then there exists a polynomial number of polyhedra P_i and lattices L_i so that $w \in B$ means that $w \in L_i \cap P_i$ for at least one i. This splits the optimization problem above into a polynomial number of problems

$$\text{maximize } \tau$$
$$\text{subject to } \tau b + Ay \leq w, \quad w \in L_i \cap P_i.$$

Each of these problems is a mixed linear program in a fixed dimension, and can be solved by an appropriate modification of Lenstra's method.

4. Neighborhood systems

This approach to integer programming was initiated by Scarf (1981, 1986). Most of the recent results to be discussed here are due to Kannan, Lovász and Scarf (1988).

Let A be a matrix with m rows and n columns. We say that two lattice points u and v are *neighboring* with respect to A, if there exists an m-vector b such that $Au \leq b$ and $Av \leq b$, but there is no lattice point w such that $Aw < b$. In other words, the polyhedron $P(A,b)$ defined by $Ax \leq b$ contains u and v (on its boundary), but no interior lattice point.

In general, if two lattice points are neighboring then there will be many choices for b establishing this fact. There is a unique *minimal* choice, which is worth considering: if a_i is any row of A then we choose $\beta_i = \max\{a_i u, a_i v\}$ as the i-th coordinate of b. It will be sometimes convenient to allow b to have coordinates $= +\infty$, which is just another way to say that we may drop some of the rows of A. Then trivially we can always choose a *maximal* lattice-free polyhedron $P(A,b)$ to establish that u and v are neighbors.

Clearly u and v are neighbors if and only if $u - v$ is a neighbor of 0. Hence from now on we shall concentrate on the neighbors of 0. It is also clear that if u is a neighbor of 0 then so is $-u$.

The motivation of this notion is the following. Assume that we want to maximize a linear objective function cx over the lattice points in a polyhedron $P(A,b)$, and that we have a solution z that we want to test whether it is optimal. Then it is easy to see that if there is a lattice point in $P(A,b)$ that "beats" z then one of the neighbors of z with respect to the matrix $\binom{A}{c}$ does so. So if for this matrix the number of neighbors is "small" in some sense then we obtain a procedure to check for optimality and to augment a non-optimal solution. For many problems, e.g. of a combinatorial nature, this could yield an efficient optimization algorithm.

Unfortunately, the number of neighbors of a lattice point may be enormous even in low dimensions. Let us consider some examples in the plane.

Example 1. Define the matrix

$$A = \begin{pmatrix} 0 & 1 \\ 0 & -1 \\ 1 & 0 \\ -1 & 0 \end{pmatrix}.$$

Then the bodies $P(A,b)$ are rectangles aligned with the axes. The neighbors of 0 with respect to A are all lattice points having at least one coordinate 0, 1 or -1. So in particular 0 has infinitely many neighbors.

This example is not very bad, however: one can come up with various ideas to "fix" it. For example, we could observe that each of these neighboring pairs can be established by an unbounded lattice-free polyhedron $P(A,b)$ (e.g. the neighbor $(1,23)$ of $(0,0)$ is established by the strip $0 \le x_1 \le 1$). Now it is clear that if u and v are neighbors in $\mathbb{R}^n$ with respect to the matrix A, then for every integer s and t, $\binom{u}{t}$ and $\binom{v}{s}$ are neighbors in $\mathbb{R}^{n+1}$ with respect to any matrix of the form

$$\begin{pmatrix} A & 0 \\ B & c \end{pmatrix}.$$

It is justified to say that if $P(A,b)$ is an unbounded lattice-free polyhedron then the fact that two lattice points on its boundary are neighbors is already established in an $(n-1)$-dimensional space (we can factor out the line contained in $P(A,b)$). We shall say that a lattice point u is an *essentially bounded* neighbor of v if the fact that they are neighbors is established only by bounded polyhedra $P(A,b)$. In the example, 0 has no essentially bounded neighbors at all. (The problem of describing the neighbors splits into two one-dimensional problems.)

Example 2. Define the matrix

$$A = \begin{pmatrix} -\alpha & 1 \\ \alpha & -1 \\ 1 & \alpha \\ -1 & -\alpha \end{pmatrix},$$

where α is an irrational number. The polyhedra $P(A,b)$ are rectangles with one pair of sides parallel to the line $y = \alpha x$. Now it follows from elementary number theory that this line contains no lattice point other than 0, but there are lattice points arbitrarily near to it. Walk along this line starting from 0; if we encounter the perpendicular projection of a lattice point on the line and the lattice point is closer to the line then any other lattice point whose projection was encountered before, mark it. Clearly, we mark infinitely many lattice points and all these lattice points are neighbors of 0. It is easy to argue that all these neighbors are essentially bounded.

The natural way to "kill" this example is to assume that A is rational (we may assume right away that A is integral). Although from the point of view of number theory this assumption may exclude the most interesting cases, it is justified if we are interested in integer programming, and we shall make it from now on.

It is easy to see that if A is integral then *the number of essentially bounded neighbors of a lattice point is finite*. In fact, it follows from Theorem 3.7 that

4.1 Proposition. *If A is integral then every essentially bounded neighbor u of 0 satisfies* $\|u - v\|_\infty \leq n^2 \Delta_{n-1}(A)$.

Unfortunately, if something behaves badly for irrational numbers it often also behaves badly for rational numbers with large numerator and denominator.

Example 3. Consider our Example 2 with $\alpha = p/q$, where p and q are "large" integers. After scaling, we get the matrix

$$A = \begin{pmatrix} -p & q \\ p & -q \\ q & p \\ -q & -p \end{pmatrix}.$$

It is not quite easy to describe the structure and number of neighbors with respect to this matrix in general (it involves the continued fraction expansion of p/q), so let us restrict our attention to the case when $p = q + 1$. Then the neighbors of 0 are:

$$(1, -1), \ (1, 0), \ (2, 1), \ \ldots, \ (p - 1, p - 2), \ (p, p - 1) \tag{1}$$

and those lattice points obtained from these by rotations by $\pm\pi/2$ and π. Note that the input size of this matrix is $O(\log p)$, while the number of neighbors is $4p + 4$, which is exponential in the input size.

One may notice, however, that all but one of the neighbors of 0 listed in (1) are placed very neatly on the line through $(1, 0)$ and $(p, p - 1)$. So all neighbors of 0 are contained in 6 lines. In Example 1, again all neighbors of 0 are contained in just 6 lines (the lines $x_i = 0$ and $x_i = \pm 1$). This is a little misleading since if we choose other values of p and q in Example 2, then the neighbors will not be contained in any bounded number of lines (e.g. let p and q be two consecutive Fibonacci numbers). However, the following can be proved (Kannan, Lovász and Scarf 1988):

4.2 Theorem. *Let A be an integral matrix with n rows and $m = n + d + 1$ columns. Then all neighbors of 0 with respect to A are contained in the union of* $O\left(n^d |\log(n\Delta_n(A))|^d\right)$ *lattice hyperplanes.*

This estimate is good if the number of rows of A is not much larger than the number of its columns. However, if m (or d) gets very large compared to n then we can use the result of Bell and Scarf (Theorem 3.4) to observe that every neighbor of 0 is established by a subsystem with at most 2^n inequalities. Hence we obtain

4.3 Corollary. *Let A be an integral matrix with n rows and m m columns. Then all neighbors of 0 with respect to A are contained in the union of* $O\left((mn|\log(n\Delta_n(A))|)^{2^n}\right)$ *lattice hyperplanes.*

The number in this corollary is enormous but *for fixed n, it is only polynomial in the input size.* Unfortunately, even for fixed n the degree of the polynomial is not best possible. For example, if $n = 2$ then Corollary 4.3 gives $O\left(m^2 \Delta_2(A)^4\right)$, although one can verify that in this case the "truth" is linear in $\Delta_2(A)$.

The proof of Theorem 4.2 depends on an approximation theorem for the family of bodies $P(A, b)$, which is perhaps of independent interest. I state the result but have to refer for the details to the forthcoming paper.

Let K and K' be two convex bodies. We define their *Banach-Mazur distance* as the least real number t with the property that for some $s > 0$, sK can be translated so that it is included in K' and $2^t sK$ can be translated so that it includes K'. In particular, the Banach-Mazur distance of two bodies is 0 if and only if they are positively homothetical. Small Banach-Mazur distance means that the bodies have "similar" shapes. The theorem roughly says that bodies defined by the same matrix can be approximated by just a polynomial number of shapes.

4.4 Theorem. *Let A be a matrix with n columns and $m = n + d + 1$ rows such that every n rows of A are linearly independent. Let $\epsilon > 0$ be given. Then there is a family $\mathcal{F}$ of bodies $P(A, b)$ such that*

$$|\mathcal{F}| \le \left(\frac{n}{\epsilon} \log(n\Delta_n(A))\right)^d$$

such that every body $P(A, b)$ has a Banach-Mazur distance at most ϵ from one member of $\mathcal{F}$.

The way this theorem can be applied is the following. Assume for simplicity that A has the property in the theorem, and apply it with $\epsilon = 1$, to get a family $\mathcal{F} = \{P(A_i, b_i) : i = 1, \ldots, N\}$. For each of the special bodies $P(A_i, b_i)$, take an integral vector w_i along which the width of $P(A_i, b_i)$ is minimal. Now each neighbor of 0 is established by some $P(A, b)$ which is lattice-free; this body can be approximated by one of the special bodies $P(A_i, b_i)$. Now it follows that the width of $P(a, b)$ along the vector w_i is $O(n^2)$. Hence each neighbor of 0 is contained in one of the hyperplanes $w_i x = k$ (const $\cdot n^2 \le k \le$ const $\cdot n^2$).

Corollary 4.2 does not describe the structure of all neighbors of 0 in full even if the dimension is bounded. We can formulate a conjecture about this structure, analogous to Conjecture 3.8:

4.5 Conjecture. Fix the dimension n. For every matrix A with n columns, there exists a family $\mathcal{F}$ of $(n-1)$-dimensional polyhedra in $\mathbb{R}^n$ such that $|\mathcal{F}|$ is polynomial in the number of bits in A (perhaps even in $\Delta_n(A)$), and the neighbors of 0 with respect to A are exactly those lattice points in the union of these polyhedra. Hopefully, the linear systems describing these polyhedra can be constructed in polynomial time.

Corollary 4.2 does lend some support to this conjecture.

We conclude with relating neighbors of the origin to Hilbert bases and successive minima. Let C be a pointed convex cone. The *Hilbert basis* of the cone is defined as the (unique) minimal set H of lattice vectors in C such that every lattice vector in C is a linear combination of members of C with non-negative integral coefficients. For a detailed treatment of Hilbert bases, see Schrijver (1986). It is easy ti verify:

4.5 Proposition. *Let C be a pointed convex cone defined by the inequalities $Ax \ge 0$. Then the elements of the Hilbert basis of the cone are just those neighbors of 0 with respect to the matrix $\left(\begin{smallmatrix} A \\ -A \end{smallmatrix}\right)$ that lie in C.*

Consider a matrix A and one of the polytopes $P = P(A, b)$. Also consider the difference body $K = P - P$, and the successive minima $\lambda_1, \ldots, \lambda_n$ of K. Let $v_1, \ldots, v_n$ be arbitrary vectors realizing these successive minima.

4.6 Proposition. *The vectors v_i are neighbors of 0 with respect to the matrix A.*

To prove this, consider any v_i. By definition, $v_i \in \lambda_i(P - P) = \lambda_i P - \lambda_i P$, and hence there is a translated copy $P' = u + \lambda_i P$ containing both 0 and v_i. Now clearly $P' = P(A, b')$ for an appropriate b', so if we show that P' contains no interior lattice point then it follows that v_i is a neighbor of 0.

Assume that P' has an interior lattice point w. By the definition of successive minima, v_i is linearly independent from $v_1, \ldots, v_{i-1}$, and hence, either w or $v_i - w$, say w, is linearly independent from $v_1, \ldots, v_{i-1}$. Then since w is an interior point of P', it follows that $w \in \lambda'(P - P)$ for some $\lambda' < \lambda_i$, contradicting the definition of successive minima.

References

L. Babai (1986), On Lovász' lattice reduction and the nearest lattice point problem, *Combinatorica* **6** 1-13.

D. E. Bell (1977), A theorem concerning the integer lattice, *Studies in Applied Math.* **56**, 187-188.

J. Bourgain and V. D. Milman (1985), Sections euclidiennes et volume des corps symétriques convexes dans $\mathbb{R}^n$, *C. R. Acad. Sci. Paris* **300**, 435-437.

J. W. S. Cassels (1959), *An Introduction to the Geometry of Numbers*, Springer.

W. Cook, A. M. H. Gerards, A. Schrijver, É. Tardos (1986), Sensitivity theorems in integer linear programming, *Math. Programming* **34**, 251-264.

P. van Emde-Boas (1981), Another NP-complete partition problem and the complexity of computing short vectors in a lattice, Report 81-04, Mathematical Institute, Univ. of Amsterdam, Amsterdam.

M. Grötschel, L. Lovász and A. Schrijver (1987), *Geometric Algorithms and Combinatorial Optimization*, Springer, Heidelberg–New York–Tokyo.

M. Grötschel, L. Lovász and A. Schrijver (1982), Geometric methods in combinatorial optimization, in: *Progress in Combinatorial Optimization*, (W. R. Pulleyblank, ed.) Academic Press, New York, 167-183.

P. M. Gruber and C. G. Lekkerkerker (1987) *Geometry of Numbers*, 2nd edition, North-Holland, Amsterdam–New York.

J. Hastad (1988), Dual vectors and lower bounds for the nearest lattice point problem, *Combinatorica* **8**, 75-81.

C. J. Hurkens (1987), Blowing up a convex body in two dimensions (preprint).

R. Kannan (1983), Improved algorithms for integer programming and related lattice problems, in: *Proc. 15th Annual Symposium on Theory of Computing*, 193-206.

R. Kannan (1987), Minkowski's convex body theorem and integer programming, *Math. of Oper. Res.* **12**, 415-440.

R. Kannan (1988), personal communication.

R. Kannan and L. Lovász (1986), Covering minima and lattice point free convex bodies, preliminary version in: *Foundations of Software Technology and Theoretical Computer Science* (K.V.Nori, ed.) Lecture Notes in Comp. Sci. **241**, Springer, 193-213; full version: *Annals of Math.* (to appear)

R. Kannan, L. Lovász and H. E. Scarf (1988), The shapes of polyhedra (preprint).

A. Khinchine (1948), A quantitative formulation of Kronecker's theory of approximation, *Izv. Acad. Nauk SSSR*, Ser. Math. **12**, 113-122 (in Russian).

J. Lagarias, H. W. Lenstra, jr. and C. P. Schnorr (1988), Korkine-Zolotarev bases and successive minima of a lattice and its reciprocal lattice, *Combinatorica* (to appear).

H. W. Lenstra, Jr. (1983), Integer programming with a fixed number of variables, *Oper. Res.* **8**, 538-548.

A. K. Lenstra, H. W. Lenstra, Jr. and L. Lovász (1982), Factoring polynomials with integral coefficients, *Math. Annalen* **261**, 515-534.

H. W. Lenstra, Jr (1988), personal communication.

L. Lovász (1986), *An Algorithmic Theory of Numbers, Graphs and Convexity*, CBMS-NSF Regional Conference Series in Applied Math. **50**, SIAM, Philadelphia.

W. R. Pulleyblank (1983), Polyhedral Combinatorics, in: *Mathematical Programming, the State of the Art* (A. Bachem, M. Grötschel, B. Korte, eds.) Springer, 312-345.

L. A. Santaló (1949), Un invariante afin pasa los cuerpos convexos del espacio de n dimensiones, *Portugal Math.* **8**, 155-161.

H. E. Scarf (1977), An observation on the structure of production sets with indivisibilities, *Proc. Nat. Acad. Sciences USA* **74**, 3637-3641.

H. E. Scarf (1981), Production sets with indivisibles, Part I: Generalities, Part II: the case of two activities, *Econometrica* **49**, 1-32 and 395-423.

H. E. Scarf (1986), Neighborhood systems for production sets with indivisibilities, *Econometrica* **54**, 507-532.

A. Schrijver (1983), Min-max results in combinatorial optimization, in: *Mathematical Programming, the State of the Art* (A. Bachem, M. Grötschel, B. Korte, eds.) Springer, 439-500.

A. Schrijver (1986), *Theory of Linear and Integer Programming*, Wiley.

Linear and Nonlinear Optimization Problems with Submodular Constraints

Satoru FUJISHIGE

Institute of Socio-Economic Planning, University of Tsukuba, Tsukuba, Ibaraki 305, Japan

Abstract. The author introduced the concepts of submodular/supermodular system and its associated base polyhedron, which give a useful mathematical framework for treating combinatorial optimization problems related to submodular/supermodular functions. From the point of view of this framework we survey a class of linear/nonlinear and continuous/discrete optimization problems with constraints described by submodular functions.

1. Introduction

The theory of matroids has successfully been applied to many practical engineering problems, where a fundamental role is played by submodular functions (see, e.g., [24], [25], [30]). Submodular and supermodular functions arise typically in matroids as rank functions [4], in network flows as cut functions [8], in the Shannon information theory as entropy functions [13], in convex games as characteristic functions [33], and in many other combinatorial systems.

The author introduced the concepts of submodular/supermodular system and its associated base polyhedron ([12], [16]), which give a useful mathematical framework for treating combinatorial optimization problems related to submodular/supermodular functions. From the point of view of this framework we survey a class of linear/nonlinear and continuous/discrete optimization problems with constraints described by submodular functions.

M. Iri and K. Tanabe (eds.), Mathematical Programming, 203–225.

For general information on submodular and supermodular functions readers should be referred, e.g., to [4], [6], [10], [16], and [28].

Propositions without any references in this paper seem to be well known to (poly-)matroid theorists or are easy corollaries.

2. Submodular/Supermodular Systems and Base Polyhedra

We give the definitions of submodular and supermodular systems, base polyhedron etc. and show their fundamental properties ([12], [16]).

2.1. Definitions

Let E be a finite nonempty set and D be a collection of subsets of E closed with respect to set union and intersection. Such D is a *distributive lattice* with set union and intersection as the lattice operations, join and meet.

Let R be the set of reals. Throughout the present paper, unless otherwise explicitly stated, R can be any totally ordered additive group, such as the set Z of integers, the set Q of rationals etc.

A function f from D to R is called a *submodular function* on D if for each pair of $X, Y \in D$

$$f(X) + f(Y) \geq f(X \cup Y) + f(X \cap Y). \tag{2.1}$$

We assume throughout the present paper that $\emptyset, E \in D$ and $f(\emptyset) = 0$. We call the pair (D, f) a *submodular system* on E and f the *rank function* of (D, f).

For a submodular system (D, f) we define the polyhedron

$$P(f) = \{\, x \mid x \in R^E, \ \forall X \in D : x(X) \leq f(X) \,\}, \tag{2.2}$$

where R^E is the set of all the vectors $x = (x(e) : e \in E)$ with coordinates indexed by E and $x(e) \in R \ (e \in E)$ and for each $x \in R^E$ and $X \in D$ $x(X)$ is defined by

$$x(X) = \sum_{e \in X} x(e). \tag{2.3}$$

We call $P(f)$ the *submodular polyhedron* associated with submodular system (D, f). We also define the polyhedron

$$B(f) = \{\, x \mid x \in P(f), \ x(E) = f(E) \,\} \tag{2.4}$$

which is called the *base polyhedron* associated with (D, f). The base polyhedron $B(f)$ consists of all the maximal vectors in $P(f)$, where the order among vectors is the one in the ordinary vector lattice, i.e., $x \leq y$ if and only if $x(e) \leq y(e)$ for each $e \in E$. A vector in $B(f)$ is called a *base* of (D, f).

Let $\leq^*$ be the dual order of the ordinary order $\leq$ in R, i.e., $a \leq^* b$ if and only if $b \leq a$, and consider the dual totally ordered additive group $(R, \leq^*)$ in stead of $(R, \leq)$ in the above definitions of submodular function, submodular system, submodular polyhedron and base polyhedron. Then replacing f by g, we have a submodular function $g : D \to R$, a submodular system (D,g), and the submodular polyhedron $P(g)$ and the base polyhedron $B(g)$ associated with the submodular system (D,g) with respect to the totally ordered additive group $(R,\leq^*)$ and they are, respectively, called, with respect to the original underlying totally ordered additive group $(R,\leq)$, a *supermodular function*, a *supermodular system*, and the *supermodular polyhedron* and the *base polyhedron associated with the supermodular system* (D,g).

Let (D,f) be a submodular system on E and define

$$\bar{D} = \{ E - X \mid X \in D \}, \tag{2.5}$$
$$f^{\#}(E - X) = f(E) - f(X) \qquad (X \in D). \tag{2.6}$$

Here, $\bar{D}$ is the dual distributive lattice of D. We call $f^{\#} : \bar{D} \to R$ the *dual supermodular function* of $f : D \to R$ and $(\bar{D}, f^{\#})$ the *dual supermodular sytem* of (D,f) ([12], [34]). Similarly we define the *dual submodular function* $g^{\#} : \bar{D} \to R$ of a supermodular function $g : D \to R$ and the *dual submodular system* $(\bar{D}, g^{\#})$ of (D,g).

The following duality between submodular functions and supermodular functions is fundamental.

Proposition 2.1 ([12], [34]):   For a submodular function $f : D \to R$ and a supermodular function $g : D \to R$ we have

$$(f^{\#})^{\#} = f, \qquad (g^{\#})^{\#} = g, \tag{2.7}$$
$$B(f) = B(f^{\#}), \qquad B(g) = B(g^{\#}). \tag{2.8}$$

$\square$

It should be noted that Proposition 2.1 holds for any set functions f and g if we formally adopt the definitions of $B(f)$ and $B(g)$. The duality of (2.7) and (2.8) may also be useful for a more general class of set functions.

We say subsets X and Y of E *intersect* if $X \cap Y$ is nonempty. A family $\mathcal{F}_1$ of subsets of E is called an *intersecting family* if for each intersecting pair of $X, Y \in \mathcal{F}_1$ we have $X \cup Y, X \cap Y \in \mathcal{F}_1$. A function $f : \mathcal{F}_1 \to R$ is called a *submodular function on the intersecting family* $\mathcal{F}_1$ if for each intersecting pair of $X, Y \in \mathcal{F}_1$ we have

$$f(X) + f(Y) \geq f(X \cup Y) + f(X \cap Y) \tag{2.9}$$

(cf.[4], [9]).

Moreover, we say subsets X and Y of E *cross* if the four sets $X \cap Y$, $(E-X) \cap Y$, $X \cap (E - Y)$ and $(E - X) \cap (E - Y)$ are nonempty. A family $\mathcal{F}_2$ of subsets of E is called a *crossing family* if for each crossing pair of $X, Y \in \mathcal{F}_2$ we have $X \cup Y$, $X \cap Y \in \mathcal{F}_2$. Also a function $f : \mathcal{F}_2 \to R$ is called a *submodular function on the crossing family* $\mathcal{F}_2$ if for each crossing pair of $X, Y \in \mathcal{F}_2$ we have the inequality (2.9) ([5]).

It should be noted that if $\mathcal{F}_2$ is a cross-free family of subsets of E, then $\mathcal{F}_2$ is a crossing family and any function $f \colon \mathcal{F}_2 \to R$ is a submodular function on the crossing family $\mathcal{F}_2$.

We see from the definitions that a distributive lattice $\mathcal{D} \subseteq 2^E$ is an intersecting family and that an intersecting family $\mathcal{F}_1 \subseteq 2^E$ is a crossing family. Therefore, the degree of generality increases from submodular functions on distributive lattices to those on intersecting families and from submodular functions on intersecting families to those on crossing families. Howerever, we have the following

Proposition 2.2 [15]:

(1) Let f be a submodular function on an intersecting family $\mathcal{F}_1$ with $\emptyset, E \in \mathcal{F}_1$ and $f(\emptyset) = 0$ and define

$$P(f) = \{\, x \mid x \in R^E, \ \forall X \in \mathcal{F}_1 : x(X) \leq f(X) \,\}. \tag{2.10}$$

Then there exists a unique submodular system $(\mathcal{D}_1, f_1)$ on E such that

$$P(f_1) = P(f). \tag{2.11}$$

(Hence $B(f_1) = B(f)$ if $B(f) \neq \emptyset$.) Moreover, if f is integer-valued, so is f_1.

(2) Let f be a submodular function on a crossing family $\mathcal{F}_2$ with $\emptyset, E \in \mathcal{F}_2$ and define

$$B(f) = \{\, x \mid x \in R^E, \forall X \in \mathcal{F}_2 : x(X) \leq f(X), x(E) = f(E) \,\}. \tag{2.12}$$

If $B(f) \neq \emptyset$, then there exists a unique submodular system $(\mathcal{D}_2, f_2)$ on E such that

$$B(f_2) = B(f). \tag{2.13}$$

Moreover, if f is integer-valued, so is f_2. $\qquad\qquad\square$

We see from this proposition that considering submodular functions f on intersecting or crossing families does not extend the class of associated polyhedra $P(f)$ and $B(f)$. It should, however, be noted that for a submodular function f on a crossing family and the submodular system $(\mathcal{D}_2, \mathcal{F}_2)$ in (2) of Proposition 2.2, we do not have $P(f) = P(f_2)$ in general.

2.2. Fundamental operations on submodular systems

Consider a submodular system $(\mathcal{D}, f)$. For any vector $x \in P(f^\#)$ the polyhedron

$$B(f)^x = \{\, y \mid y \in B(f), \forall e \in E : y(e) \leq x(e) \,\} \tag{2.14}$$

is the base polyhedron $B(f^x)$ associated with a submodular system $(2^E, f^x)$, where the rank function f^x is given by

$$f^x(X) = \min\{\, f(Y) + x(X - Y) \mid X \supseteq Y \in \mathcal{D} \,\} \tag{2.15}$$

for each $X \subseteq E$. The submodular system $(2^E, f^x)$ is called the *reduction* of $(\mathcal{D}, f)$ by vector x. Note that if f is integer-valued and x is integral, then f^x is integer-valued. Moreover, for any vector $x \in P(f)$ the polyhedron

$$B(f)_x = \{ y \mid y \in B(f), \forall e \in E : y(e) \geq x(e) \} \tag{2.16}$$

is the base polyhedron $B(f_x)$ associated with a submodular system $(2^E, f_x)$, where the rank function f_x is given by

$$f_x(X) = \min\{ f(Y) - x(Y - X) \mid X \subseteq Y \in \mathcal{D} \} \tag{2.17}$$

for each $X \subseteq E$. The submodular system $(2^E, f_x)$ is called the *contraction* of $(\mathcal{D}, f)$ by x. Note that if f is integer-valued and x is integral, then f_x is integer-valued. A submodular system obtained by repeated reductions and/or contractions is called a *(vector) minor* of $(\mathcal{D}, f)$.

Proposition 2.3: If vectors $x, y \in R^E$ satisfy (i) $x \leq y$, (ii) $B(f)_x \neq \emptyset$ and (iii) $B(f)^y \neq \emptyset$, then we have $B(f)^y_x (= (B(f)_x)^y = (B(f)^y)_x) \neq \emptyset$. □

We see from (2.14), (2.16) and the duality (2.8) that the contraction corresponds to the reduction of the dual supermodular system $(\bar{\mathcal{D}}, f^\#)$.
   For any $A \in \mathcal{D}$ define

$$\mathcal{D}^A = \{ X \mid X \subseteq A, \quad X \in \mathcal{D} \} \tag{2.18}$$

and a submodular function $f^A : \mathcal{D}^A \to R$ by

$$f^A(X) = f(X) \qquad (X \in \mathcal{D}^A). \tag{2.19}$$

The submodular system $(\mathcal{D}^A, f^A)$ on A is called the *reduction* (or *restriction*) of $(\mathcal{D}, f)$ *to* A and denoted by $(\mathcal{D}, f) \cdot A$ or $(\mathcal{D}, f) - (E - A)$. Also define

$$\mathcal{D}_A = \{ X - A \mid X \supseteq A, \quad X \in \mathcal{D} \} \tag{2.20}$$

and a submodular function $f_A : \mathcal{D}_A \to R$ by

$$f_A(X) = f(X \cup A) - f(A) \qquad (X \in \mathcal{D}_A). \tag{2.21}$$

The submodular system $(\mathcal{D}_A, f_A)$ on $E - A$ is called the *contraction of $(\mathcal{D}, f)$ by A* and denoted by $(\mathcal{D}, f)/A$ or $(\mathcal{D}, f) \times (E - A)$.
   Any vector $x \in R^E$ can be considered as a modular function on 2^E through (2.3). The submodular system $(\mathcal{D}, f + x)$ is called the *translation* of $(\mathcal{D}, f)$ by $x \in R^E$. We have

$$P(f + x) = P(f) + \{x\}, \qquad B(f + x) = B(f) + \{x\}, \tag{2.22}$$

where the sum in the right-hand side of each equation of (2.22) denotes the vector sum. It should be noted that the combinatorial structures of the submodular polyhedron and the base polyhedron are invariant under translation, whereas the monotonicity of the rank function is not invariant and any rank function can be made monotone increasing by an appropriate translation. Therefore, translation-invariant results in polymatroid theory such as the polymatroid intersection theorem [4] can easily be extended to submodular system (see [16]).

Consider a submodular system (D, f). Recall that $D \subseteq 2^E$ is a distributive lattice with $\emptyset, E \in D$. There uniquely exist a partition $\Pi = \{A_1, A_2, \ldots, A_k\}$ of E and a partial order $\preceq$ on Π such that $X \in D$ if and only if X is expressed as

$$X = \bigcup \{A_i \mid A_i \in I\} \tag{2.23}$$

for some ideal I of the partially ordered set (poset) $(\Pi, \preceq)$ ([1]). (An ideal I of the poset $(\Pi, \preceq)$ is a subset of Π such that for any A_i, $A_j \in \Pi$ with $A_i \preceq A_j \in I$ we have $A_i \in I$.) Let E' be a k-element set $\{a_1, a_2, \ldots, a_k\}$ and make a_i correspond to A_i for each $i = 1, 2, \ldots, k$. By this correspondance we obtain a distributive lattice $D' \subseteq 2^{E'}$ from D and a submodular function $f' : D' \to R$ from f. We call the pair (D', f') the *simplification* of (D, f). When the partition Π consists of singletons A_i, i.e., $|A_i| = 1$ $(i = 1, 2, \ldots, k)$, we call D and (D, f) *simple*. A simple distributive lattice $D \subseteq 2^E$ (with $\emptyset, E \in D$) is formed by the set of ideals of a poset $P = (E, \preceq)$ and is denoted by 2^P.

An example of a base polyhedron is given as follows. Suppose that we are given a real c and vectors $\ell, u \in R^E$ such that $\ell \leq u$ and $\ell(E) \leq c \leq u(E)$. Then the polyhedron $B_0 = \{ x \mid x \in R^E, \quad x(E) = c, \quad \ell \leq x \leq u \}$ is a base polyhedron. For B_0 is the minor $B(f)_\ell^u$ of $B(f)$ such that $D = \{\emptyset, E\}$ and $f(\emptyset) = 0$, $f(E) = c$. If c is an integer and ℓ, u are integral vectors, B_0 is an integral base polyhedron. Polyhedron B_0 is a typical set of feasible solutions of a resource allocation problem (see [22]).

2.3. Fundamental properties of base polyhedra

The following propositions show fundamental structural properties of base polyhedra.

Proposition 2.4 [16]:   For a submodular system (D, f) we have:
(1) The base polyhedron $B(f)$ is pointed if and only if D is simple, i.e., $D = 2^P$ for some poset $P = (E, \preceq)$.
(2) The base polyhedron $B(f)$ is bounded if and only if D is simple and complemented, i.e., $D = 2^E$.   □

Proposition 2.5 (The extreme point theorem) [19] (also see [4], [33], [28], for $D = 2^E$):   For a simple submodular system (D, f) a base $x \in B(f)$ is an extreme point of $B(f)$ if and only if for a maximal chain

$$C: \quad \emptyset = S_0 \subsetneq S_1 \subsetneq \cdots \subsetneq S_n = E \tag{2.24}$$

in D we have

$$x(S_i - S_{i-1}) = f(S_i) - f(S_{i-1}) \qquad (i = 1, 2, \ldots, n). \qquad (2.25)$$

$\square$

For a simple submodular system (D, f) the base polyhedron $B(f)$ is expressed as the vector sum of the convex hull $Q(f)$ of the extreme points of $B(f)$ and the recession cone (or the characteristic cone) $C(f)$ of $B(f)$:

$$B(f) = Q(f) + C(f), \qquad (2.26)$$

where

$$C(f) = \{\, x \mid x \in R^E, \ \forall X \in D : x(X) \le 0, \ x(E) = 0\,\}. \qquad (2.27)$$

Let $P = (E, \preceq)$ be the poset such that $D = 2^P$. Also let $G = (E, A(P))$ be the (directed) graph representing the Hasse diagram of $P = (E, \preceq)$, i.e., E is the vertex set of G and $A(P)$ is the arc set of G such that (e, e') is an arc in $A(P)$ if and only if $e' \prec e$ and there is no $e'' \in E$ such that $e' \prec e'' \prec e$.

Proposition 2.6 (The extreme ray theorem) [36]:   The set of all the extreme rays of the cone $C(f)$ in (2.26) is given by

$$\{\, \chi_e - \chi_{e'} \mid (e, e') \in A(P)\,\}, \qquad (2.28)$$

where $A(P)$ is the arc set of the Hasse diagram $G = (E, A(P))$ of the poset $P = (E, \preceq)$, and χ_e is the unit vector in R^E such that $\chi_e(e) = 1$. $\square$

For any base $x \in B(f)$ and $e \in E$ we define

$$\mathrm{dep}(x, e) = \bigcap \{\, X \mid e \in X \in D, \ x(X) = f(X)\,\}. \qquad (2.29)$$

The function dep: $B(f) \times E \to D$ is called the *dependence function* associated with (D, f) [11]. For any $e' \in \mathrm{dep}(x, e) - \{e\}$ we also define

$$\tilde{c}(x, e, e') = \min\{\, f(X) - x(X) \mid e \in X \in D, \ e' \notin X\,\}. \qquad (2.30)$$

We can easily see from the definition that $\tilde{c}(x, e, e') > 0$ for $e' \in \mathrm{dep}(x, e) - \{e\}$, and $\tilde{c}(x, e, e')$ is called the *exchange capacity* associated with $x \in B(f)$, $e \in E$ and $e' \in \mathrm{dep}(x, e) - \{e\}$. It should be noted that for any $d \in R$ with $0 \le d \le \tilde{c}(x, e, e')$ we have

$$x + d(\chi_e - \chi_{e'}) \in B(f). \qquad (2.31)$$

The transformation of $x \in B(f)$ into $x + d(\chi_e - \chi_{e'}) \in B(f)$ is called an *elementary transformation* of base x.

Proposition 2.7:   Given any two bases x, $y \in B(f)$, x can be transformed into y by (at most $\lceil \frac{1}{4}|E|^2 \rceil$) repeated elementary transformations so that $x(e)$ with $x(e) > y(e)$ monotonically decreases and $x(e)$ with $x(e) < y(e)$ monotonically increases.   □

For a base $x \in B(f)$ define a graph $G_x = (E, A_x)$ with vertex set E and arc set A_x as follows.

$$A_x = \{ (e, e') \mid e \in E, \quad e' \in \mathrm{dep}(x, e) - \{e\} \}. \tag{2.32}$$

We call G_x the *exchangeability graph* associated with base x. Note that graph G_x with selfloops appropriately attached to vertices is transitively closed.

For any base $x \in B(f)$ the *tangent cone* of $B(f)$ at x, denoted by $\mathrm{TC}(B(f), x)$, is defined by

$$\mathrm{TC}(B(f), x) = \{ \lambda y \mid \lambda \geq 0, \ y \in R^E, \ x + y \in B(f) \} \tag{2.33}$$

The following proposition is essentially subsumed in [11, Lemma 9].

Proposition 2.8:   The tangent cone $\mathrm{TC}(B(f), x)$ is generated by the set of the following vectors:

$$\chi_e - \chi_{e'} \qquad ((e, e') \in A_x), \tag{2.34}$$

where A_x is the arc set of the exchangeability graph G_x. In other words, for any vector $y \in \mathrm{TC}(B(f), x)$ there exist some nonnegative coefficients $\lambda(e, e')$ $((e, e') \in A_x)$ such that

$$y = \sum \{ \lambda(e, e')(\chi_e - \chi_{e'}) \mid (e, e') \in A_x \}. \tag{2.35}$$

□

The above characterization of the tangent cone plays an important role in the optimality conditions for a certain class of optimization problems on base polyhedra, which will be discussed in the subsequent sections.

Let $G_x^0 = (E, A_x^0)$ be the graph whose arc set A_x^0 is minimal with the property that G_x is the transitive closure of G_x^0 with possible selfloops deleted. Proposition 2.8 is apparently strengthened by replacing A_x appearing in Proposition 2.8 by A_x^0. When (D, f) is simple and x is an extreme point of $B(f)$, G_x is acyclic and this strengthened version of Proposition 2.8 implies that $\chi_e - \chi_{e'}$ $((e, e') \in A_x^0)$ are exactly the extreme rays of the tangent cone $\mathrm{TC}(B(f), x)$ (cf. [2], [36], [37]).

3. Linear Optimization

Let (D, f) be a simple submodular system with $D = 2^P$ and $P = (E, \preceq)$ and let $w : E \to R$ be an arbitrary function on E. Consider the following linear optimization problem

$$P_0: \quad \text{Minimize} \quad \sum_{e \in E} w(e)x(e)$$

$$\text{subject to} \quad x \in B(f). \tag{3.1}$$

Proposition 3.1 [19]: Problem P_0 has a finite optimal solution if and only if w is a monotone nondecreasing function from $\mathcal{P} = (E, \preceq)$ to $(R, \leq)$, where $\mathcal{P}$ is the poset such that $\mathcal{D} = 2^{\mathcal{P}}$.   □

Suppose that Problem P_0 has a finite optimal solution and let the distinct values of $w(e)$ $(e \in E)$ be given by

$$w_1 < w_2 < \cdots < w_p. \tag{3.2}$$

Also define

$$A_i = \{ e \mid e \in E, \quad w(e) \leq w_i \} \qquad (i = 1, 2, \ldots, p), \tag{3.3}$$
$$A_0 = \emptyset. \tag{3.4}$$

Note that from the assumption and Proposition 3.1 we have $A_i \in \mathcal{D}$ for each $i = 0, 1, \ldots, p$.

Proposition 3.2 (Greedy algorithm) [19] (also see [4], [28] for $\mathcal{D} = 2^E$): Let

$$\mathcal{C}: \quad S_0 = \emptyset \subsetneq S_1 \subsetneq \cdots \subsetneq S_n = E \tag{3.5}$$

be a maximal chain in $\mathcal{D}$ which contains A_i $(i = 0, 1, \ldots, p)$ in it, and define a vector $x \in R^E$ by

$$x(S_i - S_{i-1}) = f(S_i) - f(S_{i-1}) \qquad (i = 1, 2, \ldots, n). \tag{3.6}$$

Then x is an optimal solution of Problem P_0.   □

It should be noted that the optimal solution x given in Proposition 3.2 is an extreme point of base polyhedron $B(f)$ and that any optimal extreme-point solution is given by (3.5) and (3.6) by appropriately choosing a maximal chain $\mathcal{C}$ containing A_i $(i = 0, 1, \ldots, p)$. When f is integer-valued, the optimal solution x given in Proposition 3.2 is also integral. Therefore, the greedy algorithm finds an optimal solution of both continuous and discrete linear optimization problems on base polyhedra. It should, however, be noted that if f is real- or rational-valued and x is restricted to integral vectors in (3.1), then the problem becomes difficult. The structure of the set of integral points in a real or rational base polyhedron has not been elucidated at all. The difficulty is that f does not remain submodular by rounding. The rounding problems for submodular functions and base polyhedra are left for future research.

4. Continuous Nonlinear Optimization

In this section we consider a submodular system $(\mathcal{D}, f)$ with a real-valued (or rational-valued) submodular function f. The underlying totally ordered additive group is assumed to be the set R of reals (or the set Q of rationals). The problems to be considered in Sections 4 and 5 include as special cases most of the so-called *resource allocation problems* hitherto investigated in the literature. The readers should be referred to the book [22] by T. Ibaraki and N. Katoh for resource allocation problems and related topics.

4.1. Separable convex optimization

For each $e \in E$ let $w_e : R \to R$ be a real-valued convex function on R, and consider the following problem

$$P_1 : \quad \text{Minimize} \quad \sum_{e \in E} w_e(x(e))$$
$$\text{subject to} \quad x \in B(f). \tag{4.1}$$

The author [14] considered Problem P_1 where for each $e \in E$ $w_e(x(e))$ is a quadratic function given by $x(e)^2/w(e)$ with a positive real weight $w(e)$ and f is a polymatroid rank function. H. Groenevelt [20] also considered Problem P_1 where f is a rank function of a polymatroid. It is almost straightforward to generalize the result of [14] and [20] to Problem P_1 for a general submodular system.

Proposition 4.1 ([20] (also see [14])): A base $x \in B(f)$ is an optimal solution of Problem P_1 if and only if for each $e \in E$ and $e' \in \text{dep}(x, e) - \{e\}$ we have

$$w_e{}^+(x(e)) \geq w_{e'}{}^-(x(e')), \tag{4.2}$$

where $w_e{}^+$ denotes the right derivative of w_e and $w_{e'}{}^-$ the left derivative of $w_{e'}$.
(Proof) "If" part : Suppose that (4.2) holds for each $e \in E$ and $e' \in \text{dep}(x, e) - \{e\}$. From Proposition 2.8, for any base $z \in B(f)$ there exist some nonnegative coefficients $\lambda(e, e')$ $((e, e') \in A_x)$ such that

$$z = x + \sum \{ \lambda(e, e')(\chi_e - \chi_{e'}) \mid (e, e') \in A_x \}. \tag{4.3}$$

For each $e \in E$ define

$$\bar{w}_e(x(e)) = \max\{ w_{e'}{}^-(x(e')) \mid e' \in \text{dep}(x, e) \}. \tag{4.4}$$

We see from (4.2) and (4.4) that

$$w_e{}^-(x(e)) \leq \bar{w}_e(x(e)) \leq w_e{}^+(x(e)) \qquad (e \in E), \tag{4.5}$$
$$\bar{w}_e(x(e)) \geq \bar{w}_{e'}(x(e')) \qquad ((e, e') \in A_x). \tag{4.6}$$

From $(4.3) \sim (4.6)$ and the convexity of w_e $(e \in E)$ we have

$$
\begin{aligned}
\sum_{e \in E} w_e(z(e)) &= \sum_{e \in E} w_e(x(e) + \partial\lambda(e)) \\
&\geq \sum_{e \in E} \{w_e(x(e)) + \partial\lambda(e)\bar{w}_e(x(e))\} \\
&= \sum_{e \in E} w_e(x(e)) + \sum_{(e,e') \in A_x} \lambda(e,e')(\bar{w}_e(x(e)) - \bar{w}_{e'}(x(e'))) \\
&\geq \sum_{e \in E} w_e(x(e)),
\end{aligned}
\tag{4.7}
$$

where $\partial\lambda : E \to R$ is defined by

$$
\partial\lambda(e) = \sum_{(e,e') \in A_x} \lambda(e,e') - \sum_{(e',e) \in A_x} \lambda(e',e) \qquad (e \in E).
\tag{4.8}
$$

This implies the optimality of x.

"Only if" part : Suppose that for a base $x \in B(f)$ there exist $e \in E$ and $e' \in \text{dep}(x,e) - \{e\}$ such that

$$
w_e^+(x(e)) < w_{e'}^-(x(e')).
\tag{4.9}
$$

Then for a sufficiently small positive number ε we have

$$
w_e(x(e)) + w_{e'}(x(e')) > w_e(x(e) + \varepsilon) + w_{e'}(x(e') - \varepsilon),
\tag{4.10}
$$

$$
x + \varepsilon(\chi_e - \chi_{e'}) \in B(f).
\tag{4.11}
$$

Therefore, x is not an optimal solution. □

For each $e \in E$ and $\xi \in R$ define the interval

$$
J_e(\xi) = [w_e^-(\xi), w_e^+(\xi)].
\tag{4.12}
$$

$J_e(\xi)$ is the subdifferential of w_e at ξ. Conversely, for each $e \in E$ and $\eta \in R$ define

$$
I_e(\eta) = \{\, \xi \mid \xi \in R, \quad \eta \in J_e(\xi) \,\}.
\tag{4.13}
$$

Because of the convexity of w_e, $I_e(\eta)$, if nonempty, is an interval in R and we express it as

$$
I_e(\eta) = [i_e^-(\eta), i_e^+(\eta)].
\tag{4.14}
$$

In the following we assume for simplicity that $I_e(\eta) \neq \emptyset$ for every $\eta \in R$, which guarantees the existence of an optimal solution even if $B(f)$ is unbounded. When $B(f)$ is bounded, there is no loss of generality with this assumption.

By adapting the algorithm in [14] and [20] an efficient algorithm for solving Problem P_1 is given as follows, where x^* is the output vector giving an optimal solution.

Algorithm A1 by Decomposition

Step 1 : Choose $\eta \in R$ such that

$$\sum_{e \in E} i_e^-(\eta) \leq f(E) \leq \sum_{e \in E} i_e^+(\eta). \tag{4.15}$$

Step 2 : Find a base $x \in B(f)$ such that for each $e, e' \in E$

(1) if $w_e^+(x(e)) < \eta$ and $w_{e'}^-(x(e')) > \eta$, then we have $e' \notin \text{dep}(x, e)$,

(2) if $w_e^+(x(e)) < \eta$, $w_{e'}^-(x(e')) = \eta$ and $e' \in \text{dep}(x, e)$, then for any $\varepsilon > 0$ we have $w_{e'}^-(x(e') - \varepsilon) < \eta$, i.e., $x(e') = i_{e'}^-(\eta)$,

(3) if $w_e^+(x(e)) = \eta$, $w_{e'}^-(x(e')) > \eta$ and $e' \in \text{dep}(x, e)$, then for any $\varepsilon > 0$ we have $w_e^+(x(e) + \varepsilon) > \eta$, i.e., $x(e) = i_e^+(\eta)$.

Put

$$E_- = \bigcup \{ \text{dep}(x, e) \mid e \in E, \quad w_e^+(x(e)) < \eta \}, \tag{4.16}$$

$$E_+ = \bigcup \{ \text{dep}^\#(x, e) \mid e \in E, \quad w_e^-(x(e)) > \eta \}, \tag{4.17}$$

$$E_0 = E - (E_+ \cup E_-), \tag{4.18}$$

where

$$\text{dep}^\#(x, e) = \bigcap \{ X \mid e \in X \in \bar{D}, \quad x(X) = f^\#(X) \}. \tag{4.19}$$

Put $x^*(e) = x(e)$ for each $e \in E_0$.

Step 3: If $E_- \neq \emptyset$, then apply the present algorithm recursively to the problem with E and f, respectively, replaced by E_- and f^{E_-} and the base polyhedron associated with the reduction $(D, f) \cdot E_-$. Also, if $E_+ \neq \emptyset$, then apply the present algorithm recursively to the problem with E and f, respectively, replaced by E_+ and f_{E_+} and the base polyhedron associated with the contraction $(D, f) \times E_+$. (End)

The validity of the algorithm easily follows from Proposition 4.1.

It should be noted that if w_e is strictly convex for each $e \in E$, then conditions (2) and (3) in Step 2 are always satisfied, so that we have only to consider condition (1).

The above algorithm lays a basis for the algorithms for the other problems to be considered in the subsequent sections.

4.2. Lexicographically optimal base

For each $e \in E$ let h_e be a continuous and monotone increasing function from R onto R. For any vector $x \in R^E$ we denote by $T(x)$ the sequence of the components $x(e)$ ($e \in E$) of x arranged in order of increasing magnitude, i.e., $T(x) = (x(e_1), x(e_2), \ldots, x(e_n))$ with $x(e_1) \le x(e_2) \le \cdots \le x(e_n)$, where $\{e_1, e_2, \ldots, e_n\} = E$ and $|E| = n$.

Consider the following problem

$$P_2: \quad \text{Lexicographically maximize} \quad T((h_e(x(e)) : e \in E))$$
$$\text{subject to} \quad x \in B(f). \tag{4.20}$$

We call an optimal solution of Problem P_2 a *lexicographically optimal base* of (D, f) *with respect to functions* h_e $(e \in E)$. When $h_e(x(e))$ is a linear function expressed as $x(e)/w(e)$ with $w(e) > 0$ for each $e \in E$, such a lexicographically optimal base is called a *lexicographically optimal base with respect to the weight vector* $w = (w(e) : e \in E)$ [14]. It is a generalization of the concept of (lexicographically) optimal flow introduced by N. Megiddo [29] concerning multiple-source multiple-sink networks.

Proposition 4.2 (cf. [14]):   Let x be a base in $B(f)$. Define a vector $\eta \in R^E$ by

$$\eta(e) = h_e(x(e)) \qquad (e \in E) \tag{4.21}$$

and let the distinct numbers of $\eta(e)$ $(e \in E)$ be given by

$$\eta_1 < \eta_2 < \cdots < \eta_p. \tag{4.22}$$

Also, define

$$S_i = \{ e \mid e \in E, \quad \eta(e) \le \eta_i \} \qquad (i = 1, 2, \ldots, p). \tag{4.23}$$

Then the following are equivalent:

   (i) x is a lexicographically optimal base of (D, f) with respect to h_e $(e \in E)$.
   (ii) $S_i \in D$ and $x(S_i) = f(S_i)$ $(i = 1, 2, \ldots, p)$.
   (iii) $\text{dep}(x, e) \subseteq S_i$ $(e \in S_i, i = 1, 2, \ldots, p)$.
   (iv) x is an optimal solution of Problem P_1 where for each $e \in E$ the derivative of w_e coincides with h_e.

(Proof) The equivalence, (ii) $\Longleftrightarrow$ (iii), can be proved by the direct adaptation of the proof in [14]. Also, the equivalence, (ii), (iii) $\Longleftrightarrow$ (iv), follows from Proposition 4.1. We show the equivalence, (i) $\Longleftrightarrow$ (ii) $\sim$ (iv).

   (i) $\Longrightarrow$ (iii): Suppose (i). If for some $i \in \{1, 2, \ldots, p\}$ and $e \in S_i$ there exists $e' \in \text{dep}(x, e) - S_i$, then for a sufficiently small positive number ε the vector

$$y = x + \varepsilon(\chi_e - \chi_{e'}) \tag{4.24}$$

is a base in $B(f)$ and $T((h_e(y(e)) : e \in E))$ is lexicographically greater than $T((h_e(x(e)) : e \in E))$. This contradicts (i). So, (iii) holds.

   (ii), (iii) $\Longrightarrow$ (i): Suppose (ii) (and (iii)). Let $\bar{x}$ be an arbitrary base such that $T((h_e(\bar{x}(e)) : e \in E))$ is lexicographically greater than or equal to $T((h_e(x(e)) : e \in E))$. Define a vector $\bar{\eta} \in R^E$ by

$$\bar{\eta}(e) = h_e(\bar{x}(e)) \qquad (e \in E). \tag{4.25}$$

Also define $S_0 = \emptyset$. We show by induction on i that

$$x(e) = \bar{x}(e) \qquad (e \in S_i) \tag{4.26}$$

for $i = 0, 1, \ldots, p$, from which the optimality of x follows. For $i = 0$ (4.26) trivially holds. So, suppose that (4.26) holds for some $i = i_0 < p$. Since $T(\bar{\eta})$ is lexicographically greater than or equal to $T(\eta)$, we have from (4.22) and (4.23)

$$\bar{\eta}(e) \geq \eta(e) = \eta_{i_0+1} \qquad (e \in S_{i_0+1} - S_{i_0}). \tag{4.27}$$

From (4.27) and the monotone increasingness of h_e $(e \in E)$,

$$\bar{x}(e) \geq x(e) \qquad (e \in S_{i_0+1} - S_{i_0}). \tag{4.28}$$

Since $\bar{x} \in B(f)$, it follows from (4.26) with $i = i_0$, (4.28) and assumption (ii) that

$$f(S_{i_0+1}) \geq \bar{x}(S_{i_0+1}) \geq x(S_{i_0+1}) = f(S_{i_0+1}). \tag{4.29}$$

From (4.28) and (4.29) we have $\bar{x}(e) = x(e)$ $(e \in S_{i_0+1})$. □

From Proposition 4.2 we can find a lexicographically optimal base by using Algorithm A1 given in Section 4.1.

For a vector $x \in R^E$ define $T^*(x)$ to be the sequence of the components $x(e)$ $(e \in E)$ of x arranged in order of decreasing magnitude. We call a base $x \in B(f)$ which lexicographically minimizes $T^*((h_e(x(e)) : e \in E))$ a *co-lexicographically optimal base of* $(\mathcal{D}, f)$ with respect to h_e $(e \in E)$.

Proposition 4.3: x is a lexicographically optimal base of $(\mathcal{D}, f)$ with respect to h_e $(e \in E)$ if and only if it is a co-lexicographically optimal base of $(\mathcal{D}, f)$ with respect to h_e $(e \in E)$.
(Proof) Using $\eta(e)$ $(e \in E)$ and η_i $(i = 1, 2, \ldots, p)$ appearing in Proposition 4.2, define

$$S_i^* = \{ e \mid e \in E, \quad \eta(e) \geq \eta_{p-i+1} \} \qquad (i = 1, 2, \ldots, p). \tag{4.30}$$

Also define $S_0 = \emptyset = S_0^*$. Since $S_i^* = E - S_{p-i}$ $(i = 0, 1, \ldots, p)$ and $x(E) = f(E)$, we can easily see that for a base $x \in B(f)$ x satisfies (ii) of Proposition 4.2 if and only if x satisfies

(ii*) $S_i^* \in \bar{\mathcal{D}}$ and $x(S_i^*) = f^{\#}(S_i^*)$ $(i = 1, 2, \ldots, p)$.

Consequently, the present proposition follows from Proposition 4.2 and the duality shown in Proposition 2.1. □

4.3. Weighted max-min/min-max problems

For each $e \in E$ let $h_e : R \to R$ be a right-continuous and monotone non-decreasing function such that $\lim_{\xi \to +\infty} h_e(\xi) = +\infty$ and $\lim_{\xi \to -\infty} h_e(\xi) = -\infty$.

Consider the following max-min problem with nonlinear weight functions h_e $(e \in E)$:

$$P_* : \quad \text{Maximize} \quad \min_{e \in E} h_e(x(e))$$

$$\text{subject to} \quad x \in B(f). \tag{4.31}$$

For each $e \in E$ let $w_e : R \to R$ be a convex function whose right derivative w_e^+ is given by h_e.

Proposition 4.4: Consider Problem P_1 of (4.1) with w_e $(e \in E)$ defined as above. Let x be an optimal solution of Problem P_1. Then x is an optimal solution of Problem P_*.
(Proof) Define

$$\eta_1 = \min\{\, h_e(x(e)) \mid e \in E \,\}, \tag{4.32}$$
$$S_1 = \{\, e \mid e \in E, \quad h_e(x(e)) = \eta_1 \,\}, \tag{4.33}$$
$$S_1^* = \bigcup\{\, \text{dep}(x, e) \mid e \in S_1 \,\}. \tag{4.34}$$

We have from (4.34)

$$x(S_1^*) = f(S_1^*). \tag{4.35}$$

It follows from Proposition 4.1 that

$$w_e^-(x(e)) \leq \eta_1 \qquad (e \in S_1^*). \tag{4.36}$$

If there were a base $y \in B(f)$ such that

$$\eta_1 < \min\{\, h_e(y(e)) \mid e \in E \,\}. \tag{4.37}$$

Then, from (4.32) $\sim$ (4.37) we would have

$$x(e) < y(e) \quad (e \in S_1), \qquad x(e) \leq y(e) \quad (e \in S_1^* - S_1), \tag{4.38}$$

since $h_e = w_e^+$. Hence, from (4.35) and (4.38), $f(S_1^*) = x(S_1^*) < y(S_1^*)$, which contradicts the fact that $y \in B(f)$. $\qquad \square$

We see from the above proof of Proposition 4.4 that Algorithm A1 can be simplified for solving Problem P_* as follows. We may put $x^*(e) = x(e)$ for $e \in E_0 \cup E_+$ in Step 2 and apply Algorithm A1 recursively to the problem on E_- but not to the one on E_+ in Step 3(cf. [23]).

Moreover, consider the following min-max problem

$$P^* : \quad \text{Minimize} \quad \max_{e \in E} h_e(x(e))$$

$$\text{subject to} \quad x \in B(f). \tag{4.39}$$

Here, we assume that h_e is left continuous rather than right continuous for each $e \in E$.

We see from Proposition 4.4 and the duality shown in Proposition 2.1 that an optimal solution of Problem P_1 obtained by Algorithm A1, where the left derivative w_e^- of w is given by h_e, is also an optimal solution of Problem P^*.

Problems P_* and P^* are sometimes called the *sharing problems* in the literature ([3], [23]). The shairing problems with more general objective functions and sets of feasible solutions are considered by U. Zimmermann [38], [39].

4.4. The continuous fair resource allocation problem

Let $g : R^2 \to R$ be a function such that $g(u, v)$ is monotone nondecreasing in u and monotone nonincreasing in v. Also, for each $e \in E$ let h_e be a continuous monotone nondecreasing function from R onto R. Consider

$$P_3 : \quad \text{Minimize} \quad g(\max_{e \in E} h_e(x(e)), \min_{e \in E} h_e(x(e)))$$
$$\text{subject to} \quad x \in B(f). \tag{4.40}$$

We call Problem P_3 the *continuous fair resource allocation problem with submodular constraints*.

Using the same functions h_e ($e \in E$) appearing in (4.40), let us consider Problem P_* and P^* described by (4.31) and (4.39), respectively. Denote the optimal values of the objective functions of Problems P_* and P^* by v_* and v^*, respectively, and define vectors ℓ, $u \in (R \cup \{-\infty, +\infty\})^E$ by

$$\ell(e) = \min\{\, \alpha \mid \alpha \in R, \quad h_e(\alpha) \geq v_*\} \quad (e \in E), \tag{4.41}$$
$$u(e) = \max\{\, \alpha \mid \alpha \in R, \quad h_e(\alpha) \leq v^*\} \quad (e \in E). \tag{4.42}$$

Proposition 4.5: Suppose values v_* and v^* and vectors ℓ and u are defined as above. Then we have $v_* \leq v^*$ and $\ell \leq u$. Moreover, $B(f)_\ell^u$ is nonempty and any $x \in B(f)_\ell^u$ is an optimal solution of Problem P_3.
(Proof) Let x^* and x_* be, respectively, optimal solutions of Problem P^* and P_*. If $v_* > v^*$, then $x^*(e) \leq u(e) < \ell(e) \leq x_*(e)$ ($e \in E$), which contradicts the fact that $x^*(E) = f(E) = x_*(E)$. Therefore, we have $v_* \leq v^*$. This implies $\ell \leq u$. Moreover, since $x_* \in B(f)_\ell$, $x^* \in B(f)^u$ and $\ell \leq u$, from Proposition 2.3 we have $B(f)_\ell^u \neq \emptyset$. For any $x \in B(f)_\ell^u$ and $y \in B(f)$ we have

$$g(\max_{e \in E} h_e(y(e)), \min_{e \in E}(y(e)))$$
$$\geq g(v^*, v_*)$$
$$= g(\max_{e \in E} h_e(x(e)), \min_{e \in E} h_e(x(e))), \tag{4.43}$$

due to the monotonicity of g. This shows that any $x \in B(f)_\ell^u$ is an optimal solution of Problem P_3. $\qquad \square$

The continuous fair resource allocation problem P_3 can thus be solved by using Algorithm A1 given in Section 4.1.

5. Discrete Nonlinear Optimization

In this section we consider the optimization problems treated in the preceding section in the case where the values of the variables are restricted to integers. We suppose that the underlying totally ordered additive group is the set Z of integers, the rank function f of submodular system $(\mathcal{D}, f)$ is integer-valued, and the base polyhedron $B(f)$ is given by

$$B(f) = \{ x \mid x \in Z^E, \quad \forall X \in \mathcal{D} : x(X) \leq f(X), x(E) = f(E) \}. \qquad (5.1)$$

We denote the base polyhedron (5.1) by $B_Z(f)$, emphasizing that the underlying totally ordered additive group is Z. We also define for the set R of reals

$$B_R(f) = \{ x \mid x \in R^E, \quad \forall X \in \mathcal{D} : x(X) \leq f(X), x(E) = f(E) \}. \qquad (5.2)$$

Here, $B_R(f)$ is the base polyhedron associated with $(\mathcal{D}, f)$ when we consider R as the underlying totally ordered additive group. It can be seen from Propositions 2.5 and 2.6 that polyhedron $B_R(f)$ is the convex hull, in R^E, of $B_Z(f)$.

5.1. Separable convex optimization

For each $e \in E$ let $\hat{w}_e$ be a real-valued function on Z such that the piecewise linear extension, denoted by w_e, of $\hat{w}_e$ on R is a convex function, where $w_e(\xi) = \hat{w}_e(\xi)$ for $\xi \in Z$ and w_e restricted on each unit interval $[\xi, \xi+1]$ ($\xi \in Z$) is a linear function. Let us consider

$$IP_1 : \qquad \text{Minimize} \quad \sum_{e \in E} \hat{w}_e(x(e))$$
$$\text{subject to} \quad x \in B_Z(f). \qquad (5.3)$$

Also consider the continuous version of IP_1:

$$P_1 : \qquad \text{Minimize} \quad \sum_{e \in E} w_e(x(e))$$
$$\text{subject to} \quad x \in B_R(f). \qquad (5.4)$$

Proposition 5.1(cf. [20]):   There exists an integral optimal solution for Problem P_1 described by (5.4).
(Proof) Problem P_1 can be solved by Algorithm A1. Because of the definition of w_e ($e \in E$) and the integrality of $B_R(f)$, an integral optimal solution of P_1 can be obtained through Algorithm A1 by choosing an integral base x in Step 2 of Algorithm A1.   □

An integral optimal solution of P_1 is also an optimal solution of Problem IP_1 (see [7], [20]). An incremental algorithm is also given in [7].

5.2. Weighted max-min/min-max problems

For each $e \in E$ let $\hat{h}_e : Z \to R$ be a monotone nondecreasing function on Z such that $\lim_{\xi \to +\infty} \hat{h}_e(\xi) = +\infty$ and $\lim_{\xi \to -\infty} \hat{h}_e(\xi) = -\infty$ for each $e \in E$. Consider

$$IP_* : \quad \text{Maximize} \quad \min_{e \in E} \hat{h}_e(x(e))$$

$$\text{subject to} \quad x \in B_Z(f). \tag{5.5}$$

For each $e \in E$ let $w_e : R \to R$ be a piecewise-linear convex function such that its right derivative w_e^+ satisfies

$$w_e^+(\xi) = \hat{h}_e(\xi) \qquad (\xi \in Z) \tag{5.6}$$

and w_e is linear on each unit interval $[\xi, \xi + 1]$ $(\xi \in Z)$.

Proposition 5.2: Let x_* be an integral optimal solution of Problem P_1 with w_e $(e \in E)$ defined as above. Then x_* is an optimal solution of Problem IP_* of (5.5). (Proof) For each $e \in E$ let $h_e : R \to R$ be a right-continuous piecewise-constant nondecreasing function such that $h_e(\xi) = \hat{h}_e(\xi)$ $(\xi \in Z)$ and $h_e(\eta) = \hat{h}_e(\xi)$ $(\eta \in [\xi, \xi + 1), \xi \in Z)$. It follows from Proposition 4.4 that an integral optimal solution of Problem P_1 with w_e $(e \in E)$ defined by (5.6) is an integral optimal solution of Problem P_* of (4.31) with h_e $(e \in E)$ defined as above. Therefore, x_* is an optimal solution of Problem IP_*. □

The reduction of Problem IP_* to Problem P_1 was also communicated by N. Katoh [26]. A direct algorithm for Problem IP_* is given in [18].

The weighted min-max problem

$$IP^* : \quad \text{Minimize} \quad \max_{e \in E} \hat{h}_e(x(e))$$

$$\text{subject to} \quad x \in B_Z(f) \tag{5.7}$$

can be solved similarly in a dual form.

5.3. The discrete fair resource allocation problem [18]

We consider the discrete version of the continuous fair resource allocation problem P_3 treated in Section 4.5.

Let $g : R^2 \to R$ be a function such that $g(u, v)$ is monotone nondecreasing in u and monotone nonincreasing in v. Also, for each $e \in E$ let $\hat{h}_e : Z \to R$ be a monotone nondecreasing function. We assume for simplicity $\lim_{\xi \to +\infty} \hat{h}_e(\xi) = +\infty$ and $\lim_{\xi \to -\infty} \hat{h}_e(\xi) = -\infty$.

Consider the problem

$$IP_3: \qquad \text{Minimize} \quad g(\max_{e \in E} \hat{h}_e(x(e)), \min_{e \in E} \hat{h}_e(x(e)))$$

$$\text{subject to} \quad x \in B_Z(f). \tag{5.8}$$

Problem IP_3 is not so easy as its continuous version P_3 because of the integer constraints.

Using the same functions $\hat{h}_e$ ($e \in E$), consider the weighted integral max-min problem IP_* and the weighted integral min-max problem IP^*, and let $\hat{v}_*$ and $\hat{v}^*$, respectively, be the optimal values of the objective functions of IP_* and IP^*. Define vectors $\hat{\ell}, \hat{u} \in (Z \cup \{-\infty, +\infty\})^E$ by

$$\hat{\ell}(e) = \min\{\, \alpha \mid \alpha \in Z, \quad \hat{h}_e(\alpha) \geq \hat{v}_* \,\}, \tag{5.9}$$

$$\hat{u}(e) = \max\{\, \alpha \mid \alpha \in Z, \quad \hat{h}_e(\alpha) \leq \hat{v}^* \,\}. \tag{5.10}$$

We have $\hat{v}_* \leq \hat{v}^*$ but, unlike the continuous version of the problem, we may not have $\hat{\ell} \leq \hat{u}$ in general. We have

$$\hat{\ell}(e) \leq \hat{u}(e) + 1 \qquad (e \in E). \tag{5.11}$$

Similarly as in the continuous version, we can show that if we have $\hat{\ell} \leq \hat{u}$, then $B_Z(f)_{\hat{\ell}}^{\hat{u}}$ is nonempty and any $x \in B_Z(f)_{\hat{\ell}}^{\hat{u}}$ is an optimal solution of IP_3.

So, let us suppose that we do not have $\hat{\ell} \leq \hat{u}$. Define

$$D = \{\, e \mid e \in E, \quad \hat{\ell}(e) > \hat{u}(e) \,\}. \tag{5.12}$$

It follows from $(5.9) \sim (5.12)$ that

$$\hat{\ell}(e) = \hat{u}(e) + 1 \qquad\qquad (e \in D), \tag{5.13}$$

$$\hat{\ell}(e) \leq \hat{u}(e) \qquad\qquad (e \in E - D), \tag{5.14}$$

$$\hat{h}_e(\hat{u}(e)) < \hat{v}_* \leq \hat{v}^* < \hat{h}_e(\hat{\ell}(e)) \qquad (e \in D), \tag{5.15}$$

$$\hat{v}_* \leq \hat{h}_e(\hat{\ell}(e)) \leq \hat{h}_e(\hat{u}(e)) \leq \hat{v}^* \qquad (e \in E - D). \tag{5.16}$$

Moreover, let the distinct values of $\hat{h}_e(\hat{u}(e))$ ($e \in D$) be given by

$$d_1 < d_2 < \cdots < d_k \tag{5.17}$$

and define

$$A_i = \{\, e \mid e \in D, \quad \hat{h}_e(\hat{u}(e)) \leq d_i \,\} \qquad (i = 1, 2, \ldots, k). \tag{5.18}$$

Then we can show [18] that the minimum values of the objective function of Problem IP_3 is equal to the minimum of the following $k + 1$ values

$$g(\hat{v}^*, d_1), \quad g(\max_{e \in A_i} \hat{h}_e(\hat{u}(e) + 1), d_{i+1}) \qquad (i = 1, 2, \ldots, k). \tag{5.19}$$

If $g(\hat{v}^*, d_1)$ is minimum, then define vectors ℓ^0, $u^0 \in R^E$ by $\ell^0 = \hat{\ell} \wedge \hat{u}$ and $u^0 = \hat{u}$, where $\hat{\ell} \wedge \hat{u} = (\min(\hat{\ell}(e), \hat{u}(e)) : e \in E)$. If $g(\max_{e \in A_{i^*}} \hat{h}_e(\hat{u}(e) + 1), d_{i^*+1})$ is minimum for some $i^* \in \{1, 2, \ldots, k\}$, then putting $w^* = \max_{e \in A_{i^*}} \hat{h}_e(\hat{u}(e) + 1)$, define vectors ℓ^0, $u^0 \in R^E$ by

$$\ell^0(e) = \min\{\, \alpha \mid \alpha \in Z, \quad \hat{h}_e(\alpha) \geq d_{i^*+1} \,\}, \tag{5.20}$$

$$u^0(e) = \max\{\, \alpha \mid \alpha \in Z, \quad \hat{h}_e(\alpha) \leq w^* \,\} \tag{5.21}$$

for $e \in E$. We can show that $B_Z(f)^{u^0}_{\ell^0}$ is nonempty and that any $x \in B_Z(f)^{u^0}_{\ell^0}$ is an optimal solution of IP_3.

6. Some Extensions

The optimization problems considered in the previous sections have base poly-hedra as their feasible regions. For these problems we may consider the intersection of two base polyhedra as each of their feasible regions. The linear optimization problems over the intersection of two base polyhedra is equivalent to the submod-ular flow problem [5], the independent flow problem [11] and the polymatroidal flow problem [21], [27]. The nonlinear optimization problem over the intersection of two base polyhedra with a separable convex objective function can be formu-lated, for example, as a submodular flow problem with a separable convex cost function. This problem can be solved by adapting the out-of-kilter method pro-posed in [17](also see [38], [39]). All the problems considered in Sections 4 and 5 have been reduced to separable convex optimization problems. However, when the feasible region is given as the intersection of two base polyhedra, such reductions may not be possible.

The concept of lexicographically optimal base is generalized by M. Nakamura [32] and N. Tomizawa [35]. Suppose we are given two (polymatroid) base polyhe-dra $B(f_i)$ $(i = 1, 2)$ such that every base of $B(f_i)$ $(i = 1, 2)$ consists of positive components. If b_1 is a lexicographically optimal base of $B(f_1)$ with respect to a weight vector $b_2 \in B(f_2)$ and b_2 is a lexicographically optimal base of $B(f_2)$ with respect to b_1, the pair (b_1, b_2) is a *universal pair of bases* (the orignal def-inition in [32], [35] is different from but equivalent to the present one). Some characterizations of universal pairs are given by K. Murota [31].

Acknowledgement: The author is grateful to Professor Kazuo Murota for his careful reading of the manuscript.

References

[1] G. Birkhoff: Rings of sets. *Duke Mathematical Journal* **3** (1937) 443–454.

[2] R. E. Bixby, W. H. Cunningham and D. M. Topkis: The partial order of a polymatroid extreme point. *Mathematics of Operations Research* **10** (1985) 367–378.

[3] R. Brown: The sharing problem. *Operations Research* **27** (1979) 324–340.

[4] J. Edmonds: Submodular functions, matroids, and certain polyhedra. *Proceedings of the Calgary International Conference on Combinatorial Structures and Their Applications* (eds., R. Guy, H. Hanani, N. Sauer and J. Schönheim, Gordon and Breach, New York, 1970), pp. 69–87.

[5] J. Edmonds and R. Giles: A min-max relation for submodular functions on graphs. *Annals of Discrete Mathematics* **1** (1977) 185–204.

[6] U. Faigle: Matroids in combinatorial optimization.*Combinatorial Geometries* (ed., N. White, Encyclopedia of Mathematics and Its Applications **29**, Cambridge University Press, 1987) pp. 161–210.

[7] A. Federgruen and H. Groenevelt: The greedy procedure for resource allocation problems – necessary and sufficient conditions for optimality. *Operations Research* **34** (1986) 909–918.

[8] L. R. Ford, Jr. and D. R. Fulkerson: *Flows in Networks*. Princeton University Press, Princeton, 1962.

[9] A. Frank: An algorithm for submodular functions on graphs. *Annals of Discrete Mathematics* **16** (1982) 189–212.

[10] A. Frank and É. Tardos: Generalized polymatroids and submodular flows. Report No.85389, Institut für Ökonometrie und Operations Research, Universität Bonn, West Germany, August 1986; to appear in *Mathematical Programming*, Series B.

[11] S. Fujishige: Algorithms for solving the independent-flow problems. *Journal of the Operations Research Society of Japan* **21** (1978) 189–204.

[12] S. Fujishige: The independent-flow problems and submodular functions (in Japanese). *Journal of the Faculty of Engineering, University of Tokyo* **A-16** (1978) 42–43.

[13] S. Fujishige: Polymatroidal dependence structure of a set of random variables. *Information and Control* **39** (1978) 55–72.

[14] S. Fujishige: Lexicographically optimal base of a polymatroid with respect to a weight vector. *Mathematics of Operations Research* **5** (1980) 186–196.

[15] S. Fujishige: Structures of polyhedra determined by submodular functions on crossing families. *Mathematical Programming* **29** (1984) 125–141.

[16] S. Fujishige: Submodular systems and related topics. *Mathematical Programming Study* **22** (1984) 113–131.

[17] S. Fujishige: An out-of-kilter method for submodular flows. *Discrete Applied Mathematics* **17** (1987) 3–16.

[18] S. Fujishige, N. Katoh and T. Ichimori: The fair resource allocation problem with submodular constraints. *Mathematics of Operations Research* **13** (1988) 164–173.

[19] S. Fujishige and N. Tomizawa: A note on submodular functions on distributive lattices. *Journal of the Operations Research Society of Japan* **26** (1983) 309–318.

[20] H. Groenevelt: Two algorithms for maximizing a separable concave function over a polymatroid feasible region. Working Paper, The Graduate School of

Management, The University of Rochester, August 1985.

[21] R. Hassin: Minimum cost flow with set-constraints. *Networks* **12** (1982) 1–21.

[22] T. Ibaraki and N. Katoh: *Resource Allocation Problems – Algorithmic Approaches*. Foundations of Computing Series, MIT Press, 1988.

[23] T. Ichimori, H. Ishii and T. Nishida: Optimal sharing. *Mathematical Programming* **23** (1982) 341–348.

[24] M. Iri: Applications of matroid theory. *Mathematical Programming – The State of the Art* (eds., A. Bachem, M. Grötschel and B. Korte, Springer, Berlin, 1983), pp. 158–201.

[25] M. Iri and S. Fujishige: Use of matroid theory in operations research, circuits and system theory. *International Journal of Systems Science* **12** (1981) 27–54.

[26] N. Katoh: Private communication, 1985.

[27] E. L. Lawler and C. U. Martel: Computing maximal polymatroidal network flows. *Mathematics of Operations Research* **7** (1982) 334–347.

[28] L. Lovász: Submodular functions and convexity. *Mathematical Programming – The State of the Art* (eds., A. Bachem, M. Grötschel and B. Korte, Springer, Berlin, 1983), pp. 235–257.

[29] N. Megiddo: Optimal flows in networks with multiple sources and sinks. *Mathematical Programming* **7** (1974) 97–107.

[30] K. Murota: *Systems Analysis by Graphs and Matroids – Structural Solvability and Controllability*. Algorithms and Combinatorics **3**, Springer, Berlin, 1987.

[31] K. Murota: Note on the universal bases of a pair of polymatroids. Research Memorandum, RMI88-05, Department of Mathematical Engineering and Information Physics, University of Tokyo, February 1988; to appear in *Journal of the Operations Research Society of Japan*.

[32] M. Nakamura: *Mathematical Analysis of Discrete Systems and Its Applications* (in Japanese). Dissertation, Department of Mathematical Engineering and Instrumentation Physics, Faculty of Engineering, University of Tokyo, 1983.

[33] L. S. Shapley: Cores of convex games. *International Journal of Game Theory* **1** (1971) 11–26.

[34] N. Tomizawa: Theory of hyperspaces (I) – supermodular functions and generalization of concept of 'base' (in Japanese). Papers of the Technical Group on Circuits and Systems, Institute of Electronics and Communication Engineers of Japan, CAS 80-72 (1980).

[35] N. Tomizawa: Theory of hyperspaces (IV) – principal partitions of hypermatroids (in Japanese). Papers of the Technical Group on Circuits and Systems, Institute of Electronics and Communication Engineers of Japan, CAS 80-85 (1980).

[36] N. Tomizawa: Theory of hyperspaces (XVI) – on the structures of hedrons (in Japanese). Papers of the Technical Group on Circuits and Systems, Institute of Electronics and Communication Engineers of Japan, CAS 82-174 (1983).

[37] D. M. Topkis: Adjacency on polymatroids. *Mathematical Programming* **30** (1984) 229–237.

[38] U. Zimmermann: Linear and combinatorial sharing problems. *Discrete Applied Mathematics* **15** (1986) 85–105.

[39] U. Zimmermann: Sharing problems. *Optimization* **17** (1986) 31-47.

Sequential and Parallel Methods for Unconstrained Optimization

Robert B. SCHNABEL

October 1988

Department of Computer Science, Campus Box 430, University of Colorado, Boulder, Colorado 80309, U.S.A.

Abstract

This paper reviews some interesting recent developments in the field of unconstrained optimization. First we discuss some recent research regarding secant (quasi-Newton) methods. This includes analysis that has led to an improved understanding of the comparative behavior of the BFGS, DFP, and other updates in the Broyden class, as well as computational and theoretical work that has led to a revival of interest in the symmetric rank one update. Second we discuss recent research in methods that utilize second derivatives. We describe tensor methods for unconstrained optimization, which have achieved considerable gains in efficiency by augmenting the standard quadratic model with low rank third and fourth order terms, in order to allow the model to interpolate some function and gradient information from previous iterations. Finally, we will review some work that has been done in constructing general purpose methods for solving unconstrained optimization problems on parallel computers. This research has led to a renewed interest in various ways of performing the linear algebra computations in secant methods, and to new algorithms that make use of multiple concurrent function evaluations.

This research was supported by AFOSR grant AFOSR-85-0251, ARO grant DAAL 03-88-K-0086, and NSF cooperative agreement CCR-870243.

M. Iri and K. Tanabe (eds.), Mathematical Programming, 227–261.
© *1989 by KTK Scientific Publishers, Tokyo.*

1. Introduction

This paper reviews some interesting developments in the solution of unconstrained optimizatio
problems over the last few years. The unconstrained optimization problem is

$$\text{given } f : R^n \rightarrow R \quad , \text{ find } x_* \text{ for which } f(x_*) \leq f(x) \text{ for all } x \in D, \tag{1.1}$$

where D is an open neighborhood containing x_*. We assume that the function f is at least twice continu
ously differentiable, even though the analytic derivatives may not be readily available. Our orientation i
towards problems where the number of variables, n, is not too large, say $n \leq 100$. Even if n is not large,
is often the case in practice that the evaluation of $f(x)$ is very expensive, so it is important that alg
rithms make as few evaluations of $f(x)$ and its derivatives as possible.

In Section 2 we give a very brief and superficial summary of the standard methods for solvin
unconstrained optimization problems, and their relative advantages and disadvantages. Readers familia
with unconstrained optimization may wish to skip this section. More extensive references can be foun
in various books, including Fletcher [1980], Gill, Murray, and Wright [1981], and Dennis and Schnabe
[1983].

Section 3 discusses a number of recent interesting research developments in *secant methods* fo
unconstrained optimization. By secant methods, we mean methods that update an approximation to th
Hessian matrix at each iteration, using only values of the gradient at current and previous iterates. Th
developments we discuss include experiments and analysis that have led to a better understanding of th
comparative behavior of the BFGS, DFP, and other updates in the Broyden class, as well as new theoret
cal and experimental research related to the symmetric rank one update.

In Section 4 we review some recent research into *second derivative methods*, methods that utiliz
the analytic or finite difference value of the Hessian matrix at each iteration. We concentrate on tens
methods, a new class of methods that augment the standard quadratic model with low rank approxima
tions to higher derivatives at each iteration. We briefly describe these methods in the contexts of bo
nonlinear equations and unconstrained optimization.

Finally, Section 5 summarizes some research in the fairly new field of parallel methods for uncon
strained optimization. We concentrate on parallel secant methods. We briefly discuss both some ways t
parallelize the linear algebra computations in these methods, and some approaches that make use of mu
tiple concurrent function evaluations.

2. Background

Algorithms for solving the unconstrained optimization problem (1.1) are iterative. The basic framework of an iteration of the methods that are used when the number of variables is not too large is shown in Algorithm 2.1.

At the highest level, two aspects of Algorithm 2.1 require elaboration. The first is the method for calculating or approximating the Hessian matrix $\nabla^2 f(x_+)$. The second is the method for chosing the new iterate x_+. In this section, we give a very brief description of the main alternatives that are used in practice, and how they compare.

Methods for calculating or approximating the Hessian matrix can be divided into two classes, second derivative methods and secant methods. In second derivative methods, one calculatesthe Hessian matrix $\nabla^2 f(x)$, analytically or by finite differences, at each iteration. If finite differences are used, this costs either n additional evaluations of the gradient $\nabla f(x)$ or $n^2 + \frac{3n}{2}$ additional evaluations of the function $f(x)$ at each iteration.

In secant methods, one forms an approximation H_+ to the Hessian $\nabla^2 f(x_+)$ using only values of the gradient at the current and previous iterates. This approximation is chosen so that the model of $f(x)$ around the new iterate x_+,

---

Algorithm 2.1 -- Basic Unconstrained Optimization Iteration

given current iterate x_c , $f(x_c)$,
$\quad g_c = \nabla f(x_c)$ or finite difference approximation,
$\quad H_c = \nabla^2 f(x_c)$ or finite difference approximation
$\qquad$ or secant approximation

select new iterate x_+ by a line search or trust region method
$\quad$ (often $x_+ = x_c - H_c^{-1} g_c$)
evaluate $g_+ = \nabla f(x_+)$ or finite difference approximation if not done in previous step
$\quad$ ($f(x_+)$ is evaluated in previous step)
decide whether to stop; if not
calculate $H_+ = \nabla^2 f(x_+)$ or finite difference approximation
$\qquad$ or secant approximation

$$m(x_+ + d) = f(x_+) + g(x_+)^T d + \frac{1}{2}d^T H_+ d,\qquad(2.2)$$

interpolates not only the function and gradient values at x_+, but also the gradient value at the previous iterate x_c. This interpolation condition is satisfied if the new Hessian approximation H_+ satisfys the secant equation

$$H_+ s = y\qquad(2.3)$$

where $s = x_+ - x_c$, $y = g(x_+) - g(x_c)$.

The most commonly used Hessian approximation is the BFGS update (Broyden [1970], Fletcher [1970] Goldfarb [1970], Shanno [1970])

$$H_+ = H_c - \frac{H_c ss^T H_c}{s^T H_c s} + \frac{yy^T}{y^T s} \ .\qquad(2.4)$$

This update makes a symmetric, rank two change to the previous Hessian approximation H_c, and if H_c is positive definite and $s^T y > 0$, then H_+ is positive definite. In practice, the initial approximation is sym metric and positive definite, and usually $s^T y$ is greater than 0, otherwise the update is skipped. Thus each Hessian approximation in a BFGS method is symmetric and positive definite.

A brief comparison between the costs and theoretical properties of second derivative and secant methods is given in Table 2.5. It shows three main differences between these two classes of methods First, second derivative methods require an evaluation of the Hessian matrix at each iteration, while secant methods do not. Second, second derivative methods require $O(n^3)$ arithmetic operations at each iteration, because they must factorize a symmetric matrix, while secant methods can be implemented in $O(n^2)$ operations at each iteration, because they can use techniques to update the factorization of H_c into the factorization of H_+ (see e.g. Gill, Murray, and Wright [1981] or Dennis and Schnabel [1983], or Section 5.1). Third, if the Hessian matrix at the solution is nonsingular, the eventual rate of convergence of the second derivative methods is quadratic, while secant methods such as the BFGS converge at a slower superlinear rate.

Thus secant methods cost less per iteration than second derivative methods, but can be expected to require more iterations. Computational experience confirms that this is usually the case. In experiments by Schnabel, Koontz, and Weiss [1985] on a set of standard test problems, it was found that second derivative methods and secant methods had roughly the same reliability (ability to find the solution) , and that the number of iterations required by secant methods was usually a relatively small multiple (less than

Table 2.5 -- Comparison, Second Derivative Methods vs Secant Methods

		2nd Derivative	Secant
Evaluations	$f(x)$	Usually 1 or 2	
per	$\nabla f(x)$	Usually 1	
Iteration	$\nabla^2 f(x)$	1	0
Arithmetic Operations per Iteration		$\frac{n^3}{6} + 0(n^2)$	$(2-6)\,n^2$
Storage		$n^2/2$ or n^2	
Rate of Local Convergence		Quadratic	Superlinear

$\frac{n}{2}$) times the number of iterations required by second derivative methods. This implies that if the cost of function evaluation is dominant, and the cost of an analytic or finite difference Hessian evaluation is at least $\frac{n}{2}$ times the cost of a gradient evaluation, then it is probably preferable to use secant methods. This is the case when the Hessian is approximated by finite differences. If function evaluation is not expensive, or if the cost of Hessian evaluation is not much more than the cost of gradient evaluation, then either method is probably satisfactory with second derivative methods possibly having a slight advantage in their reliability. It is our understanding that other computational studies have come to similar conclusions.

The other aspect of Algorithm 2.1 that we discuss briefly is the method for choosing the new iterate x_+. Two important classes of methods, line search methods and trust region methods, are used in practice. Roughly speaking, the objective of either type of method is that close to the solution, the new iterate should be chosen to be the minimizer $-H_c^{-1} g(x_c)$ of the model

$$m(x_c + d) = f(x_c) + g(x_c)^T d + \tfrac{1}{2} d^T H_c\, d \tag{2.6}$$

and that otherwise the new iterate $x_+ = x_c + d$ should be some value for which $f(x_+) < f(x_c)$.

In a standard line search algorithm, the new iterate x_+ is chosen to be

$$x_+ = x_c - \lambda_c (H_c + D_c)^{-1} g_c \tag{2.7}$$

where $\lambda_c > 0$ is the step length and $(H_c + D_c)^{-1} g_c$ is the search direction. Here the diagonal matrix D_c is if H_c is safely positive definite, and is a non-negative matrix such that $H_c + D_c$ is positive definite otherwise (see Gill, Murray, and Wright [1981], Dennis and Schnabel [1983]). Thus the search direction is

always a descent direction for $f(x)$, and close to the solution, the search direction is the Newton direction. The step length λ_c generally is chosen so that two conditions, slightly stronger than $f(x_+) < f(x_c$
and $s^T y > 0$, are satisfied, and so that $\lambda_c = 1$ is used if it is acceptable. It has been shown that it is always
possible to choose λ_c to satisfy such conditions, and that many such strategies for choosing the step
length and the search direction cause line search algorithms to be both globally convergent, and to retain
fast local convergence.

The trust region approach is somewhat different. Rather than first choosing a search direction and
then a step length, the trust region algorithm first chooses an approximate step length Δ_c, and then
chooses the next iterate x_+ to be an approximate solution to the problem

$$\text{minimize } f(x_c) + g(x_c)^T d + \frac{1}{2} d^T H_c d \quad \text{subject to } \| d \| \leq \Delta_c \ . \tag{2.8}$$

If this step does not result in a satisfactory decrease in $f(x)$, the trust region Δ_c is reduced and problem
(2.8) is solved again. If x_+ is satisfactory, the trust region is adjusted for the next iteration.

The solution to the trust region problem (2.8) is $d = -H_c^{-1} g_c$ if H_c is positive definite and
$\| H_c^{-1} g_c \| \leq \Delta_c$, otherwise it is a pair (d, μ_c) for which

$$(H_c + \mu_c^2 I) d = -g_c \ , \tag{2.9}$$

$H_c + \mu_c^2 I$ is at least positive semi-definite, and $\| d \| = \Delta_c$. Since this step is expensive to calculate
trust region algorithms usually approximate it in one of two ways. Either they calculate a d which
satisfies (2.9) for some μ_c and has length approximately equal to the trust radius, or they restrict the
choice of d to a two-dimensional subspace, as in the "dogleg" algorithms. Many strategies for approximately solving the trust region problem (2.8), and for adjusting the trust radius Δ_c, have been shown to
lead to algorithms that are globally convergent and retain fast local convergence. For more information
on trust regions methods, see for example Dennis and Schnabel [1983], Moré and Sorensen [1983], or
Shultz, Schnabel, and Byrd [1985].

In our computational experience (see e.g., Schnabel, Koontz, and Weiss [1985]), we have found no
systematic differences between line search and trust region algorithms. This is true both in the case
where H_c is the analytic or finite difference Hessian, and where it is a secant approximation. Sometimes
however, there are substantial differences between the efficiency of line search and trust region algorithms on specific problems, so that it may be useful to have both options available in software. Line
search methods may enjoy a small advantage when used in conjunction with secant approximations to the
Hessian, because they can assure that the necessary and sufficient condition for a positive definite secant

ipdate, $s^T y > 0$, is satisfied at each iteration, while trust region methods cannot do this. Conversely, trust region methods may enjoy a small advantage when used with analytic or finite difference Hessians, because they seem to deal more directly with indefinite Hessians.

3. Recent Research on Secant Methods

In this section, we review some interesting recent developments concerning secant methods for unconstrained optimization. These developments fall into two categories. First, in Section 3.1, we discuss several research contributions that have helped explain the differences between the two best known secant updates, the BFGS and the DFP. Some of this work has also considered other updates in the Broyden class, which consists of all linear combinations of the BFGS and DFP. Secondly, in Section 3.2, we discuss some research that has caused a revival of interest in the symmetric rank one update.

3.1 Understanding the difference in performance between the BFGS, DFP, and other updates in the Broyden class

As we mentioned in Section 2, secant methods are commonly used to solve unconstrained optimization problems when function evaluation is expensive and analytic values of the Hessian are not available. These methods update an approximation H_c to $\nabla^2 f(x_c)$ into an approximation H_+ to $\nabla^2 f(x_+)$, using values of the gradient at x_c and x_+.

The most commonly used secant update for unconstrained optimization is the BFGS update (2.4). Among its important properties are that it obeys the secant equation (2.3), that H_+ differs from H_c by a symmetric matrix of rank two, and that H_+ is positive definite as long as H_c is positive definite and $s^T y > 0$. It has long been known that many other updates have these same algebraic properties. In particular, the DFP update

$$H_+ = H_c + \frac{(y - H_c s)y^T + y(y - H_c s)}{y^T s} - \frac{yy^T (y - H_c s)^T s}{(y^T s)^2} \tag{3.1}$$

Davidon [1959], Fletcher and Powell [1963]), the oldest known secant update for unconstrained optimization, is another rank two update that obeys the secant equation and preserves symmetry and positive definiteness under the same conditions as the BFGS. In addition, any update in the Broyden class, which consists of all linear combinations of the BFGS and DFP

$$H_+(\phi) = (1-\phi)H_+^{BFGS} + \phi H_+^{DFP} \tag{3.2}$$

also is a symmetric rank two update that obeys the secant equation. Furthermore, if $\phi \in [0,1]$, then $H_+(\phi)$ also is positive definite if H_c is positive definite and $s^T y > 0$. All the updates in the Broyden class also share the important property that they are invariant under linear transformations of the variable space (see e.g. Dennis and Schnabel [1983]).

For some time, the conventional wisdom has been that the BFGS is the best update in the Broyden class in practice, and that it is significantly superior to the DFP. There has been little theoretical analysis, however, that helps explain this superiority. Two types of convergence analysis have been successfully applied to secant methods. One type, which originated with the work of Broyden, Dennis, and Moré [1973], proves local superlinear convergence for a direct prediction algorithm (where each iterate $x_{k+1} = x_k - H_k^{-1}\nabla f(x_k)$) under the initial assumptions that $\| x_0 - x_* \| \leq \epsilon$ and $\| H_0 - \nabla^2 f(x_0) \| \leq \delta$ for ϵ and δ sufficiently small. Under these assumptions, Broyden, Dennis, and Moré [1973] proved the superlinear convergence of the iterates produced by either the BFGS or the DFP update. Stachurski [1981] and Griewank and Toint [1982] proved analogous results for any $\phi \in [0,1]$ in (3.2) under the same assumptions. Thus this convergence theory does not differentiate at all between the BFGS, DFP, and their convex combinations.

A second type of convergence result has been proven by Powell [1976]. Under the assumption that $f(x)$ is convex, he establishes both global and local superlinear convergence of a BFGS algorithm that uses a standard line search. While he does not establish this result for the DFP or any other update in the Broyden class, until recently there was no clear understanding of whether or not the result could be extended to other updates. Therefore it was not clear whether this result helped distinguish between various secant updates.

The first recent contribution that we discuss towards understanding the difference between the BFGS and other updates was provided by Powell [1986]. He compares the behavior of the BFGS and the DFP when solving the problem

$$f(x) = x^T x, \quad n = 2, \quad H_0 = \begin{bmatrix} 1 & 0 \\ 0 & \lambda \end{bmatrix} \tag{3.3}$$

under the assumption that the iterates are chosen by $x_{k+1} = x_k - H_k^{-1}\nabla f(x_k)$. (Note that due to the scale invariance of these methods, (3.4) is equivalent to $f(x) = x_1^2 + \frac{x_2^2}{\lambda}, H_0 = I$.) Powell shows that if λ is much greater than 1, and if a particularly difficult starting point,

$$x_0 = \begin{bmatrix} \cos \psi \\ \sin \psi \end{bmatrix}, \quad \psi = \arctan \sqrt{\lambda} \tag{3.4}$$

is chosen, then the BFGS will require about $2.4 \log_{10}\lambda$ iterations to achieve convergence, whereas the DFP will require about λ iterations. If λ is much less than 1, either method requires at most about 10 iterations.

This analysis shows a potentially huge difference between the performance of the BFGS and the DFP in some situations. Table 3.5 summarizes some computations performed by Powell [1986] to confirm this analysis. The first two lines show the number of iterations required by both methods to reduce $\| x \|$ by a factor of 10^4, using x_0 given in (3.4). They correspond very closely to the estimates from Powell's analysis. The last two rows show the number of iterations required to reduce $\| x \|$ by a factor of 10^4 when $x_0 = \begin{bmatrix} \cos 40^o \\ \sin 40^o \end{bmatrix}$. While they indicate that the worst case is somewhat extreme, they still show a clear superiority of the BFGS.

Thus the analysis by Powell [1986] of the behavior of the BFGS and DFP on the problem (3.3) gives some indication of a fundamental difference between these two updates. A second recent paper, by Byrd, Nocedal, and Yuan [1987], sheds additional light on the subject.

Byrd, Nocedal, and Yuan extend the global and local superlinear convergence of Powell [1976] for the BFGS update to any update in the Broyden class (3.2) with $\phi \in [0,1)$, that is any convex combination

---

Table 3.5 -- Iterations Required for $\| x_k \| \le 10^{-4} \| x_0 \|$ on Problem (3.4)
(from Powell [1986])

	λ	10^1	10^2	10^3	10^4	10^6
Bad x_o	BFGS	8	10	-	15	20
	DFP	16	107	1006	-	-
Average x_o	BFGS	6	7	-	7	7
	DFP	10	15	-	19	24

of the BFGS and the DFP except for the DFP itself. The techniques they use to prove this result give some interesting insight into the behavior of these methods. A key quantity in their analysis is $\cos(\theta_k)$, where θ_k is the angle between the step direction and the negative gradient direction at the kth iteration. If this angle is less than any $\sigma < 90°$ infinitely often, then the iterates produced by any standard line search method will be globally convergent (Wolfe [1969, 1971]). Byrd, Nocedal, and Yuan show that, if $f(x)$ is uniformly convex and $x_{k+1} = x_k - \lambda_k H_k^{-1} \nabla f(x_k)$, then

$$\cos(\theta_k) \geq \frac{\lambda_k}{c \cdot \text{Trace}(H_k)} \tag{3.6}$$

for some constant c. This indicates that the method can fail to have the desired global convergence properties only if the step lengths λ_k become arbitrarily close to 0, or if the trace of H_k becomes arbitrarily large. Next they show that for any $\phi \in [0,1]$, the geometric mean of the step lengths $\{\lambda_k\}$ produced by the algorithm is bounded below. This means that the step lengths do not converge to 0 for any $\phi \in [0,1]$, so that the only possible impediment to convergence for any such update is $\text{Trace}(H_k)$. Finally, Byrd, Nocedal, and Yuan show that

$$\text{Trace}(H_{k+1}) \leq \text{Trace}(H_k) - \frac{(1-\phi)\lambda_k}{c \cdot (cos(\theta_k))^2} + t(\phi_k, \lambda_k, \theta_k, H_k) \tag{3.7}$$

where $t(\phi_k, \lambda_k, \theta_k, H_k)$ are some additional terms that are less crucial to the analysis. Equation (3.7) indicates that if the method takes a bad step (i.e., $\cos(\theta_k)$ close to 0), then the trace of H_{k+1} will be significantly less than the trace of H_k, as long as $\phi < 1$. This in turn can be used to show that for any method with $\phi \in [0,1]$, there cannot be too many bad steps, which leads to both global and local super-linear convergence.

An interesting aspect of the convergence of Byrd, Nocedal, and Yuan [1987] is that it shows that secant methods with $\phi < 1$ have a "self-correcting" property with respect to $\text{Trace}(H_k)$ that becomes less strong as ϕ gets closer to 1, and is not present for $\phi = 1$, the DFP. This analysis does not show that the DFP fails to possess the same global and local convergence properties, but it does seem to point out a fundamental deficiency of the DFP update.

Byrd, Nocedal, and Yuan [1987] also provide a simple example that shows the deterioration of the computational performance of secant methods as ϕ goes from 0 to 1. They consider the function

$$f(x) = \frac{1}{2} x^T x + (0.1)(\frac{1}{2} x^T \begin{bmatrix} 5 & 1 \\ 1 & 3 \end{bmatrix} x)^2, \quad n = 2 \tag{3.8}$$

with the starting values $x_o = \begin{bmatrix} \cos 70° \\ \sin 70° \end{bmatrix}$, $H_o = \begin{bmatrix} 1 & 0 \\ 0 & 10^4 \end{bmatrix}$. Table 3.9 shows the number of iterations required by an unconstrained optimization method with a modern line search to achieve $\| x_k \| \leq 10^{-4} \| x_o \|$ for various values of ϕ. This example clearly exhibits a deterioration in efficiency as ϕ goes from 0 to 1, but shows that this deterioration is most marked very close to the DFP ($\phi=1$).

The analysis and computational example of Byrd, Nocedal, and Yuan [1987] also naturally suggest that one might try values of $\phi < 0$. This possibility has been investigated in a paper by Zhang and Tewarson [1986]. They suggest a heuristic for chosing a value of $\phi < 0$ in (3.2), and show that on a set of test problems, their method is about 10% to 15% more efficient than the BFGS on the average. They are able to prove global and r-linear convergence as long as $\theta_k \geq c_k$, where $c_k < 0$ is a quantity that is computable in practice, but can only show superlinear convergence if $\theta_k \geq \bar{c}_k$, where $\bar{c}_k < 0$ is not computable in practice. Thus, their computational and theoretical results are very interesting, but given the heuristic nature of the choice of ϕ_k and the lack of a fully satisfactory superlinear convergence result, there is probably not yet strong enough computational evidence to warrant switching from the BFGS to their new method.

It will be seen that the same contrast, between slightly improved computational performance and the lack of fully satisfactory superlinear convergence results, exists for some methods to be discussed in Section 3.2.

3.2 Recent Research on the Symmetric Rank One Update

It has long been known that there is one rank one update that satisfies the secant equation (2.3) and preserves symmetry. This update is known as the symmetric rank one (SR1) update,

---

Table 3.9 -- Iterations required for $\| x_k \| \leq 10^{-4} \| x_o \|$ on Problem 3.8
(from Byrd, Nocedal, Yuan [1987])

ϕ	0	0.2	0.4	0.6	0.8	0.9	0.99	0.999	1
iterations	15	21	26	32	66	115	630	2223	4041

$$H_+ = H_c + \frac{(y-H_c s)(y-H_c s)^T}{(y-H_c s)^T s} \ . \tag{3.10}$$

While this update has not been used much in practice, some recent research is leading to a revival of interest in it. In this section we briefly review the properties of this update, and then discuss this recent research.

It is straightforward to show that the SR1 update is a member of the Broyden class (3.2), with

$$\phi = \frac{y^T s}{(y-H_c s)^T s} \ . \tag{3.11}$$

This value of ϕ is always outside the range [0,1], as long as H_c is positive definite and $s^T y > 0$. Thus the SR1 update is not covered by the convergence theory mentioned in Section 2 or 3.1, nor is it guaranteed to be positive definite even if $s^T y > 0$. These properties figure prominently in the main advantages and disadvantages of the update.

The SR1 update has two main advantages. First, it is a rank one modification whereas all the other members of the Broyden class are rank two modifications; this may make it cheaper to implement. Second, it is well known that the SR1 possesses *quadratic termination,* meaning that if it is applied to a quadratic function $f(x)$ and the step $-H_c^{-1} \nabla f(x_c)$ is used at each iteration, then in exact arithmetic, the minimizer will be found exactly in $n+1$ or fewer iterations. Furthermore, if $n+1$ iterations are required, then the final Hessian approximation will equal the exact Hessian $\nabla^2 f(x)$. It can be shown that no update in the Broyden class that always preserves positive definiteness has this quadratic termination property. This at least raises the possibility that the SR1 update may produce more accurate Hessian approximations than other updates on general functions, and that it may have attractive local convergence properties.

On the other hand, the SR1 update has several disadvantages. First, there is no reason why the denominator $(y-H_c)^T s$ cannot be zero or nearly zero, even close to the solution. This indicates a potential instability in the update. Secondly, aside from the quadratic termination result mentioned above, no global or local convergence results analogous to those mentioned in Sections 2 and 3.1 have been established for the SR1.

Finally, we have already said that the SR1 will not necessarily yield a positive definite H_+ even when $s^T y > 0$. This could be an advantage if it allows the update to better model the actual Hessian when it is indefinite, or it could be a disadvantage if it leads to an indefinite approximation in a region where the actual Hessian is positive definite.

The revival of interest in the SR1 update was started by the research of Conn, Gould, and Toint 1986, 1987, 1988]. The main focus of their research is somewhat different than that considered in this paper. Conn, Gould, and Toint consider the bound-constrained optimization problem

$$\text{minimize } f(x) \quad \text{subject to } l_i \le x_i \le u_i \, , \, i = 1, ..., n \, . \tag{3.12}$$

Their research has many interesting and novel aspects, including the generalization of the notion of a Cauchy point for unconstrained optimization to the problem (3.12), the use of inexact Newton methods methods that solve the linear system of equations associated with each iteration inexactly) in the context of problem (3.12), the introduction of new techniques that allow large changes in the set of active constraints at each iteration, and the extension of the known global convergence theory for trust region methods for unconstrained optimization to problem (3.12). We will not discuss these aspects of their research further since they are outside the scope of this paper.

The part of the research of Conn, Gould, and Toint that interests us most from the perspective of this paper is one of their computational experiments. Conn, Gould, and Toint [1986] ran their algorithm for problem (3.12) (a trust region method using secant updates and an inexact Newton method) on 50 problems, of which 15 are unconstrained. They tried both the BFGS and the SR1 updates. A summary of their results is given in Table 3.13.

The results of Conn, Gould, and Toint [1986] show a large overall advantage for the SR1 update in comparison to the BFGS. The advantage is great on problems where bounds are present, while the two updates appear similar on unconstrained problems.

These unconstrained optimization results interested us considerably, because if the SR1 is even competitive with the BFGS in general, then there are situations where it may be preferable due to its simpler form, quadratic termination, and its ability to reflect indefiniteness. Thus we decided to experiment with using the SR1 update instead of the BFGS update in the UNCMIN code (Schnabel, Koontz, and Weiss [1985]), a fairly standard unconstrained optimization method. The results of running UNCMIN, using the BFGS and the SR1, on the same unconstrained problems as were used by Conn, Gould, and Toint [1986] are shown in Table 3.14.

The UNCMIN results in Table 3.14 are very different than the unconstrained optimization results in Table 3.13, with the UNCMIN results strongly favoring the BFGS over the SR1. The reasons for the difference in the comparative performance of the BFGS and SR1 updates within the algorithms of Conn, Gould, and Toint [1986] and in UNCMIN are not clear. It should be stressed that these two algorithm are

Table 3.13 -- Computational Results from Conn, Gould, and Toint [1986]

Problems	BFGS Better	SR1 Better	BFGS, SR1 Similar	total SR1 iterations / total BFGS iterations
All (50)	9	30	11	0.71
Unconstrained (15)	6	7	2	1.12
Bounds Present (35)	3	23	9	0.56

Table 3.14 -- UNCMIN Results on the Unconstrained Problems from Table 3.13

Global Method	BFGS Better	SR1 Better	BFGS, SR1 Similar	total SR1 iterations / total BFGS iterations
Trust Region	10	1	3	2.66
Line Search	12	0	2	1.93

considerably different. Most importantly, Conn, Gould, and Toint [1986] use an inexact Newton strategy while UNCMIN finds the minimizer of the quadratic model exactly. It is also important to mention that the performances of the BFGS versions of the two algorithms are fairly similar; the big difference between the unconstrained optimization results in Tables 3.13 and 3.14 stems from the fact that the Conn, Gould, and Toint algorithm performs much better using the SR1 than UNCMIN does using the SR1. We do not yet understand why this is so, nor why the comparative advantage of the SR1 over the BFGS in Conn, Gould, and Toint's tests is so much bigger for problems with simple bounds. However all these results do seem to indicate that the SR1 update in general, and the above questions in particular, warrant additional research.

Another interesting aspect of the research of Conn, Gould, and Toint [1987] is an examination of the convergence of the sequence of matrices generated by the SR1 update. They show that if the sequence of iterates $\{x_k\}$ converges to a strong local minimizer x_*, if each set of n consecutive steps $\{x_{i+1}-x_i\}$, $i=k,\cdots,k+n-1$ is uniformly linearly independent, and if the denominators of (3.10) are bounded below in the sense that $|(y_k-H_k s_k)^T s_k| \geq c \, \|y_k-H_k s_k\| \, \|s_k\|$ for all k, then the equence of Hessian approximations $\{H_k\}$ generated by the SR1 algorithm converges to $\nabla^2 f(x_*)$. This in urn implies that the rate of local convergence is at least superlinear. While these assumptions are strong, and can probably not be guaranteed to be satisfied in theory, this convergence result still gives an indication of what might often happen in practice. Indeed, Conn, Gould, and Toint [1987] conduct some experiments, using a fourth order polynomial for $f(x)$ and running to a very tight convergence tolerance, where they show that the SR1 produces final Hessian approximations that agree with the actual Hessian at the solution to within between 10^{-8} and 10^{-13}, whereas the final approximations produced by the BFGS only agree to about 10^{-3}. This research supports the hypothesis that the quadratic termination property of the SR1 might lead to better final Hessian approximations in practice. It also seems to further indicate a need for continued research on the role of the SR1 update in unconstrained optimization.

We conclude this section by discussing a second, somewhat different, new algorithm involving the SR1 update that was recently proposed by Osborne and Sun [1988], motivated in part by the work of Conn, Gould, and Toint. Osborne and Sun's approach is to use the SR1 in such a way that it always produces positive definite Hessian approximations. They do this by first multiplying the current Hessian approximation H_c by a scale factor $\gamma > 0$, and then applying the SR1 update to γH_c. That is

$$H_+ = \gamma H_c + \frac{(y - \gamma H_c s)(y - \gamma H_c s)^T}{(y - \gamma H_c s)^T s} . \tag{3.15}$$

It is fairly easy to see that if H_c is positive definite and $s^T y > 0$, then H_+ given by (3.15) will be positive definite if γ is either sufficiently large or sufficiently close to 0. In fact, Osborne and Sun [1988] show that H_+ is positive definite if $s^T y > 0$ and

$$\gamma \in (0, \frac{s^T y}{s^T H_c s}) \text{ or } \gamma \in (\frac{y^T H_c^{-1} y}{s^T y}, \infty). \tag{3.16}$$

Note that the standard SR1 update, $\gamma = 1$, may or may not be contained in one of the two intervals in (3.16).

Osborne and Sun [1988] propose using the standard SR1 update, $\gamma = 1$ in (3.15), if it satisfies (3.16). Otherwise, they propose choosing the value of γ that satisfies (3.16) and that leads to the optimally conditioned update in the sense proposed by Davidon [1975], namely that it minimizes the l_2 condition number of $H_c^{-1/2} H_+ H_c^{-1/2}$ among all the updates of the form (3.15). Osborne and Sun derive a closed form for this optimal value of γ; actually there are two values of γ that yield equally optimal solutions, one in each of the intervals in (3.16).

Osborne and Sun report promising computational results using this scaled SR1 method on a small set of test problems. We have tried using their update in the UNCMIN code, and have found that on the average it leads to a 10-15% improvement over the BFGS update on a standard set of test problems. Therefore, since the Osborne and Sun algorithm has the additional advantage in comparison to the BFGS that it only requires a rank one update, it appears to merit further consideration. On the other hand, the scaled SR1 update (3.15) shares with the standard SR1 update (3.10) the apparent disadvantage that no global or local convergence results have been proven for it under the standard assumptions that are used in the convergence analysis of many secant methods, including the BFGS (see Section 3.1). Thus more research seems necessary to understand both the practical and theoretical properties of all the SR1 methods discussed in this section.

4. Recent Research on Second Derivative Methods -- Tensor Methods

In this section we turn our attention to second derivative methods, methods where the analytic or finite difference Hessian is available at each iteration. We discuss one recent development, the development of *tensor methods*. This is a class of methods that bases each iteration upon a higher order model than is used by standard methods. The higher order terms in this model are chosen so that the model is hardly more expensive to form, store, and solve than the standard model.

Tensor methods were first developed by Schnabel and Frank [1984] in the context of solving systems of nonlinear equations. These methods base each iteration upon a quadratic model, rather than the linear model that is standard for solving systems of nonlinear equations. Since an understanding of tensor methods for nonlinear equations is helpful in understanding the more complicated tensor methods for unconstrained optimization, we review tensor methods for nonlinear equations in Section 4.1. Then in Section 4.2 we describe tensor methods for unconstrained optimization. These methods base each iteration upon a fourth order model, rather than the standard quadratic model (2.6).

We note that another recent body of research has considered the use of higher ordered models when the objective function is a "factorable function." See for example Jackson [1983], Jackson and McCormick [1986], and McCormick [1983].

4.1 Tensor Methods for Nonlinear Equations

The nonlinear equations problem is

$$\text{given } F : R^n \rightarrow R^n, \text{ find } x_* \text{ for which } F(x_*) = 0. \tag{4.1}$$

When the Jacobian matrix $F'(x_c)$ is available, algorithms for solving (4.1) generally base each iteration upon the linear model

$$M(x_c + d) = F(x_c) + F'(x_c)d. \tag{4.2}$$

This model requires n^2 storage locations, and $\frac{n^3}{3}$ arithmetic operations to solve it at each iteration. Its use leads to quadratic convergence for problems where $F'(x_*)$ is non-singular. but at best linear convergence for problems where $F'(x_*)$ is singular (see for example Decker and Kelly [1980a, b], Griewank 1980]).

The tensor method proposed by Schnabel and Frank [1984] instead bases each iteration upon the model

$$M(x_c + d) = F(x_c) + F'(x_c)d + \tfrac{1}{2}T_c dd \tag{4.3}$$

where $T_c \in R^{n \times n \times n}$ is a three-dimensional object often referred to as a tensor. If $T_c = F''(x_c)$, then (4.3) is just a second order Taylor series model. However using (4.3) with $T_c = F''(x_c)$ is not practical, as it leads to huge increases in the costs to form, store, and solve the model. Instead, Schnabel and Frank chose T_c in (4.3) to be a very low rank approximation to $F''(x_c)$. We now briefly summarize how they make this choice, and some of its consequences.

Schnabel and Frank [1984] choose T_c in (4.3) by requiring the model to interpolate the values of $F(x)$ at p (not necessarily consecutive) previous iterates $x_{-1}, \cdots, x_{-p}$. They impose the limit $p \leq \sqrt{n}$, and also require that the steps $s_i = x_c - x_{-i}$ from x_c to these p previous iterates be strongly linearly independent. The latter condition is usually much more restrictive than the limit $p \leq \sqrt{n}$, and most often results in the choice $p=1$, meaning that only information from the most recent previous iterate is used. Schnabel and Frank then choose the smallest T_c, in the Frobenius norm, that satisfies the p interpolation conditions $M(x_c - s_i) = F(x_{-i})$, $i=1, \cdots, p$. The result is a rank p tensor T_c of the form $T_c = \sum_{i=1}^{p} a_i s_i s_i$, for some $a_i \in R^n$, $i = 1, \cdots, p$. Thus the model (4.3) becomes

$$M(x_c + d) = F(x_c) + F'(x_c)d + \tfrac{1}{2} \sum_{i=1}^{p} a_i (s_i^T d)^2. \tag{4.4}$$

The additional cost of forming this model is about $n^2 p$ arithmetic operations, while the additional storage

244   Robert B. Schnabel

cost is about $4np$ locations. Both of these are small in comparison to the basic $O(n^3)$ arithmetic per
iteration and n^2 storage costs of the linear model.

Due to the special form of the model (4.4), the problem of finding its root (or of minimizing
$\| M(x_c + d) \|_2^2$ if there is no root) can be reduced to solving p quadratic equations in p unknowns plus
$n-p$ linear equations in $n-p$ unknowns. This is hardly more expensive than finding the root of the linear
model (4.2), requiring only about n^2p additional arithmetic operations (recall that usually $p=1$). Furthermore, Schnabel and Frank [1984] show that the solution of (4.4) is usually well posed as long as the rank
of $F'(x_c)$ is at least $n-p$, whereas the solution of the linear model (4.2) only is well posed if rank $(F'(x_c))$
$= n$.

The above simply shows that, by utilizing a low rank quadratic term, it is possible to add a small
amount of information to the linear model at a small cost. What is perhaps surprising is that this small
amount of information seems to lead to fairly large improvements in the cost of solving problem (4.1) in
practice. Schnabel and Frank [1984] compare an implementation of their tensor method, using a standard
line search, to a standard method for nonlinear equations that uses the linear model (4.2) and the same
line search, on a set of problems from Moré, Garbow, and Hillstrom [1981]. Their results are summarized in Table 4.5; for more details, see Schnabel and Frank [1984].

Table 4.5 indicates that the tensor method leads to consistent and rather substantial improvements
in efficiency over a standard derivative-based method for solving systems of nonlinear equations.

Table 4.5 -- Comparison of Tensor and Standard Methods for Nonlinear Equations
(from Schnabel and Frank [1984])

Rank $F'(x_*)$	Problem Set	Tensor Better	Standard Better	Two Methods similar	Average Ratio of Tensor Iterations/Standard Iterations
n	All	21	1	6	0.77
	Harder Only*	14	0	0	0.61
$n-1$	All	15	0	2	0.58
	Harder Only*	9	0	0	0.39
$n-2$	all	11	2	0	0.63
	Harder Only*	7	0	0	0.50

*Only those problems where slower method required at least 10 iterations.

Furthermore, in these tests the number of past points, p, used in the tensor model was generally 1 and never more than 3, so that the additional cost of using the tensor method was low. A Fortran code for solving systems of nonlinear equations, and nonlinear least squares problems, by tensor methods is available from the author. Our positive experience with tensor methods for nonlinear equations also has motivated us to consider using tensor methods for unconstrained optimization, which are described in the next section.

4.2 Tensor Methods for Unconstrained Optimization

The extension of the tensor method for nonlinear equations that we have just described into a tensor method for unconstrained optimization brings up an interesting general issue. In some cases, such as the basic Newton method, methods for nonlinear equations and for unconstrained optimization are very closely related. In some other cases, such as secant methods, methods for nonlinear equations and unconstrained optimization are considerably different. These differences are generally due to the considerations of symmetry, convexity, and positive definiteness which are present for unconstrained optimization but not for nonlinear equations. In the case of tensor methods, the differences between unconstrained optimization and nonlinear equations cause the tensor methods for the two problems to differ significantly.

The obvious extension of the tensor method for nonlinear equations to the unconstrained optimization problem would be to base each iteration upon the quadratic model of the gradient

$$\nabla m \ (x_c + d) = \nabla f \ (x_c) + \nabla^2 f \ (x_c) d + \tfrac{1}{2} T_c \ dd \ . \tag{4.6}$$

that would result from simply substituting $\nabla f (x)$ for $F(x)$ in the method of Section 4.1. This model is unappealing for unconstrained optimization for several reasons. First, the tensor T_c produced by the method described in Section 4.1 is not symmetric, but for optimization it should be. Second, we will want to interpolate past values of $f(x)$ as well as $\nabla f(x)$, so a procedure based solely on modeling $\nabla f(x)$ is too restrictive. Most importantly, the model (4.6) corresponds to using a third order model of $f(x_c)$, and a third order model has two basic deficiencies for unconstrained optimization. One is that it does not have a global minimizer. A second is that it does supply enough information to lead to faster than linear convergence for problems where $\nabla^2 f(x_*)$ is singular. For this, fourth order information is necessary as well.

For these reasons, Schnabel and Chow [1988] propose using a fourth order model of $f(x)$

$$M(x_c + d) = f(x_c) + \nabla f(x_c)^T d + \frac{1}{2} d^T \nabla^2 f(x_c) d + \frac{1}{6} T_c \, ddd + \frac{1}{24} V_c \, dddd \tag{4.7}$$

where $T_c \in R^{n \times n \times n}$ is a symmetric approximation to $\nabla^3 f(x_c)$ and $V_c \in R^{n \times n \times n \times n}$ is a symmetric approximation to $\nabla^4 f(x_c)$. While this model may appear complicated, the crucial point is that T_c and V_c will again be low rank tensors. We will see that this again makes the model hardly more expensive to form, store, and solve than the standard quadratic model.

To form the tensor model (4.7), Schnabel and Chow require it to interpolate the values of $f(x)$ and $\nabla f(x)$ at p previous iterates x_{-i}, $i=1, \cdots, p$. These iterates are chosen by the same criteria as in the tensor method for nonlinear equations, namely that $p \leq \sqrt{n}$ and that the steps $s_i = x_c - x_{-i}$ are strongly linearly independent. In practice, usually the second criterion limits p to 1, meaning that only information from the most recent previous iteration is used. Next Schnabel and Chow choose the smallest V_c, in the Frobenius norm, that is consistent with these interpolation conditions. The result is a symmetric rank p tensor V_c of the form

$$V_c = \sum_{i=1}^{p} \alpha_i \, s_i s_i s_i s_i \, , \tag{4.8}$$

where each α_i is a scalar. Finally they choose T_c to be the smallest symmetric tensor, in the Frobenius norm, that is consistent with the choice of the V_c and the interpolation conditions. This yields a rank $2p$ tensor T_c of the form

$$T_c = \sum_{i=1}^{p} (b_i s_i s_i + s_i b_i s_i + s_i s_i b_i \,) \tag{4.9}$$

where each $b_i \in R^n$. This order of choosing V_c and T_c causes the model to interpolate as much information as possible with a third order model, and to use the fourth order term only where a third order model is insufficient.

Even though T_c is a rank $2p$ tensor, Schnabel and Chow [1988] show that the minimizer of the resultant tensor model (4.7) can still be found by solving p cubic equations in p unknowns and $n-p$ linear equations in $n-p$ unknowns. Thus the size of the system of nonlinear equations that must be solved to minimize the tensor model for unconstrained optimization is the same as for nonlinear equations, with the difference being that the p nonlinear equations are cubic rather than quadratic. Since p generally is 1, this is not significant. The total additional costs of using the tensor model rather than the standard quadratic model are again $O(np)$ storage locations, and $O(n^2 p)$ arithmetic operations per iteration. This is again minor in comparison to the n^2 storage locations and at least $\frac{n^3}{6}$ arithmetic operations per iteration that are required by the standard method.

Schnabel and Chow [1988] compare an implementation of their tensor method for unconstrained optimization to a standard method based on a quadratic model, on the problems from Moré, Garbow, and

Iillstrom [1981]. Both algorithms use the same, fairly standard trust region strategy. A summary of their computational results is given in Table 4.10. The lines labeled "$p \geq 1$" show the results when p, the number of previous iterates whose function and gradient is interpolated is interpolated, is allowed to xceed one, while the lines labeled "p=1" show the results when only information from the most recent iterate is used.

These results indicate that once again, the tensor method seems to obtain a considerable improvement in efficiency from using a rather small additional amount of information. Indeed, the tensor method or unconstrained optimization rarely chooses $p > 1$ even when we allow this possibility, and Table 4.10 hows if we require $p = 1$, then the results do not change appreciably. Thus it appears that we can achieve ubstantial savings at a very low cost.

It is not yet clear to us why tensor methods for both unconstrained optimization and nonlinear quations achieve rather large improvements in efficiency from utilizing a rather small additional amount f information. For one class of problems, nonlinear equations with rank($F'(x_*)$) = $n-1$, the observed mprovement is probably due to a faster local convergence rate, since Frank [1984] shows that the tensor method achieves 3-step superlinear convergence as opposed to the linear rate of standard methods. This attern of convergence is observed in practice, and may also occur in the analogous case for unconstrained optimization. But the rate of convergence of the tensor methods is almost certainly not superior o standard methods when $F'(x_*)$, or $\nabla^2 f(x_*)$, is non-singular, and yet the tensor methods usually are onsiderably faster in these cases too. Our conjecture is that this improvement occurs because the

---

Table 4.10 -- Comparison of Tensor and Standard Methods for Unconstrained Optimization
(from Schnabel and Chow [1988])

Rank $\nabla^2 f(x_*)$	Value of p in Tensor Method	Tensor Better	Standard Better	Two Methods Similar	Average Ratio of Tensor Iterations to Standard Iterations
n	$p \geq 1$	27	4	5	0.65
	$p = 1$	27	1	7	0.62
$n-1$	$p \geq 1$	27	2	4	0.56
	$p = 1$	30	3	2	0.55
$n-2$	$p \geq 1$	26	2	4	0.59
	$p = 1$	28	5	3	0.58

directions of two consecutive steps are often rather similar in practice, so that the additional information provided by the tensor model, which is along the previous step direction, turns out to be especially useful. In any case, it appears to us that the use of higher order models seems fruitful, and that additional research on such approaches is warranted.

5. Parallel Methods for Unconstrained Optimization

It is becoming clear that the fastest computers in the future will be parallel computers. Therefore, it is natural that there has been increased interest recently in designing parallel optimization algorithms. In this section we discuss one aspect of this work, parallel methods for the general unconstrained optimization problem. Some of this research will be seen to consist of the development of new algorithms motivated by the consideration of parallel computers, while other parts simply consist of determining good ways to parallelize existing sequential methods.

Before we begin, we need to clarify which of the various types of parallel computers we are concerned with. At a high level, currently available parallel computers can be divided into three categories, vector computers, processor arrays, and multiprocessors. Vector computers can perform pairwise addition or multiplication of long vectors of numbers quickly. They are best suited to a low level, fine grain type of parallelism. Processor arrays (SIMD computers) are computers with many processors that can perform the same operation in lock step on multiple data at the same time. Thus they are also suited to a fairly low level, fine to medium grain type of parallelism. By multiprocessors (MIMD computers), we mean computers that can perform entirely different operations on different data at the same time. These include both shared memory multiprocessors, where all the processors share access to some of the memory, and distributed memory multiprocessors, where each processor has its own memory and there is no shared memory. These computers support a higher level, medium to coarse grain type of parallelism.

Among these various types of parallel architectures, the type of optimization problems we consider in this paper seem best suited to solution on (MIMD) multiprocessors. This is because parallel algorithms for general unconstrained optimization problems generally seem to entail a high level, coarse grain type of parallelism. Largely this is because the evaluation of the objective function $f(x)$, an atomic operation in these algorithms, is itself often a lengthy calculation. Furthermore, if we wish to perform two or more computations of $f(x)$ concurrently, this will often require an MIMD computer since for two different values of x, the program that evaluates $f(x)$ may well execute different sequences of instructions. Thus the discussion in the remainder of this section is mainly oriented towards parallel computation on multiprocessors, though the specific type of multiprocessor, for example shared or distributed memory, is relatively unimportant.

The remainder of this section will discuss parallel methods that are related to the most commonly used general purpose unconstrained optimization method, the BFGS method with a line search. A high level description of this method is given in Algorithm (5.1).

Since presumably we are interested in parallel optimization in order to solve expensive problems, we need to focus on why the BFGS method can be expensive. There are two main reasons. One is that the function and derivative evaluations are expensive. The second is that the linear algebra computations, namely updating the Hessian approximation and calculating the search direction, are expensive because the number of variables is large.

There are several obvious possibilities for adapting these potentially expensive portions of the BFGS algorithm to parallel computers. The first is to parallelize the individual calculations of the objective function $f(x)$. This may or may not be feasible, depending upon the form of $f(x)$, the availability of pre-existing parallel routines to calculate it, or the interest of the user in devising new parallel routines to evaluate it. In any case, it is outside the scope of basic optimization research. Thus for the remainder of this section we will assume that $f(x)$ is evaluated on one processor.

Algorithm 5.1 -- Iteration of the BFGS Method with Line Search

given current iterate x_c , $f(x_c)$,
   $g_c = \nabla f(x_c)$ or finite difference approximation,
   H_c = symmetric positive definite Hessian approximation

calculate search direction d_c :
   solve $H_c\, d_c = -g_c$
line search:
   find $\lambda > 0$ for which $x_+ = x_c + \lambda_c\, d_c$ is satisfactory next iterate
evaluate $g_+ = \nabla f(x_+)$ or finite difference approximation if not done in previous step
   ($f(x_+)$ is evaluated in previous step)
decide whether to stop; if not
update Hessian approximation by BFGS formula (2.4):
   $H_+ = H_c$ + rank two matrix

The second possibility is to parallelize the dominant linear algebraic computations of the BFGS method, mainly the Hessian updates and the calculations of the step directions. We consider this topic in Section 5.1. While it basically consists of ways to parallelize the existing method, it brings up some interesting fundamental issues in unconstrained optimization.

The third possibility is to perform multiple evaluations of the objective function $f(x)$ concurrently. The possibilities here range from calculating several components of the finite difference gradient concurrently to new algorithms that make use of concurrent function evaluations. We consider both of these possibilities in Section 5.2.

The material in this section is taken largely from Schnabel [1987] and Byrd, Schnabel, and Shultz [1988a, b].

5.1 Parallel Methods for the Linear Algebra Calculations in Secant Methods

In this section, we consider the parallelization of the linear algebra calculations in the standard BFGS method. This topic is interesting because it leads us to re-examine various ways for implementing these calculations. It also leads us to interpret a recent proposal of Han [1986] in a different light.

It is convenient to view the linear algebra calculations involved in an iteration of the BFGS method as the update of the Hessian approximation followed by the calculation of the next search direction. Table 5.2 summarizes four different ways of implementing these calculations, along with their costs. These four options arise from choosing between two possibilities each for two orthogonal attributes of how the Hessian approximation may be stored. First, one can store either an approximation to the Hessian itself or an approximation to the inverse of the Hessian. Second, this approximation can either be kept in the straightforward, unfactored, form, or one can store a matrix M from a factorization MM^T of the Hessian or the inverse Hessian approximation.

The upper left variant in Table 5.2 is the most obvious way to implement the BFGS method. It simply consists of performing the update (2.4) to obtain H_+, and then performing a Cholesky factorization of H_+ to solve $H_+d_+=-g_+$ for d_+. This is the only variant that requires $O(n^3)$ arithmetic operations; all the others require only $O(n^2)$.

The upper right and lower left variants in Table 5.2 are the two well-known ways for performing the linear algebra calculations in a BFGS algorithm in $O(n^2)$ operations. The lower left variant was the first known $O(n^2)$ implementation of the BFGS; it consists of using the formula that gives the effect of

Table 5.2 -- Four Possible Implementations of the Linear Algebra Calculations :

$$H_{k+1} = H_k + \text{rank-two-matrix}$$
$$\text{solve } H_{k+1} d_{k+1} = -g_{k+1} \text{ for } d_{k+1}$$

	Matrix Stored Unfactored	Matrix Stored Factored
	$(H_k$ stored, updated to $H_{k+1})$	$(L_k$ lower triangular stored, for which $H_k = L_k L_k^T$, updated to L_{k+1} lower triangular for which $H_{k+1} = L_{k+1} L_{k+1}^T)$
Direct (H_k) Update	$H_{k+1} = H_k + \text{rank-two}$ Cholesky factor H_{k+1} 2 triangular solves to find d_{k+1}	$J_{k+1} = L_k + \text{rank-one}$ $J_{k+1} = Q_{k+1} L_{k+1}$ by Givens rotations 2 triangular solves to find d_{k+1}
	$\dfrac{n^3}{6} + 2n^2$	$6n^2$ $(2.5n^2)$
	$(H_k^{-1}$ stored, updated to $H_{k+1}^{-1})$	$(M_k$ stored for which $H_k^{-1} = M_k M_k^T$, updated to M_{k+1} for which $H_{k+1}^{-1} = M_{k+1}M_{k+1}^T)$
Inverse (H_k^{-1}) Update	$H_{k+1}^{-1} = H_k^{-1} + \text{rank-two}$ Matrix-vector multiply to find d_{k+1}	$M_{k+1} = M_k + \text{rank-one}$ 2 Matrix-vector multiples to find d_{k+1}
	$2n^2$	$4n^2$

the BFGS update on the inverses of H_c and H_+,

$$H_+^{-1} = H_c^{-1} + \frac{(s-H_c^{-1}y)s^T + s(s-H_c^{-1}y)^T}{s^T y} - \frac{s s^T (s-H_c^{-1}y)^T y}{(s^T y)^2}, \tag{5.3}$$

and then multiplying $-g_+$ by H_+^{-1} to get d_+. The upper right variant was subsequently discovered (Gill and Murray [1972], Goldfarb [1976]), and has become the favored $O(n^2)$ method for implementing the BFGS. It consists of directly updating the Cholesky factorization LL^T of H_c into the Cholesky factorization $L_+L_+^T$ of H_+. This is accomplished by expressing the BFGS update as a rank one change to the Cholesky factor L, resulting in a non-triangular matrix J_+, and then transforming J_+ into a lower triangular matrix L_+ through a series of $2n-2$ Given's rotations. Then d_+ is found by using two backsolves to solve $L_+L_+^T d_+ = -g_+$ for d_+. An advantage of this implementation is that by always keeping a Cholesky factorization, it implicitly guarantees that the Hessian approximation is numerically positive definite.

(For details on the operation counts, see Byrd, Schnabel, and Shultz [1988b].)

The fourth, lower right, variant is closely related to the upper right variant. It consists of updating a factorization MM^T of H_c^{-1} directly into a factorization $M_+ M_+^T$ of H_+^{-1} by making a rank one change to M, and then multiplying $-g_+$ by $M_+ M_+^T$ to obtain d_+. Since the savings that would occur in these matrix vector multiplications if M_+ were triangular would be exceeded by the cost of restoring M_+ to triangularity at each iteration, it is preferable to omit the Given's rotations that are used in the upper right variant and simply store M as a full matrix.

To our knowledge, the lower right variant has hardly been considered in the form in which we have just described it, but it is equivalent to a suggestion of Han [1986]. Han proposes implementing the linear algebra of the BFGS on parallel computers by keeping a matrix Z_c for which

$$Z_c^T H_c Z_c = I, \tag{5.4}$$

and then updating it to a matrix Z_+ for which

$$Z_+^T H_+ Z_+ = I \tag{5.5}$$

by making a rank one change to Z_c to obtain Z_+. But since equations (5.4) and (5.5) are equivalent to $H_c^{-1} = Z_c Z_c^T$ and $H_+^{-1} = Z_+ Z_+^T$, this is simply another way of interpreting the lower right variant in Table 5.2, and it is easy to confirm that the calculations in Han's method and the lower right variant in Table 5.2 are identical.

These four possible ways of implementing the linear algebra calculations of the BFGS method produce identical results in exact arithmetic. They differ in the number of arithmetic operations they require, in their suitability to parallel computation, and perhaps in their accuracy when implemented in computer arithmetic. None appears to be best in all these regards. There has long been a belief that the two left hand variants in Table 5.2 might have problems in finite precision arithmetic because they might produce numerically non-positive definite Hessian approximations. Powell [1987] has also shown that the two bottom variants may have difficulties in finite precision arithmetic handling very badly scaled problems. The combination of these two observations thus leads one to favor the upper right variant and this is the conventional wisdom.

In several computational tests that have compared the use of these four variants in a BFGS method, however, the differences between them have turned out to be negligible, even on somewhat extreme test problems. These include tests performed by Grandinetti [1978], Connolly and Nocedal [1987], and tests we recently performed using all four variants of the BFGS shown in Table 5.2 in the UNCMIN code. In our tests, on no problem was there more than a 1-2% variation between the costs of solving it using these four different implementations.

Thus we consider any of the variants in Table 5.2 to be a suitable candidate for the implementation of the BFGS method on a parallel computer. Among these four, the bottom two variants clearly are the most amenable to parallelization. Both require only matrix vector multiplications and rank one updates, computations that can be easily and efficiently parallelized over a large variety of parallel architectures and problem sizes. The upper right variant appears far more difficult to parallelize efficiently due to the need to perform the $2n-2$ Given's rotations, on vectors of size 2 through n, sequentially. The upper left variant is clearly too expensive to consider since some of the much less expensive sequential variants parallelize well. The difference in efficiency between the two bottom variants on parallel computers can be expected to be about a factor of two (but only 4/3 on a distributed memory computer, see Byrd, Schnabel, and Shultz [1988b]), so we would tend to favor the lower left variant, but experiments using the two bottom variants on parallel computers need to be performed.

Thus the discussion of this section has shown that the consideration of parallel computation may lead to a different choice for implementing the linear algebraic computations in the BFGS method than is generally made on sequential computers. It also makes it clear that, to conclusively resolve this issue, optimization researchers need to carefully consider whether and when the inverse updates, factored or unfactored, have practical computational deficiencies.

5.2 Utilizing Concurrent Function Evaluations for Unconstrained Optimization

In this section we review some possibilities for utilizing concurrent function evaluations in general purpose unconstrained optimization algorithms. These possibilities include both the parallelization of standard optimization algorithms, and new algorithms motivated by the consideration of parallel computation. Our point of departure is the standard BFGS method described in Algorithm 5.1.

Recall first the pattern of function and gradient evaluations in a standard secant method like Algorithm 5.1. Each iteration performs one or more *trial point function evaluations*, evaluations of $f(x_c + \lambda d_c)$ for some values of the step length parameter λ, within the line search. The last of these is at the successful next iterate x_+, and is followed by the gradient evaluation at x_+. Generally this is the only gradient evaluation in the iteration. In our experience, the average number of trial point function evaluations per iteration over the course of the algorithm tends to be between 1.2 and 1.5.

We are most concerned with performing these function and derivative evaluations efficiently on a parallel computer if they are expensive. In our practical experience, when $f(x)$ is expensive, $\nabla f(x)$ usually is calculated by the finite difference approximation

$$\nabla f(x)_i = \frac{f(x + h_i e_i) - f(x)}{h_i}, \quad i = 1, \cdots, n \tag{5.6}$$

where e_i is the ith unit vector and h_i is an appropriately chosen finite difference step size. This requires n additional evaluations of $f(x)$. Clearly these function evaluations can be performed concurrently on a parallel computer. Thus on a machine with p processors, the finite difference gradient evaluation requires $\left\lceil \frac{n}{p} \right\rceil$ concurrent function evaluation steps, steps where each processor performs at most one evaluation of $f(x)$ concurrently.

Thus if $f(x)$ is expensive and the number of processors, p, is much less than n, then simply parallelizing the finite difference gradient calculation by performing groups of p function evaluations concurrently leads to very good speedups. If $p >= n$, however, the overall speedup for all the function and gradient evaluation steps will be no better than about half of optimal. This is because all but one of the processors will be unused during the trial point function evaluations, and there will be at least as many steps devoted to trial point function evaluations as to gradient evaluations.

This leads to the obvious question, "How can we utilize additional processors while evaluating $f(x_c + \lambda d_c)$ on one processor in the line search?" There are two obvious possibilities. The most common suggestion (see for example Dixon and Patel [1982], or Lootsma [1984]) is to supplement the line search by evaluating $f(x)$ at $p-1$ other points simultaneously. We refer to this strategy as a *multiple point search*.

A second possibility, originally suggested by Schnabel [1987], is to use the remaining $p-1$ processors to evaluate (part of) $\nabla f(x_c + \lambda d_c)$ by finite differences *before* it is known whether this gradient value will be needed. We refer to this as a *speculative* (partial) gradient evaluation. If $x_c + \lambda d_c$ is accepted as the next iterate, as it usually will be, then the function evaluations that have been performed in the speculative gradient evaluation have all turned out to be necessary ones. If $x_c + \lambda d_c$ is not accepted as the next iterate, then the speculative gradient evaluation has been unnecessary, although Byrd, Schnabel, and Shultz [1988b] describe a new method that makes some beneficial use of this information.

We believe that using the extra processors that are available, while evaluating $f(x_c + \lambda d_c)$ in the line search, to perform a speculative finite difference gradient evaluation will usually result in a more efficient parallel algorithm than utilizing the extra processors to perform a multiple point search. Indeed, in order for the multiple point search to be the superior strategy, it would need to reduce the overall number of iterations by a factor of almost $\frac{n+p}{n}$ (assuming $p \le n+1$). In this case, the multiple point search would lead to a better sequential algorithm as well. We consider this unlikely, especially since it is usually very hard to improve upon the choice $\lambda = 1$ close to the solution. We cannot make a more definite assessment because, to our knowledge, proposers of multiple point searches have not provided the data that would allow us to compare their strategies to speculative gradient evaluations.

If $p \le n+1$ and gradients are evaluated by finite differences, then we feel there is little more to say about utilizing concurrent function evaluations for unconstrained optimization. The remaining interesting cases are when $p > n+1$, or when $f(x)$ and $\nabla f(x)$ are naturally evaluated together utilizing one processor. In either of these cases, additional processors are available while $f(x)$ and $\nabla f(x)$ are being evaluated. Byrd, Schnabel, and Shultz [1988a, b] consider several ways to utilize these additional processors. In the remainder of this section we briefly summarize some of their suggestions.

If there are enough processors so that the function, gradient, and finite difference Hessian can all be evaluated at once, then the analogous idea to that discussed above is to perform speculative finite difference Hessian evaluations. (Finite difference approximation of the Hessian requires $\frac{n^2+3n}{2}$ additional evaluations of $f(x)$ or n additional evaluations of $\nabla f(x)$.) This means evaluating all of $\nabla^2 f(x_c + \lambda d_c)$, as well as $\nabla f(x_c + \lambda d_c)$, concurrently with the trial point function evaluation $f(x_c + \lambda d_c)$. If $x_c + \lambda d_c$ is accepted as the new iterate x_+, as it usually will be, all the speculative evaluations perform useful work. Note that this is basically just a way to parallelize a standard second derivative method, although new algorithmic features could be introduced to attempt to the speculative gradient and Hessian information at unsuccessful trial points.

If there are more processors than necessary to evaluate $f(x)$ and $\nabla f(x)$ simultaneously, but not enough to evaluate the entire finite difference Hessian simultaneously as well, then Byrd, Schnabel, and Shultz [1988a, b] propose utilizing the remaining processors to perform speculative evaluation of *part* of the finite difference Hessian at each iteration. This leads to the consideration of new optimization algorithms. Now we briefly review this work

Byrd, Schnabel, and Shultz [1988a] consider many alternatives for evaluating part of the Hessian at each iteration, and for incorporating this information into an optimization algorithm. The strategy for evaluating part of the Hessian which they find to be most successful is simply to evaluate some set of columns of the Hessian at each iteration, with these sets sweeping through all the columns as the iterations proceed. At any given iteration, this means one evaluates $z_i = \nabla^2 f(x) \cdot e_i$ for some values

$i \in I_c \subseteq [1,n]$. The strategy for incorporating this information that Byrd, Schnabel, and Shultz [1988a, b]
find best is to first update H to $\bar{H}$ by the standard BFGS update (2.4), utilizing the normal secant equation
(2.3). Then they update $\bar{H}$ to H_+ by performing a multiple BFGS update (Schnabel [1983]), which causes
H_+ to satisfy $H_+ e_i = z_i$ for each $i \in I_c$. The rationale for inserting the information in this order is that the
normal BFGS update gives information about the Hessian between x_c and x_+, while the finite difference
Hessian information gives values at x_+. Thus the finite difference information is inserted last.

Byrd, Schnabel, and Schultz [1988b] show that an algorithm that utilizes the above strategy for
incorporating part of the finite difference Hessian at each iteration retains the superlinear convergence
rate of the BFGS method. Of course the intent is that the new method should perform better than the
BFGS in practice, since it uses more information, but this is about the best theoretical result one can
expect.

The results of extensive computational tests utilizing the partial Hessian methods described above
are reported in Byrd, Schnabel, and Shultz [1988a, b]. Table 5.7 summarizes their results on a set of test
problems with $n=20$. The second row shows the speedup of the new method, that evaluates the function
gradient, and q columns of the finite difference Hessian at each trial point, over a parallel BFGS method
that just evaluates the function and gradient at each trial point, under the assumption that there are enough
processors to evaluate $f(x)$, $\nabla f(x)$, and q columns of the finite difference Hessian concurrently. The
third row shows the speedups of the same partial Hessian methods, under the same assumptions about the

---

**Table 5.7 -- Average Speedup of a Method Using Speculative Partial Hessian Evaluations
on a Test Set with $n=20$ (speed measured in function evaluations)**
(from Byrd, Schnabel, and Shultz [1988b])

Number of columns of Hessian calculated at each iteration	0	1	2	3	4	5	20*
Speedup over Parallel BFGS utilizing speculative gradient evaluation	(1.0)	1.8	2.3	2.9	3.0	3.2	6.0
Speedup over Sequential BFGS with finite difference gradients	17.5	31.5	40.3	50.8	52.5	56.0	105
Number of processors required to evaluate $f(x)$, $\nabla f(x)$, $\nabla^2 f(x)$ concurrently from function values	21	42	62	81	99	116	231

*Newton's method

number of processors, in comparison to a standard sequential BFGS method that performs finite differ-ence gradient evaluations and utilizes only one processor. These numbers in the third row are 17.5 times as large as the numbers in the second row. This reflects the fact that the parallel BFGS is 17.5 times as fast as the sequential BFGS on these problems on the average, or equivalently, that there are an average of 1.21 trial point function evaluations per iteration on these problems. The final row shows the number of processors that would be necessary to implement the parallel partial Hessian algorithms for each value of q, when the finite difference gradient and Hessian are computed from function values.

The results in Table 5.7 show that the new method that utilizes otherwise idle processors to perform speculative partial Hessian evaluations is more efficient, in terms of concurrent function evaluation steps, than a standard method that doesn't utilize these processors, but that these improvements are not propor-tional to the amount of new information that is being utilized. This is to be expected, however, because as the table shows, Newton's method, which uses $\frac{n}{2}$ times as much information as the BFGS method, is only about 6 times as fast on the average on these problems. Combining the last two rows of Table 5.7 shows that the speedups are at least 0.45 of the optimal in all cases, which is considered reasonable in parallel computation. These results also indicate, however, that there is still an opportunity to find new algorithms that make better use of partial Hessian information.

6. Concluding Remarks

Recent research in unconstrained optimization has demonstrated that even though the field has reached a fairly mature state, many interesting and potentially fruitful research possibilities still exist. The research on the BFGS, DFP, and their convex combinations show that there are still fundamental theoreti-cal issues concerning secant updates that need to be better understood. The research on the SR1 and on updates beyond the BFGS ($\phi<1$) illustrates that there may be updates that perform better than the BFGS in practice, but that we need to understand these updates better in both practice and theory. It appears increasingly unlikely, however, that any new secant update will result in a large improvement, say greater than 25%, over the BFGS.

The research on derivative tensor methods for nonlinear equations and unconstrained optimization has led to surprisingly large improvements in efficiency over standard methods, often in the range of 30-50%. Since these methods constitute just one of many possibilities for incorporating additional function or derivative information into optimization algorithms by using nonstandard models, this research indi-cates that there may be other interesting unexplored possibilities for utilizing nonstandard models in unconstrained optimization algorithms.

Research in parallel optimization methods is still in its infancy. Much of the research described in Section 5 can be considered to be fairly straightforward adaptations or generalizations of standard methods. We consider it likely that more novel parallel optimization research will result from considering parallel methods for specific classes of large scale optimization problems. Indeed, the consideration of specific classes of large scale optimization problems, which has been neglected in this paper, probably holds many of the future challenges for research in unconstrained optimization.

7. References

C. G. Broyden [1970], "The convergence of a class of double-rank minimization algorithms", Parts I and II, *Journal of the Institute of Mathematics and its Applications* 6, pp. 76-90, 222-236.

C. G. Broyden, J. E. Dennis Jr., and J. J. Moré [1973], "On the local and superlinear convergence of quasi-Newton methods", *Journal of the Institute of Mathematics and its Applications* 12, pp. 223-246.

R. H. Byrd, J. Nocedal, and Y. Yuan [1987], "Global convergence of a class of quasi-Newton methods on convex problems", *SIAM Journal on Numerical Analysis* 24, pp. 1171-1190.

R. H. Byrd, R. B. Schnabel, and G. A. Shultz [1988a], "Using parallel function evaluations to improve Hessian approximation for unconstrained optimization", *Annals of Operations Research* 14, pp. 167-193.

R. H. Byrd, R. B. Schnabel, and G. A. Shultz [1988b], "Parallel quasi-Newton methods for unconstrained optimization", Technical Report No. CU-CS-396-88, Department of Computer Science, University of Colorado at Boulder, to appear in *Mathematical Programming*.

A. R. Conn, N. I. M. Gould, and Ph. L. Toint [1986], "Testing a class of methods for solving minimization problems with simple bounds on the variables," Research Report CS-86-45, Faculty of Mathematics, University of Waterloo, Waterloo, Canada.

A. R. Conn, N. I. M. Gould, and Ph. L. Toint [1987], "Convergence of quasi-Newton matrices generated by the symmetric rank one update", Report 87/12, Department of Computer Sciences, University of Waterloo, Waterloo, Canada.

A. R. Conn, N. I. M. Gould, and Ph. L. Toint [1988], "Global convergence of a class of trust region algorithms for optimization with simple bounds", *SIAM Journal on Numerical Analysis* 25, pp. 433-460.

K. A. Connolly and J. Nocedal [1987], "Quasi-Newton methods for parallel processing", Technical Report NAM 05, Department of Electrical Engineering and Computer Science, Northwestern University.

W. C. Davidon [1959], "Variable metric methods for minimization", Argonne National Laboratory Report ANL-5990 (rev.).

W. C. Davidon [1975], "Optimally conditioned optimization algorithms without line searches", *Mathematical Programming* 9, pp. 1-30.

D. W. Decker and C. T. Kelley [1980a], "Newton's method at singular points I", *SIAM Journal on Numerical Analysis* 17, pp. 66-70.

D. W. Decker and C. T. Kelley [1980b], "Newton's method at singular points II", *SIAM Journal on Numerical Analysis* 17, pp. 465-471.

, E. Dennis Jr. and R. B. Schnabel [1983], *Numerical Methods for Nonlinear Equations and Unconstrained Optimization*, Prentice-Hall, Englewood Cliffs, New Jersey.

. C. W. Dixon and K. D. Patel [1982], "The place of parallel computation in numerical optimization IV, parallel algorithms for nonlinear optimisation", Technical Report No. 125, Numerical Optimisation Centre, The Hatfield Polytechnic.

R. Fletcher [1970], "A new approach to variable metric methods", *Computer Journal* 13, pp. 317-322.

R. Fletcher [1980], *Practical Method of Optimization, Vol. 1, Unconstrained Optimization,* John Wiley and Sons, New York.

R. Fletcher and M. J. D. Powell [1963], "A rapidly convergent descent method for minimization", *Computer Journal* 6, pp. 163-168.

P. D. Frank [1984], "Tensor methods for solving systems of nonlinear equations", Ph.D. Thesis, Department of Computer Science, University of Colorado at Boulder.

P. E. Gill and W. Murray [1972], "Quasi-Newton methods for unconstrained optimization", *Journal of the Institute of Mathematics and its Applications* 9, pp. 91-108.

P. E. Gill, W. Murray, and M. H. Wright [1981], *Practical Optimization,* Academic Press, London.

D. Goldfarb [1970], "A family of variable metric methods derived by variational means", *Mathematics of Computation* 24, pp. 23-26.

D. Goldfarb [1976], "Factorized variable metric methods for unconstrained optimization", *Mathematics of Computation* 30, pp. 796-811.

L. Grandinetti [1978], "Factorization versus nonfactorization in quasi-Newtonian methods for differentiable optimization," Report N5, Dipartimento di Sistemi, Universita della Calabria.

A. O. Griewank [1980], "Starlike domains of convergence for Newton's method at singularities", *Numerische Mathematik* 35, pp. 95-111.

A. O Griewank and Ph. L. Toint [1982], "On the unconstrained optimization of partially separable functions", in *Nonlinear Optimization 1981,* M. J. D. Powell ed., Academic Press, London, pp. 301-312.

S. P. Han [1986], "Optimization by updated conjugate subspaces," in *Numerical Analysis: Pitman Research Notes in Mathematics Series 140,* D.F. Griffiths and G.A. Watson, eds., Longman Scientific and Technical, Burnt Mill, England, pp. 82-97.

R. H. F. Jackson [1983], "Tensors, polyads, and high-order methods in factorable programming", Ph.D. Dissertaion, The George Washington University, Department of Operations Research, Washington, DC.

R. H. F. Jackson and G. P. McCormick [1986], "The polyadic structure of factorable function tensors with application to high-order minimization techniques", *Journal of Optimization Theory and Applications* 51, pp. 63-94.

F. A. Lootsma [1984], "Parallel unconstrained optimization methods," Report No. 84-30, Department of Mathematics and Informatics, Technische Hogeschool Delft.

G. P. McCormick [1983], *Nonlinear Programming: Theory, Algorithms and Applications*, John Wiley & Sons, New York.

J. J. Moré, B. S. Garbow, and K. E. Hillstrom [1981], "Testing unconstrained optimization software", *ACM Transactions on Mathematical Software* 7, pp. 17-41.

J. J. Moré and D. C. Sorensen [1983], "Computing a trust region step", *SIAM Journal on Scientific and Statistical Computing* 4, pp. 553-572.

M. R. Osborne and L. P. Sun [1988], "A new approach to the symmetric rank-one updating algorithm", Report NMO/01, Department of Statistics, Australian National University, Canberra, Australia.

M. J. D. Powell [1976], "Some global convergence properties of a variable metric method without exact line searches", in *Nonlinear Programming,* R. Cottle and C. Lemke, eds. AMS, Providence, R.I., pp. 53-72.

M. J. D. Powell [1986], "How bad are the BFGS and DFP methods when the objective function is quadratic?", *Mathematical Programming* 34, No. 1, pp. 34-47.

M. J. D. Powell [1987], "Updating conjugate directions by the BFGS method", *Mathematical Programming* 38, pp. 29-46.

R. B. Schnabel [1983], "Quasi-Newton methods using multiple secant equations," Technical Report CU-CS-247-83, Department of Computer Science, University of Colorado at Boulder.

R. B. Schnabel [1987], "Concurrent function evaluations in local and global optimization," *Computer Methods in Applied Mechanics and Engineering* 64, pp. 537-552.

R.B. Schnabel and T. Chow [1988], "Tensor methods for unconstrained optimization," (in preparation).

R. B. Schnabel and P. Frank [1984], "Tensor methods for nonlinear equations", *SIAM Journal on Numerical Analysis* 21, pp. 815-843.

R. B. Schnabel, J. E. Koontz, and B. E. Weiss [1985], "A modular system of algorithms of unconstrained minimization", *ACM Transactions on Mathematical Software* 11, pp. 419-440.

D. F. Shanno [1970], "Conditioning of quasi-Newton methods for function minimization", *Mathematics f Computation* 24, pp. 647-657.

G. A. Shultz, R. B. Schnabel, and R. H. Byrd [1985], "A family of trust region based algorithms for unconstrained minimization with strong global convergence properties", *SIAM Journal on Numerical nalysis,*22, pp. 47-67.

A. Stachurski [1981], "Superlinear convergence of Broyden's bounded Θ-class of methods", *Mathematical Programming*, 20, pp. 196-212.

. Wolfe [1969], "Convergence conditions for ascent methods", *SIAM Review*, 11, pp. 226-235.

. Wolfe [1971], "Convergence conditions for ascent methods II: some corrections", *SIAM Review*, 13, p. 185-188.

Y. Zhang and R. P. Tewarson [1986], "On the development of algorithms superior to the BFGS method in Broyden's family of updates", Department of Applied Mathematics and Statistics Report AMS 86-69, State University of New York, Stony Brook, New York.

A Survey of Bundle Methods for Nondifferentiable Optimization[1]

Krzysztof C. KIWIEL

Systems Research Institute, Polish Academy of Sciences, Newelska 6, 01-447 Warsaw, Poland

Abstract. We review descent methods for constrained nondifferentiable optimization problems that find search directions via quadratic programming subproblems based on polyhedral approximations to the problem functions derived from their accumulated subgradients. Several techniques for handling nondifferentiability have been proposed, and their extensions to constrained problems attempt to exploit some successful approaches to smooth nonlinear programming (successive QP, exact penalty function, gradient reduction, gradient projection and feasible directions methods). Recent research on these questions is described and discussed.

. Introduction

We are concerned with numerical methods for solving the following nondifferentiable optimization (NDO) problem

$$\text{minimize} \quad f(x) \quad \text{over all} \quad x \in \mathbb{R}^N \tag{1.1a}$$

$$\text{satisfying} \quad F_i(x) \leq 0 \quad \text{for} \quad i=1,\ldots,m_I , \tag{1.1b}$$

$$h_i(x) \leq 0 \quad \text{for} \quad i \in I_h , \tag{1.1c}$$

where the (possibly nonsmooth) functions f and F_i are locally Lipschitz continuous on $\mathbb{R}^N$, each h_i is affine and $|I_h| < \infty$. We assume that at each x in $S_h = \{x \in \mathbb{R}^N : h_i(x) \leq 0 \ \forall i \in I_h\}$ we can evaluate f and F_i together with their single subgradients (called generalized gradients by Clarke [C2]) $g_f(x) \in \partial f(x)$ and $g_{Fi}(x) \in \partial F_i(x)$; moreover, f and F_i should be upper subdifferentiable (see [B2]). This property is likely to hold in most applications (see [M3]). Thus the potential application area of general-purpose NDO methods is vast (see [L8]).

Our aim is to give a short review of bundle methods for problem (1.1). Such methods construct piecewise linear (polyhedr-

---

[1] This research was supported by Project CPBP.02.15.

M. Iri and K. Tanabe (eds.), *Mathematical Programming*, 263–282.
© 1989 by KTK Scientific Publishers, Tokyo.

al) models of the problem functions by accumulating their subgra
dients and use quadratic programming (QP) for search directio
finding. They were introduced by Lemarechal [L1, L2] and Wolf
[W1] for convex unconstrained problems, and by Mifflin [M2] fo
constrained nonconvex ones. Their state of the art was describe
in 1985 in [K9]. Therefore, we shall concentrate on subsequen
improvements, mainly in the constraint handling techniques. W
refer the reader to [K9, L7, L8, Z1] for more extensive intro
ductions to bundle methods and larger bibliographies.

Space limitations prevent us from discussing some othe
classes of NDO methods. Simple subgradient algorithms, subgradien
algorithms with space dilation and ellipsoid methods are studie
in several books [D2, M8, M9, S2] and many papers, e.g. [A1, A3
B4, C1, E1, G2, L8, N1, S1]. NDO methods for semi-infinite optimi
zation problems (i.e. those involving finitely many variables an
infinitely many smooth constraints) are reviewed in [P2]. We mus
stress that particular classes of NDO problems (e.g. minima
problems and composite NDO problems with $f(x)=f_0(x)+\phi(\psi(x))$
where $f_0:\mathbb{R}^N\to\mathbb{R}$ and $\psi:\mathbb{R}^N\to\mathbb{R}^M$ are smooth, and $\phi:\mathbb{R}^M\to\mathbb{R}$ is a pointwis
maximum of a finite number of linear functions) can be solved mor
efficiently by specialized methods (see, e.g. [F1, P5, B1, B5, P2
P3]). There are problems (e.g. with $f(x)=\max\{0,\min(x_1,x_2)\}$) whic
are not composite NDO problems and for which the bundle method
may fail to find an approximate solution; such problems may b
solved by the quasidifferentiable optimization methods in [D1
K10, K15, K20, M1].

Our review is organized as follows. We start in Section
with methods for unconstrained problems, concentrating on the ne
promising technique of [K23]. Extensions to linearly constraine
problems are discussed in Section 3, and to nonlinear constraint
in Section 4. In Section 5 we mention some interesting works tha
cannot be reviewed here because of lack of space.

We use the following notation. We denote by $\langle\cdot,\cdot\rangle$ and $|\cdot|$
respectively, the usual inner product and norm in $\mathbb{R}^N$. We use x
to denote the i-th component of the vector x. Superscripts ar
used to denote different vectors. For simplicity, from now on w
assume that the problem functions of (1.1) are convex (extension
to nonconvex problems are mentioned in Section 5). For $\varepsilon\geq 0$, th
ε-subdifferential of f at x is defined by

$$\partial_{\varepsilon} f(x) = \{ p \in \mathbb{R}^N : f(y) \geq f(x) + \langle p, y-x \rangle - \varepsilon \quad \forall \; y \in \mathbb{R}^N \}.$$

e denote by ∂f the ordinary subdifferential $\partial_0 f$.

. **Unconstrained convex minimization**

e start by reviewing the proximal bundle method of [K23] for inimizing a convex function f. The algorithm generates a sequence $\{x^k\}_{k=1}^{\infty} \subset \mathbb{R}^N$ that should converge to a minimizer of f, and a equence of trial points $\{y^k\} \subset \mathbb{R}^N$ at which the linearizations f f

$$\overline{f}(x;y) = f(y) + \langle g_f(y), x-y \rangle \quad \forall \; x \in \mathbb{R}^N \tag{2.1}$$

re calculated, where $x^1 = y^1$ is a given starting point.

At the k-th iteration the polyhedral approximation to f

$$\hat{f}^k(x) = \max\{ \overline{f}(x;y^j) : j \in J_f^k \} \quad \text{for all} \quad x \tag{2.2}$$

ith $J_f^k \subset \{1, \ldots, k\}$ satisfies $\hat{f}^k \leq f$ and $\hat{f}^k(y^j) = f(y^j)$ for all $j \in$ J_f^k. The next trial point is chosen as

$$y^{k+1} = \text{argmin} \{ \hat{f}^k(x) + u^k |x-x^k|^2/2 : x \in \mathbb{R}^N \}, \tag{2.3}$$

here $u^k > 0$ is intended to keep y^{k+1} in the region where $\hat{f}^k$ hould be close to f. A *serious step* from x^k to $x^{k+1} = y^{k+1}$ ccurs if y^{k+1} is significantly better than x^k in the sense that

$$f(y^{k+1}) \leq f(x^k) + m_L v^k, \tag{2.4}$$

here $m_L \in (0, 0.5)$ is a parameter and

$$v^k = \hat{f}^k(y^{k+1}) - f(x^k) \tag{2.5}$$

s the predicted descent. Otherwise, a *null step* $x^{k+1} = x^k$ nproves the polyhedral approximation $\hat{f}^{k+1}$. Namely, $\hat{f}^{k+1}$ is elected with $J_f^{k+1} = \{k+1\} \cup \tilde{J}_f^k$ and $\tilde{J}_f^k \subset J_f^k$ so that *a posteriori*

$$\hat{f}_s^k(x) = \max\{ \overline{f}(x;y^j) : j \in \tilde{J}_f^k \} \quad \text{for all} \quad x \tag{2.6}$$

ay replace $\hat{f}^k$ in (2.3) and (2.5), i.e. $\hat{f}_s^k$ incorporates all the tive linearizations, and the inactive ones may be dropped to ave storage without impairing convergence to a solution.

More specifically, the direction $d^k = y^{k+1} - x^k$ and v^k solve the subproblem

$$\text{minimize} \quad u^k |d|^2/2 + v \quad \text{cver all} \; (d,v) \in \mathbb{R}^{N+1} \tag{2.7a}$$

satisfying $-\alpha^k_{f,j} + \langle g^j_f, d \rangle \leq v$ for $j \in J^k_f$,   (2.7)

where $g^j_f = g_f(y^j)$, $f^k_j = \bar{f}(x^k, y^j)$ and $\alpha^k_{f,j} = f(x^k) - f^k_j = \alpha_f(x^k, y^j)$, and
where $\alpha_f(x,y) = f(x) - \bar{f}(x;y) \geq 0$ is the linearization error. The
corresponding Lagrange multipliers $\lambda^k_j \geq 0$, $j \in J^k_f$, sum up to 1 and
can be chosen so that $\hat{J}^k_f = \{ j : \lambda^k_j \neq 0 \}$ satisfies $|\hat{J}^k_f| \leq N+1$ (since
(2.7) involves N+1 variables; see [K12]). By the Kuhn-Tucker
optimality conditions, it suffices to choose any $\tilde{J}^k_f \supseteq \hat{J}^k_f$.

We note that $v^k \leq 0$ yields the optimality estimate

$$f(x) \geq f(x^k) - |u^k v^k|^{1/2} |x - x^k| + v^k \quad \forall x \in \mathbb{R}^N.   (2.8)$$

so that the method may terminate if v^k is sufficiently small.

For practical efficiency it is crucial to choose the weights
u^k in a way that reduces sensitivity to the scaling of
(multiplication of f by a positive constant). Although in theory
any constant $u^k \equiv \bar{u} > 0$ suffices for global convergence [K9
Chapter 2.5], if $\bar{u}$ is too large for a given f then we have
small $|v^k|$ and $|d^k|$, almost all steps serious and slow descent
On the other hand, a small $\bar{u}$ produces large $|v^k|$ and $|d^k|$, and
each serious step is followed by many null steps. The case of
too large u^k may be detected by

$$f(y^{k+1}) \leq f(x^k) + m_R v^k   (2.9)$$

with $m_R \in (m_L, 1)$ (cf. (2.4)), i.e. u^k may be decreased if $\hat{f}$
is close to f at y^{k+1}. To update u^k, assume temporarily that
N=1, f is quadratic and strictly convex, and k=1, so tha
$v^k = \langle g(x^k), d^k \rangle = -u^k |d^k|^2$. Then simple calculations show that th
Hessian of f equals

$$u^{k+1}_{int} = 2 u^k (1 - [f(y^{k+1}) - f(x^k)] / v^k) ,   (2.10)$$

and if $u^2 = u^2_{int}$ then $\hat{f}^2 + u^2 | . - x^2 |^2 / 2 = f$, and $x^3 = y^3$ is optimal
For a general f, (2.9) with $m_R \in (1/2, 1)$ ensures that u^{k+1}_{int}/u^k
$2(1 - m_R) < 1$. However, it is useful to safeguard our quadrati
interpolation by letting

$$u^{k+1} = \max\{ u^{k+1}_{int}, u^k/10, u_{min} \} ,   (2.11)$$

where u_{min} is a small positive constant. A safeguarded versio
of (2.10) is also used for increasing u^{k+1} after a null step whe
the error $\alpha_f(x^k, y^{k+1})$ is sufficiently large (see [K23]).

To sum up, the main idea is to choose the weight u^{k+1} by safeguarded quadratic interpolation so that it estimates the curvature of f between x^k and y^{k+1}. Our preliminary computational experience indicates that this technique can significantly decrease the number of objective evaluations required to reach a desired accuracy in the optimal f-value (this measure of efficiency is suitable for typical applications: see Lemarechal [L6]).

The algorithm of [K23] is globally convergent in the sense that either $\{x^k\}$ converges to a point in $X = \text{Argmin } f$, or $X=\emptyset$ and $\lim|x^k|=\infty$. We may add that Auslender [A4] studied convergence of modifications of the algorithms in [K9, Chapter 2] with bounded $\{u^k\}$ for an inf-compact f, but he neither established global convergence nor gave practical rules for choosing u^k (see also Fukushima [F2]). Moreover, he also observed that the results of [R2] on proximal point methods, which might ensure global convergence, can hardly be applied in practice when f is neither smooth nor polyhedral. The connection with proximal point methods stems from the fact that for $\phi^k(x)=f(x)+u^k|x-x^k|^2/2$ we have $\phi^k(y^{k+1})\leq\min\phi^k+\varepsilon^k$ with $\varepsilon^k=f(y^{k+1})-\hat{f}^k(y^{k+1})\leq-(1-m_L)v^k$ and $|y^{k+1}-\text{argmin }\phi^k|\leq(\varepsilon^k/u^k)^{1/2}$ if $x^{k+1}\neq x^k$ (see [A4]). Incidentally, this explains why we say that relations (2.1)-(2.5) define a proximal bundle method.

We shall now comment on the conjugate subgradient methods (see [L2, W1, M2, B3, P4] for their different versions, and [K9] for globally convergent modifications). They use in (2.3) $u^k\equiv 1$ and a simplified version of $\hat{f}^k$ that corresponds to assuming arbitrarily that all the linearization errors $\alpha^k_{f,j}$ are zero. This loss of accuracy in $\hat{f}^k$ requires special resets for dropping obsolete linearizations and slows down convergence in practice [L6]. Similar comments apply to the method of [G1], which uses (2.3) with $u^k\equiv 1$ and $\alpha^k_{f,j}$ replaced by $f(y^j)-f(x^k)$.

The ε-steepest descent methods, which were introduced in [L3] and later modified in [L11] and [K9], are based on the following observation. If $\text{Nr}\partial_\varepsilon f(x^k)$ is the shortest element of $\partial_\varepsilon f(x^k)$ denoted by p^k for some $\varepsilon\geq 0$, then either $p^k=0$ and $f(x^k)\leq\inf f+\varepsilon$, or $p^k\neq 0$ and there exists a stepsize $t>0$ along the direction $d^k=-p^k$ from x^k such that $f(x^k+td^k)\leq f(x^k)-\varepsilon$ [L8]. In practice the algorithm uses the inner approximation $\partial_\varepsilon\hat{f}^k(x^k)\subseteq$

$\partial_\varepsilon f(x^k)$ with $\varepsilon = \varepsilon^k > 0$ and computes

$$-d^k = p^k = \sum_{j \in J_f^k} \lambda_j^k g_f^j \qquad (2.12$$

by finding multipliers λ_j^k that

$$\text{minimize} \quad | \sum_{j \in J_f^k} \lambda_j g_f^j |^2 / 2 , \qquad (2.13)$$

$$\text{subject to } \lambda \geq 0, \sum_{j \in J_f^k} \lambda_j = 1, \sum_{j \in J_f^k} \lambda_j \alpha_{f,j}^k \leq \varepsilon^k.$$

Several techniques for choosing ε^k were proposed in [L3, L6], whereas [K9,Chapter 7] gave ones that ensure global convergence However, the resulting algorithms seem to be less efficient than the proximal bundle method of [K23]. In particular, their line searches usually require more than one f-evaluation per iteration, in contrast to the simple acceptance test (2.4).

We may add that the subgradient selection strategy (i.e. choosing $\tilde{J}_f^{k+1} = \{k+1\} \cup \tilde{J}_f^k$ with $\tilde{J}_f^k$ containing $\{j: \lambda_j^k \neq 0\}$) can also be used in the conjugate subgradient methods and the ε-steepest descent methods [K9]. This requires storing at least N+2 subgradients (the points y^j need not be stored because the linearization values f_j^k may be updated). One may also employ the subgradient aggregation strategy of [K9] to decrease storage and work per iteration at the cost of slower convergence.

It is worth adding that the proximal bundle method will find an optimal solution in a finite number of iterations when f is polyhedral and satisfies a mild regularity condition; moreover, such conditions are not required for establishing finite convergence of specialized versions of the method (see [K16] and [R3]). The other methods do not have this finite termination property in the polyhedral case.

Specialized QP subroutines are required for implementing the above-mentioned methods, since their search direction finding subproblems are not strictly convex and have a particular structure that makes general-purpose QP methods rather inefficient in terms of additional workspace, stability and work per iteration (see [K21]). Subproblem (2.13) may be solved by the method of [M4], whereas its version with $\varepsilon^k = \infty$ (employed by the conjugate subgradient methods) may be solved by the simpler method of [W2]. Subproblem (2.7) may be handled by the dual QP methods of [K12] nd [R3]. These methods may loose accuracy when the weigth u^k ecomes small, and then the dual method of [K21] is recommended. owever, the latter method requires workspace of order $1.5N^2$, hereas the former ones need only $N^2/2$.

Linearly constrained convex minimization

ll the three classes of bundle methods of the preceding section an be extended to handle the linearly constrained convex problem

$$\text{minimize} \quad f(x) \quad \text{over all} \quad x \in S_h. \tag{3.1}$$

ormally, this amounts to replacing f in the various preceding ormulae by the extended objective $f + \delta$, where δ is the indi- ator function of S_h ($\delta(x)=0$ if $x \in S_h$, $\delta(x)=\infty$ otherwise). This rick turns the algorithms into feasible point methods. i.e. with $x^k \in S_h$ and $\{y^k\} \in S_h$ (where, of course, $x^1 = y^1 \in S_h$), so that f need ot be evaluated outside S_h.

The resulting extension of the proximal bundle method of K5, K18, K23] is, perhaps, the most natural one. Namely, subprob- em (2.3) becomes

$$y^{k+1} = \text{argmin} \{ \hat{f}^k(x) + u^k|x-x^k|^2/2 : x \in S_h^N \} \tag{3.2}$$

cf. (3.1)), and (2.7) is augmented with the constraints

$$h_i(x^k) + \langle \nabla h_i, d \rangle \leq 0 \quad \forall \ i \in I_h, \tag{3.3}$$

hereas the optimality estimate (2.8) takes the form

$$f(x) \geq f(x^k) - |u^k v^k|^{1/2} |x - x^k| + v^k \quad \forall \ x \in S_h^N. \tag{3.4}$$

t is worth observing that (in theory) the method is insensitive o the constraint scaling (as well as to the scaling of f when he weigth updating technique of (2.10) and (2.11) is used; see K23]). Moreover, the method retains both global convergence and he finite termination property in the polyhedral case.

Subproblem (2.7), (3.3) may be solved by the QP subroutines f [K12, R3]. These subroutines are quite efficient. Still when here are many linear constraints some work could be saved at irection finding by considering only almost active constraints. .e. those indexed by

$$I_h(x^k, \varepsilon_a^k) = \{ \ i \in I_h : \ h_i(x^k) \geq -\varepsilon_a^k \ \}$$

for some activity tolerance $\varepsilon_a^k > 0$. A reasonable choice of ε that does not impair global convergence is $\varepsilon_a^k = -v^k$ or $\varepsilon_a^k = \max\{-v^k$ $10^{-3}\}$. However, the gain in effort at direction finding could b outweigted by an increase in the number of iterations required t find an acceptable solution (cf. [N2]).

The same extensions can be introduced in the versions of th conjugate subgradient methods given in [K9], and the proofs o [K5] and [K16] may be used to show that these extensions do no impair their global convergence. Such extensions have not bee described in the literature, probably because they would be les efficient than the proximal bundle method sketched above.

We now pass to extensions of the ε-steepest descent methods The one in [S4] uses $d^k = -\mathrm{Nr} \partial_{\varepsilon^k}[\hat{f}^k + \delta](x^k)$, which boils down t finding multipliers λ_j^k and ν_i^k that

$$\text{minimize} \quad | \ \sum_{j \ \in \ J_f^k} \lambda_j \ g_f^j + \sum_{i \ \in \ I_h} \nu_i \ \nabla h_i \ |^2 / 2 \ , \tag{3.5a}$$

$$\text{subject to} \quad \lambda \geq 0, \ \nu \geq 0, \ \sum_{j \ \in \ J_f^k} \lambda_j + \sum_{i \ \in \ I_h} \nu_i = 1. \tag{3.5b}$$

$$\sum_{j \ \in \ J_f^k} \lambda_j \ \alpha_{f,j}^k - \sum_{i \ \in \ I_h} \nu_i \ h_i(x^k) \leq \varepsilon^k \tag{3.5c}$$

and setting

$$-d^k = p^k = \sum_{j \ \in \ J_f^k} \lambda_j^k \ g^j + \sum_{i \ \in \ I_h} \nu_i^k \ \nabla h_i. \tag{3.6}$$

The usual line search criteria are simply modified to ensur feasibility by testing feasible stepsizes only. The convergenc results of [S4] are quite weak and require both unbounded storag and the Slater constraint qualification ($\exists x: h_i(x) < 0 \ \forall i \in I_h$), whic eliminates equality constraints. These deficiencies can be remove by combining the results of [K9, Chapter 7] and [K18]. However even with such improvements the algorithm would probably be les efficient than the proximal bundle method of [K23]. In particular (3.5c) implies sensitivity to the scaling of h_i. A possible reme dy (which does not seem to have been considered in the literature is to consider subproblems of the form

$$\text{minimize} \quad \left| \sum_{j \in J_f^k} \lambda_j \, g_f^j \; + \; \sum_{i \in I_h} \nu_i \nabla h_i \right|^2 / 2 \; - \; \sum_{i \in I_h} \nu_i \, h_i(x^k)$$

$$\text{subject to} \quad \lambda \geq 0, \; \nu \geq 0, \quad \sum_{j \in J_f^k} \lambda_j = 1, \quad \sum_{j \in J_f^k} \lambda_j \, \alpha_{f,j}^k \leq \varepsilon^k.$$

The convergence analysis would still go through, but such subproblems would require a special QP subroutine (e.g. an extension of [K21]).

The ε-steepest descent method was extended in [B3] to problems of the form

$$\text{minimize} \quad f(x), \; \text{subject to} \quad Ax = b, \; x \geq 0,$$

where the matrix A is $M \times N$, $b \in \mathbb{R}^M$ and $\text{rank}(A) = M$. Following the classical reduced gradient motivation, partition A into an $M \times M$ nonsingular basis matrix B and the remaining $M \times (N-M)$ matrix $\mathbb{N}$. This partitions the variables into $(N-M)$ nonbasics x_N and M basics $x_B = B^{-1}(b - \mathbb{N} x_N)$, e.g. $x = (x_B, x_N)$ when $A = [B, \mathbb{N}]$. Hence the reduced objective

$$\phi(x_N) = f(\, B^{-1}b - B^{-1}\mathbb{N}x_N, \; x_N)$$

should be minimized subject to $x_N \geq 0$ and $x_B \geq 0$. If x^k is feasible and $x_B^k > 0$, we may temporarily ignore the constraint $x_B \geq 0$ and consider finding a feasible descent direction for ϕ at x_N^k for the problem

$$\text{minimize} \quad \phi(x_N), \; \text{subject to} \quad x_N \geq 0.$$

To this end, observe that the linearization of ϕ at y_N

$$\overline{\phi}(x_N; y_N) = \phi(y_N) + g_\phi(y_N)^T (x_N - y_N)$$

involves the subgradient $g_\phi(y_N) \in \partial\phi(y_N)$ given by

$$g_\phi(y_N) = g_N - \mathbb{N}^T B^{-1} g_B,$$

where $(g_B, g_N) = g_f(y)$ corresponds to the partition $y = (y_B, y_N)$ (since $\langle g, x-y \rangle = g_N^T(x_N - y_N) + g_B^T(x_B - y_B) = g_N^T(x_N - y_N) - g_B^T B^{-1}\mathbb{N}(x_N - y_N)$). Hence we may use the polyhedral approximation to ϕ (cf. (2.2))

$$\hat{\phi}^k(x) = \max\{\, \overline{\phi}(x_N; y_N^j) \; : \; j \in J_f^k \,\}$$

and the indicator function δ_N of $\{x_N: x_N \geq 0\}$ for finding the nonbasic component $d_N^k = -\text{Nr}\partial_{\varepsilon_k}[\hat{\phi}^k + \delta_N](x_N^k)$ of the direction $d^k = (d_B^k, d_N^k)$

with $d_B^k = -B^{-1}Nd_N^k$. (This involves replacing g_f^j by $g_\phi(y_N^j)$ and $h_i(x)$ by $-(x_N)_i$ for $i=1,\ldots,N-M$ in (3.5), and in (3.6) for calculating d_N^k.) The line search from x^k along d^k may reduce some basic component of x^{k+1} to zero, and then a new basis must be found. Under a standard nondegeneracy assumption, a suitable anti-jamming rule for basis changes was given in [B3].

It is worth observing that the subgradient reduction strategy of [B3] may also be employed in the framework of proximal bundle methods. This involves replacing (3.2) by

$$d_N^k = \text{argmin} \{ \hat\phi^k(x_N^k + d_N) + u^k|d_N|^2/2 : x_N^k + d_N \geq 0 \}$$

and setting $d_B^k = -B^{-1}Nd_N^k$. The QP formulation of the above sub-problem involves fewer variables than (2.7),(3.3), and hence may require less work. Yet this version should, in general, be less efficient than the former one in terms of the number of iterations because it neglects certain constraints at direction finding. In particular, since to maintain feasibility it has to use $y^{k+1}=x^k + t_R^kd^k$ with $t_R^k=\min\{t_{max}^k,1\}$ and $t_{max}^k=\max\{t\geq0: x^k+td^k\in S_h\}$, the step lengths $|x^{k+1}-x^k|$ may be smaller. Similar comments apply to the method of [B3] and the algorithm of [P1], which combines the ε-steepest descent method with (sub)gradient projection techniques for constraints of the form $Ax \leq b$. Of course, more numerical experience is needed to clarify the merits of all those methods.

4. Nonlinearly constrained convex minimization

Consider the convex program

$$\text{minimize} \quad f(x), \text{ subject to } F(x) \leq 0, \quad x \in S_h, \qquad (4.1)$$

where f and F are convex (e.g. $F=\max\{F_i: i=1,\ldots,m_1\}$). Suppose that the Slater constraint qualification holds (i.e. $F(x)<0$ for some $x\in S_h$) and S_h is bounded. Let $S_F=\{x: F(x)\leq0\}$ and $S=S_F\cap S_h$.

One way of solving (4.1) consists in minimizing the (partial) exact penalty function

$$e(x;c) = f(x) + c F_+(x)$$

over all $x\in S_h$, where $F_+(x)=\max\{F(x),0\}$ and $c>0$ is a sufficiently large penalty coefficient (not smaller than the Lagrange multiplier of (4.1) [B1]). Since a suitable value of c is seldom

nown in advance, we may try to choose $c = c^k$ by increasing it
.uring the minimization of $e(.,c)$. For instance, if we employ the
.ethod based on (3.2) (with f replaced by $e(.,c^k)$) then we may
.pdate c^k as follows [K7]. If $v^k \geq -\varepsilon_c^k$ and $F(x^k) > -v^k$, set $c^{k+1} =$
.c^k and $\varepsilon_c^{k+1} = \varepsilon_c^k/2$; otherwise, set $c^{k+1} = c^k$ and $\varepsilon_c^{k+1} = \varepsilon_c^k$, where
.$^1 > 0$ and $\varepsilon_c^1 > 0$ are arbitrary. Thus c^k is increased only if
.$(.,c^k)$ has been approximately minimized, as indicated by (3.4)
.pplied to $e(.,c^k)$ and $|v^k| \leq \varepsilon_c^k$ (with progressively smaller mini-
.ization tolerances ε_c^k), but x^k is significantly infeasible.

Moreover, (3.4) holds with with S_h replaced by S, so the
.ethod may terminate if both v^k and $F_+(x^k)$ are small.) The
.lternative "parameter-free" rule of [K19], which reads: "If $F(x^k)$
.$-c^k v^k$ set $c^{k+1} = c^k$, otherwise $c^{k+1} = 2c^k$ ", has a similar motiva-
.ion. Both rules ensure a finite growth of c^k (c^k=const for large
.), and global convergence to a solution of (4.1), including
.inite convergence in the polyhedral case [K16]. Other more abstr-
.ct penalty updating schemes are given in [A4], but it is not
.lear how to implement them.

A more efficient approach of [K8,K19] takes advantage of the
.act that the subgradients of f and F are available separately.
.lence we may use separate approximations to f and F (cf. (2.2))
.or constructing a more accurate model of $e(.,c^k)$ of the form

$$\hat{e}^k(x;c^k) = \hat{f}^k(x) + c^k \max\{ \hat{F}^k(x), 0 \}.$$

.or example, with $N/2$ linearizations of f and $N/2$ of F it has
.$^2/4$ pieces, whereas the preceding one would have only N (for
.torage of order N; the QP subproblems can still be solved by the
.ubroutines of [K12,R3], which produce at most $N+2$ nonzero Lagra-
.ge multipliers for subgradient selection). Our numerical experi-
.ents [K19] indicate that this exploitation of the structure of
.$(.,c^k)$ does increase the speed of convergence, sometimes very si-
.nificantly.

Observe that the preceding methods do not recognize the rela-
.ionship of $e(.,x^k)$ to problem (4.1), e.g. the fact that the
.second term of $e(.,x^k)$ is zero at the solution of (4.1). More-
.over, they are sensitive to the scaling of F. To mitigate these
.rawbacks, the constraint linearization method of [K16] chooses

$$y^{k+1} = \text{argmin}\{ \; \hat{f}^k(x) + |x-x^k|^2/2 : \hat{F}^k(x) \leq 0, \; x \in S_h \; \}, \qquad (4.2$$

and employs the exact penalty function only for testing whethe y^{k+1} is better than x^k. More specifically, it updates the penal ty by the rule: "If $c^k < \tilde{c}^k$ then $c^{k+1} = \max\{ \tilde{c}^k, 2c^k \}$. otherwis $c^{k+1}=c^k$ ", where $\tilde{c}^k$ is the Lagrange multiplier of the first con straint in (4.2), and $c^1=0$. Then $c^{k+1} \geq \tilde{c}^k$ implies that also

$$y^{k+1} = \text{argmin}\{ \; \hat{e}^k(x;c^{k+1}) + |x-x^k|^2/2 : \; x \in S_h \; \},$$

so that $v^k = \hat{e}^k(y^{k+1};c^{k+1})-e(x^k;c^{k+1})$ is the predicted descent fo an improvement test related to (2.4). In other words, we choos c^{k+1} so that the second term of $\hat{e}^k(y^{k+1};c^{k+1})$ vanishes. Th global convergence properties of this method are the same as thos of the preceding ones, but I prefer it in practice because of it insensitivity to constraint scaling.

We may add that our penalty updating schemes can be grafte onto the ε-steepest descent methods (especially those in [K9]) although this has not been described in the literature. A schem for one conjugate subgradient method was given in [P4].

Problem (4.1) may also be solved by the feasible point metho of [K9]. Briefly speaking, its k-th iteration is just one itera tion of the proximal bundle method applied to the minimizatio over S_h of the improvement function $H(x;x^k)=\max\{f(x)-f(x^k),F(x)$ approximated by $\hat{H}^k(x)=\max\{\hat{f}^k(x)-f(x^k),\hat{F}^k(x)\}$ in (3.2). Thus, i $F(x^k)<0$, we wish to find a feasible $(\hat{F}^k(y^{k+1}) \leq 0)$ direction d^k $y^{k+1}-x^k$ of descent $(\hat{f}^k(y^{k+1}) \leq f(x^k))$, whereas for $F(x^k)>0$, d should be a descent direction for F $(\hat{F}^k(y^{k+1}) \leq F(x^k))$, since the we would like to decrease the constraint violation. Feasible poin extensions of other methods may be derived in a similar manne [A3, M2, M5, P4, P5]. We note that such methods are usually les efficient than the exact penalty function methods because the cannot approach the boundary of the feasible set S at a fast rate.

5. Modifications and extensions

In this section we mention some important ideas that cannot b described in detail from lack of space.

First, suppose that we wish to minimize

$$f(x) = \sum_{i=1}^{n} f_i(x) \quad \text{for all} \quad x \in \mathbb{R}^N,$$

here $f_i:\mathbb{R}^N \to \mathbb{R}$ are convex functions with subgradients $g_{fi}(x) \in$ $f_i(x)$, for $i=1,\ldots,n$. By exploiting the structure of f one may ncrease the speed of convergence at the cost of more storage and ork per iteration. To this end, replace $\hat{f}^k$ in subproblem (2.3) y the approximations

$$\hat{f}^k(x) = \sum_{i=1}^n \hat{f}^k_i(x) , \qquad (5.1)$$

$$\hat{f}^k_i(x) = \max\{ \overline{f}_i(x;y^j) : j \in J^k_i \}$$

onstructed from the linearizations of f_i

$$\overline{f}_i(x;y) = f_i(y) + \langle g_{fi}(y),x-y \rangle$$

s in (2.1), where the sets J^k_i satisfying $\sum_{i=1}^n |J^k_i| \le M_g$ with $g \ge N+2n$ are selected by finding at most $N+n$ nonzero Lagrange ultipliers λ^k_{ij}, $j \in J^k_i$, $i=1,\ldots,n$, of the corresponding exten- ion of (2.7) (see Kiwiel [K21]). Thus $J^{k+1}_i = J^k_{si} \cup \{k+1\}$, where $j \in J^k_i : \lambda^k_{ij} \ne 0\} \subset J^k_{si} \subset J^k_i$, $i=1,\ldots,n$. This is just an exten- ion of our subgradient selection strategy of [K23]. Moreover, our echnique for the additional constraint of (3.1) can be employed. e note that Ruszczyński [R3] gives encouraging numerical results or a similar method (with $u^k \equiv 1$) for polyhedral problems.In act, when this method is applied to the decomposition of linear tochastic programs, it can be two or three times faster than pecialized simplex algorithms [R6]. It also leads in a natural ay to a new augmented Lagrangian decomposition method for angular inear programs [R5], and to a highly promissing parallel decompo- ition method for multistage stochastic programs [R4]. (See [R1] or another application of bundle methods to decomposition.)

The idea of exploiting an additive structure of the objective n the context of bundle NDO methods seems to have been used for he first time in [K8], and is applied in [F3,K3,K11,K17] to esign specialized methods for various problems. In particular, if $(x)=f_1(x)+f_2(x)$ with f_1 convex and f_2 nonconvex, but smooth (so hat $g_{f2}=\nabla f_2$), then it may be more efficient to use $\hat{f}^k_2(x)=\overline{f}_2(x;x^k)$ n (5.1) and a suitable line search [K11], rather than resort to eneral-purpose NDO methods for nonconvex programs (see below), ince there is no need to accumulate subgradient information about he smooth term f_2.

In some applications it may be difficult to evaluate f and g_f

accurately, but for any x and $\varepsilon>0$ one may calculate an ε-subgra-
dient of f at x and a number $f_\varepsilon(x)$ satisfying $|f_\varepsilon(x)-f(x)|\leq\varepsilon$. Then
the method of [K6] may be used (or that of [A4] if f(x) is avail-
able). An extension of [K6] to constrained problems will be reported elsewhere.

We would like to add that Nurminski [N3, N4, N5] has recently
proposed an interesting alternative approach to the derivation of
bundle methods for convex problems.

From lack of space we cannot describe extensions of bundle
methods to nonconvex problems with upper semidifferentiable
functions. Detailed descriptions may be found in [K4,K9,K13,K14,
K19,M5] for the proximal bundle methods, in [K9,L11] for the
ε-steepest descent methods, and in [B2,K9,M2,P4] for the conjugate
subgradient methods. In particular, [K19] and [P4] give exact
penalty methods for problems with additional smooth equality
constraints. Broadly speaking, two techniques are used to handle
nonconvexity: either subgradient locality measures, or subgradient
deletion rules for dropping obsolete information (see [K9]). It is
not clear which of them is preferable, and this requires further
research.

This review would be quite incomplete without at least men-
tioning some attempts to introduce variable metric techniques that
should ensure faster convergence. Lemarechal in his pioneering
work [L4] proposed an algorithm in which

$$y^{k+1} = \text{argmin} \{ \hat{f}^k(x) + \langle A_k(x-x^k), x-x^k\rangle/2 : x \in \mathbb{R}^N \}. \qquad (5.2)$$

where the N×N matrix A_k was intended to accumulate information
about the curvature of f around x^k. However, updating A_k by
the BFGS secant formula gave disappointing numerical results [L6].
Next, Kiwiel [K1] tried updating A_k as in the Shor method with
space dilation [S2] (this idea was also used in [A2]). Mifflin
[M6] proposed a more rigorous approach that could yield superlinear convergence for some functions (in which A_k is a convex
combination of some N matrices that approximate the Hessians of
smooth pieces of f); unfortunately, it is not known how to make
it implementable. We refer the reader to Lemarechal [L7, L8] for a
discussion of other attempts to construct superlinearly convergent methods, whose success so far has been restricted to the one-
dimensional case (see [M7]).

The basic question is, of course, how to define a generalized Hessian" for a nondifferentiable   f. One such notion has been made precise by Goffin [G2] in the context of solving linear inequalities via the ellipsoid method, which asymptotically generates a suitable matrix for (5.2). A similar idea appears in the method of [K22], in which

$$y^{k+1} = \text{argmin} \{ \hat{f}^k(x) + u^k \langle A_k(x-x_c^k), x-x_c^k \rangle /2 \; : \; x \in \mathbb{R}^N \},$$

where x_c^k is the center of the current ellipsoid

$$E_k = \{ x \in \mathbb{R}^N : \langle A_k(x-x_c^k), x-x_c^k \rangle \le 1 \}$$

which intersects the optimal set X and is updated by ellipsoidal techniques; see, e.g. [A1, B4]), and where u^k is a weight calculated by a trust region technique (see [M10]) so as to restrict the choice of y^{k+1} to E_k. However, the underlying ellipsoid method is quite slow in generating A_k, and it seems that it should be replaced by a Karmarkar-like projective or affine method [G3, G4]. Our common work with J.-L. Goffin and C. Lemarechal on this subject will (hopefully) be reported soon.

Finally, we would like to add that much of the progress that has taken place in the NDO bundle methods during the last decade was predicted in 1978 by C. Lemarechal [L5].

References

A1]   A. Agkül, *Topics in Relaxation and Ellipsoidal Methods*, Research Notes in Mathematics **97**, Pitman, Boston, 1984.

A2]   N.T. Aliev, "ε-Subgradient algorithms with space dilation", *Kibernetika* **4** (1986) 106–107.

A3]   E. Allen, R. Helgason and E. Shetty, "A generalization of Polyak's convergence result for subgradient optimization". *Mathematical Programming* **37** (1987) 309–317.

A4]   A. Auslender, "Numerical methods for nondifferentiable convex optimization", *Mathematical Programming Study* **30** (1986) 102– 126.

A5]   A. Auslender, J.T. Crouzeix and P. Fedit, "Penalty-proximal methods in convex programming", *Journal of Optimization Theory and Applications* **55** (1987) 1–21.

B1]   D. Bertsekas, *Constrained Optimization and Lagrange Multiplier Methods*, Academic Press, New York, 1982.

B2]   A. Bihain, "Optimization of upper semidifferentiable functions", *Journal of Optimization Theory and Applications* **44** (1984) 545–568.

B3]   A. Bihain, V.H. Nguyen and J.-J. Strodiot, "A reduced subgradient algorithm", *Mathematical Programming Study* **30** (1987) 127–149.

[B4]   R.B. Bland, D. Goldfarb and M.J. Todd, "The ellipsoid
       method: a survey", *Operations Research* **29** (1981) 1039-1091.
[B5]   J.V. Burke, "Descent methods for composite nondifferentiable
       optimization problems", *Mathematical Programming* **33** (1985)
       260-279.
[C1]   N.D. Chepurnoi, "Monotone methods with subgradient averaging
       and their stochastic finite-difference analogues", *Kiberne-
       tika* **1** (1988) 62-66 (in Russian).
[C2]   F.H. Clarke, *Optimization and Nonsmooth Analysis* (Wiley, New
       York, 1983).
[D1]   V.F. Demyanov and L.C.W. Dixon, eds., *Quasidifferentiable
       Calculus. Mathematical Programming Study* **29**, North-Holland,
       Amsterdam, 1986.
[D2]   V.F. Demyanov and L.V. Vasiliev, *Nondifferentiable Optimiza-
       tion*, Optimization Software Inc., New York, 1985.
[E1]   J.G. Ecker and A. Kupferschmidt, "A computational comparison
       of the ellipsoid method with several nonlinear programming
       algorithms", *SIAM Journal on Control and Optimization* **5**
       (1985) 657-674.
[F1]   R. Fletcher, *Practical Methods of Optimization, Vol.2,
       Constrained Optimization*, Wiley, New York, 1981.
[F2]   M. Fukushima, "A descent algorithm for nonsmooth convex
       programming", *Mathematical Programming* **30** (1984) 163-175.
[F3]   M. Fukushima, "A successive quadratic programming method for
       a class of constrained nondifferentiable optimization prob-
       lems", Technical Report #87024, Department of Applied Mathe-
       matics and Physics, Kyoto University, Kyoto, 1987.
[G1]   M. Gaudioso and M.F. Monaco, "A bundle type approach to the
       unconstrained minimization of convex nonsmooth functions",
       Mathematical Programming **23** (1982) 216-226.
[G2]   J.-L. Goffin, "Variable metric relaxation methods. part II:
       the ellipsoid method", *Mathematical Programming* **30** (1984)
       147-162.
[G3]   J.-L. Goffin, "Affine methods in nondifferentiable optimiza-
       tion", CORE Discussion Paper 8744, Center for Operations Re-
       search and Econometrics, Louvain-la-Neuve, Belgium, 1987.
[G4]   J.-L. Goffin and J.-P. Vial, "Cutting planes and column
       generation techniques with the projective algorithm", CORE
       Discussion Paper 8744, Center for Operations Research and
       Econometrics, Louvain-la-Neuve, Belgium, 1988.
[K1]   K.C. Kiwiel, "A variable metric method of centers for
       nonsmooth minimization", CP-81-27, International Institute
       for Applied Systems Analysis, Laxenburg, Austria, 1981.
[K2]   K.C. Kiwiel, "An aggregate subgradient method for nonsmooth
       convex minimization", *Mathematical Programming* **27** (1983)
       320-341.
[K3]   K.C. Kiwiel, "A descent algorithm for large-scale linearly
       constrained convex nonsmooth minimization". CP-84-15,
       International Institute for Applied Systems Analysis,
       Laxenburg, Austria, 1984.
[K4]   K.C. Kiwiel, "A linearization algorithm for nonsmooth mini-
       mization", *Mathematics of Operations Research* **10** (1985) 185-
       194.
[K5]   K.C. Kiwiel, "An algorithm for linearly constrained convex
       nondifferentiable minimization problems", *Journal of Mathe-
       matical Analysis and Applications* **105** (1985) 452-465.
[K6]   K.C. Kiwiel, "An algorithm for nonsmooth convex minimization
       with errors", *Mathematics of Computation* **45** (1985) 173-180.

K7]   K.C. Kiwiel, "An exact penalty function method for nonsmooth
      constrained convex minimization problems", *IMA Journal of
      Numerical Analysis* 5 (1985) 111–119.
K8]   K.C. Kiwiel, "Descent methods for nonsmooth convex con-
      strained minimization", in: V.F. Demyanov and D. Pallaschke,
      eds., *Nondiffertiable Optimization: Motivations and
      Applications,* Lecture Notes in Economics and Mathematical
      Systems 255 (Springer, Berlin, 1985) pp. 203–214.
K9]   K.C. Kiwiel, *Methods of Descent for Nondifferentiable Opti-
      mization* (Lecture Notes in Mathematics 1133, Springer,
      Berlin, 1985).
K10] K.C. Kiwiel, "A linearization method for minimizing certain
      quasidifferentiable functions", *Mathematical Programming
      Study* 29 (1986) 85–94.
K11] K.C. Kiwiel, "A method for minimizing the sum of a convex
      function and a continuously differentiable function",
      Journal of Optimization Theory and Applications 48 (1986)
      437–449.
K12] K.C. Kiwiel, "A method for solving certain quadratic prog-
      ramming problems arising in nonsmooth optimization". *IMA
      Journal of Numerical Analysis* 6 (1986) 137–152.
K13] K.C. Kiwiel, "A method of linearizations for linearly
      constrained nonconvex nonsmooth optimization". *Mathematical
      Programming* 34 (1986) 175–187.
K14] K.C. Kiwiel, "An aggregate subgradient method for nonsmooth
      and nonconvex minimization", *Journal of Computational and
      Applied Mathematics* 14 (1986) 391–400.
K15] K.C. Kiwiel, "Randomized search directions in descent meth-
      ods for minimizing certain quasidifferentiable functions",
      Optimization 17 (1986) 1–11.
K16] K.C. Kiwiel, "A constraint linearization method for nondif-
      ferentiable convex minimization", *Numerische Mathematik* 51
      (1987) 395–414.
K17] K.C. Kiwiel, "A decomposition method of descent for minimi-
      zing a sum of convex functions", *Journal of Optimization
      Theory and Applications* 52 (1987) 255–271.
K18] K.C. Kiwiel, "A subgradient selection method for minimizing
      convex functions subject to linear constraints". *Computing*
      39 (1987) 293–305.
K19] K.C. Kiwiel, "An exact penalty function method for nondif-
      ferentiable constrained minimization", Prace IBS PAN 155,
      Warszawa, 1987.
K20] K.C. Kiwiel, "Descent methods for quasidifferentiable mini-
      mization", *Applied Mathematics and Optimization* 18 (1988)
      163–180.
K21] K.C. Kiwiel, "A dual method for solving certain positive se-
      midefinite quadratic programming problems", *SIAM Journal on
      Scientific and Statistical Computing* (to appear).
K22] K.C. Kiwiel, "An ellipsoid trust region bundle method for
      nonsmooth convex minimization", *SIAM Journal on Control and
      Optimization* (to appear).
K23] K.C. Kiwiel, "Proximity control in bundle methods for convex
      nondifferentiable minimization", *Mathematical Programming*
      (to appear).
L1]   C. Lemarechal, "An algorithm for minimizing convex func-
      tions", in: J.L. Rosenfeld, ed., *Proceedings IFIP '74 Con-
      gress,* North–Holland, Amsterdam, 1974, pp. 552–556.
L2]   C. Lemarechal. "An extension of Davidon methods to nondif-

ferentiable problems", *Mathematical Programming Study* 3
(1975) 95-109.

[L3]   C. Lemarechal, "Bundle methods in nonsmooth optimization",
in: C. Lemarechal and R. Mifflin, eds., *Nonsmooth Optimization*, Pergamon Press, Oxford, 1978, pp. 79-102.

[L4]   C. Lemarechal, "Nonsmooth optimization and descent methods",
Research Report RR-78-4, International Institute of Applied
Systems Analysis, Laxenburg, Austria, (1977).

[L5]   C. Lemarechal, "Nonlinear programming and nonsmooth optimi-
zation - a unification", Rapport de Recherche No. 332,
Institut de Recherche d'Informatique et d'Automatique,
Rocquencourt, Le Chesnay, (1978).

[L6]   C. Lemarechal, "Numerical experiments in nonsmooth optimization", in: E.A. Nurminski, ed., *Progress in Nondifferentiable Optimization* (CP-82-S8, International Institute for
Applied Systems Analysis, Laxenburg, Austria, 1982)
pp.61-84.

[L7]   C. Lemarechal, "Constructing bundle methods for convex optimization", in: J.B. Hiriart-Urruty, ed., *Fermat Days 85: Mathematics for Optimization* (North-Holland, Amsterdam, 1986)
pp. 201-240.

[L8]   C. Lemarechal, "Nondifferentiable optimization", in: G.L.
Nemhauser, A.H.G. Rinooy-Kan and M.J. Todd, eds., *Handbook
for Operations Research*, North-Holland, Amsterdam (to
appear)

[L9]   C. Lemarechal and R. Mifflin, eds., *Nonsmooth Optimization*
(Pergamon Press, Oxford, 1978).

[L10]   C. Lemarechal and J.-J. Strodiot, "Bundle methods, cutting
plane algoritms and σ-Newton directions", in: V.F. Demyanov
and D. Pallaschke, eds., *Nondifferentiable Optimization: Motivations and Applications*, Lecture Notes in Economics and
Mathematical Systems 255, Springer, Berlin, 1985, pp. 25-33.

[L11]   C. Lemarechal, J.-J. Strodiot and A. Bihain, "On a bundle
algorithm for nonsmooth optimization", in: O.L. Mangasarian,
R.R. Meyer and S.M. Robinson, eds., *Nonlinear Programming 3*,
Academic Press, New York, 1981, pp. 245-281.

[M1]   D.Q. Mayne and E. Polak, "Algorithms for optimization problems with exclusion constraints", *Journal of Optimization
Theory and Applications* 51 (1986) 453-473.

[M2]   R. Mifflin, "An algorithm for constrained optimization with
semismooth functions", *Mathematics of Operations Research* 2
(1977) 191-207.

[M3]   R. Mifflin, "Semismooth and semiconvex functions in constrained optimization", *SIAM Journal on Control and Optimization*
15 (1977) 959-972.

[M4]   R. Mifflin, "A stable method for solving certain constrained
least-squares problems", *Mathematical Programming 16* (1979)
141-158.

[M5]   R. Mifflin, "A modification and an extension of Lemarechal's
algorithm for nonsmooth minimization", *Mathematical Programming Study* 17 (1982) 77-90.

[M6]   R. Mifflin, "Better than linear convergence and safeguarding
in nonsmooth optimization", in: P. Thoft-Christensen, ed.,
System Modelling and Optimization, Lecture Notes in Control
and Information Sciences 59, Springer, Berlin, 1984, pp.
321- 330.

[M7]   R. Mifflin and J.-J. Strodiot, "A bracketing technique to
ensure desirable convergence in univariate minimization",

Mathematical Programming (to appear).

M8] V.S. Mikhalevich, A.M. Gupal and V.I. Norkin. *Nonconvex Optimization Methods*, Nauka, Moscow, 1987 (in Russian).

M9] V.S. Mikhalevich, V. A. Trubin and N.Z. Shor, *Optimization Problems of Production-Transportation Planning*, Nauka, Moscow, 1986 (in Russian).

M10] J.J. More, "Recent developments in algorithms and software for trust region methods", in: A. Bachem, M. Grötschel and B. Korte, eds., *Mathematical Programming, The State of the Art, Bonn 1982*, Springer, Berlin, 1983, pp. 258–287.

N1] Yu. E. Nesterov, "Minimization methods for nonsmooth convex and quasiconvex functions", *Ekonomika i Matematicheskije Metody* **XX** (1984) 519–531 (in Russian).

N2] V.H. Nguyen and J.-J. Strodiot, "A linearly constrained algorithm not requiring derivative continuity". *Engineering Structures* **6** (1984) 7–11.

N3] E.A. Nurminski, "The ε-subgradient mapping and the convex optimization problem", *Kibernetika* **6** (1985) 61–63 (in Russian).

N4] E.A. Nurminski, "Global properties of ε-subgradient mappings", *Kibernetika* **1** (1986) 120–122 (in Russian).

N5] E.A. Nurminski, "Convex optimization problems with constraints", *Kibernetika* **4** (1987) 29–31 (in Russian).

P1] E. Panier, "An active set method for solving linearly constrained nonsmooth optimization problems". *Mathematical Programming* **37** (1987) 269–292.

P2] E. Polak, "On the mathematical foundations of nondifferentiable optimization in engineering design". *SIAM Review* **29** (1987) 21–89.

P3] E. Polak and D.Q. Mayne, "Algorithm models for nondifferentiable optimization", *SIAM Journal on Control and Optimization* **23** (1985) 477–491.

P4] E. Polak, D.Q. Mayne and Y. Wardi, "On the extension of constrained optimization methods from differentiable to nondifferentiable problems", *SIAM Journal on Control and Optimization* **25** (1983) 179–203.

P5] M.J.D. Powell, "General algorithms for discrete nonlinear approximation calculations", in: C.K. Chui, L.L. Schumaker and J.D. Ward, eds., *Approximation Theory IV*. Academic Press, New York, 1983, pp. 187–218.

R1] S.M. Robinson, "Bundle-based decomposition: Description and preliminary results", in: A. Prekopa, J. Szelezsan and B. Strazicky, eds., *System Modelling and Optimization*, Lecture Notes in Control and Information Sciences, Springer, Berlin, 1986, pp. 751–756.

R2] R.T. Rockafellar, "Monotone operators and the proximal point algorithm", *SIAM Journal of Control and Optimization* **14** (1976) 877–898.

R3] A. Ruszczyński, "A regularized decomposition method for minimizing a sum of polyhedral functions", *Mathematical Programming* **35** (1986) 309–333.

R4] A. Ruszczyński, "Parallel decomposition of multistage stochastic programming problems", Technical Report, Institute of Automatic Control, Warsaw University of Technology, Warsaw, 1988.

R5] A. Ruszczyński, "Regularized decomposition and augmented Lagrangian decomposition for angular linear programming problems", WP–88–88, International Institute for Applied Systems

Analysis, Laxenburg, Austria, 1988.

[R6]   A. Ruszczyński, "Regularized decomposition of stochastic programming programs: algorithmic techniques and numerical results", *Operations Research* (to appear).

[R7]   S.V. Rzhevski, "A conditional ε-subgradient method for solving the convex programming problem", *Kibernetika* **1** (1987) 69- 72 (in Russian).

[S1]   S. Sen and H.D. Sherali, "A class of convergent primal-dual subgradient algorithms for decomposable convex programs, *Mathematical Programming* **35** (1986) 279-297.

[S2]   N.Z. Shor, *Minimization Methods for Nondifferentiable Functions* (Springer, Berlin, 1985).

[S3]   J.-J. Strodiot and V.H. Nguyen, "On the numerical treatment of the inclusion $0 \in \partial f(x)$", in: J.J. Moreau, P.D. Panagiotopoulos and G. Strang, eds., *Nonsmooth Mechanics*, Birkhauser, Basel (to appear).

[S4]   J.-J. Strodiot, V.H. Nguyen and N. Heukemes, "ε-Optimal solutions in nondifferentiable convex programming and related questions", *Mathematical Programming* **25** (1983) 307-328.

[W1]   P. Wolfe, "A method of conjugate subgradients for minimizing nondifferentiable convex functions", *Mathematical Program-ming Study* **3** (1975) 145-173.

[W2]   P. Wolfe, "Finding the nearest point in a polytope", *Mathematical Programming* **11** (1976) 128-149.

[Z1]   J. Zowe, "Nondifferentiable optimization", in: K. Schittkowski, ed., *Computational Mathematical Programming* (Springer, Berlin, 1985) pp.323-356.

Fixed Point Algorithms for Stationary Point Problems

Yoshitsugu YAMAMOTO

Institute of Socio-Economic Planning, University of Tsukuba, Tsukuba, Ibaraki 305, Japan

ABSTRACT: Over several years, we have had a new class of fixed point algorithms called variable dimension algorithms. They were originally for finding a fixed point on the unit simplex and since then have been extended to the solution of systems of nonlinear equations. Recently several algorithms were developed tailored for the stationary point problem, which could be viewed as a fixed point problem without any boundary condition. We first demonstrate that the stationary point problem affords a unifying view on problems arising from various fields such as mathematical economics, game theory and mathematical programming. Then we review the basic and common ideas of the variable dimension algorithms, path-following and the primal-dual pair of subdivided manifolds by taking two algorithms for the stationary point problem and the equilibrium problem.

1. INTRODUCTION

The aim of this paper is to provide an introductory review of variable dimension algorithms for stationary point problems. The algorithm can be traced back to the works of Kuhn [K3] and Shapley [S3] but was substantially improved by van der Laan and Talman [LT1,2,3,4]. It is distinguished by three features: no artificial dimension, arbitrary starting point and generating simplices of variable dimensions. The algorithm was originally developed for finding Brouwer's fixed point

M. Iri and K. Tanabe (eds.), Mathematical Programming, 283–307.
© *1989 by KTK Scientific Publishers, Tokyo.*

on the unit simplex and then extended to the solution of systems of
nonlinear equations on R^n. In this process of extension several new
algorithms joined in and now it is not a sole algorithm but a class of
algorithms based on the same idea. See [T3,4],[K2],[KY2] and [B] for
introduction and [F2,3],[KY1,3],[L],[LS],[LT5,6,7,8,9],[T1],[TV],[W]
and [Y1,2,3,6] for algorithms and theory.

The stationary point problem, often referred to as the
variational inequality problem, is the one embracing problems of
economic equilibrium, nonlinear complementarity, nonlinear program and
traffic assignment. The early studies were carried out mainly for
infinite-dimensional problems in the area of control theory, for
which the reader is referred to the books by Kinderlehrer and
Stanpacchia [KS] and Glowinski et al.[GLT]. The problem can indeed be
reformulated as a fixed point problem through the projection, but
solving the original problem directly sounds more reasonable than
solving the reformulated one. Mathematical equivalence does not
always imply computational equivalence. Van der Laan and Talman's
adapting their variable dimension algorithm to the linear
complementarity problem with upper and lower bounds led a series of
works of developing a tailored algorithm for the stationary point
problems. It is closely related with the unproper integer labelling
[LTV2] and also with the generalization of Sperner's Lemma [F1],[F4,5]
and [Y6].

In the following sections after citing several problems that are
reduced to the stationary point problem, we will briefly review the
primal-dual pair of subdivided manifolds in Section 3 and take two
variable dimension algorithms to explain the basic and common idea.
We confine ourselves to the variable dimension algorithm, so the reader
is referred to the survey by Allgower and Georg [AG], the monograph by
Todd [T2] and the book by Garcia and Zangwill [GZ] for other algorithms
such as homotopy method.

2. EXISTENCE OF A STATIONARY POINT AND APPLICATIONS

Let K be a closed convex subset of R^n and let f be a continuous

function from K into R^n. A point s of K is called a stationary
point of the problem (f,K) if it satisfies the so-called variational
inequality:

(2.1) $\langle x - s, f(s) \rangle \geq 0$ for any $x \in K$.

This inequality geometrically means that f(s) is an inward normal to
K when s is on the boundary of K and s is a zero point of f
when s is in the interior of K. If we define the normal cone to
K at point x of R^n as

$$N_K(x) = \emptyset \qquad\qquad\qquad\qquad\qquad\qquad\qquad \text{if} \quad x \notin K;$$
$$\qquad = \{ y : \langle y, z-x \rangle \leq 0 \ \text{ for any } \ z \in K \} \quad \text{if} \quad x \in K,$$

then (2.1) will hold if and only if s satisfies the generalized
equation (see [R])

$$0 \in f(s) + N_K(s).$$

The following theorem (2.2) states the equivalence between the
existence of a stationary point and Brouwer's Fixed Point Theorem (see
for example [I]).

(2.2) Theorem: Suppose K is a nonempty compact convex subset of
R^n. Then the following two statements are equivalent:

(2.3a)   any continuous function g from K into K has a fixed
          point;

   b)   for any continuous function f from K into R^n there
          is a stationary point of (f,K).

Proof: Let $f(x) = g(x) - x$ and let s be a stationary point
of (f,K). Then since g maps K into K, it is readily seen that
$f(s) = 0$, i.e., s is a fixed point of g. To show the converse let
$r(z) = \text{argmin} \{ \|z-x\| : x \in K \}$ for each $z \in R^n$ and consider g(x)
$= r(x-f(x))$, where $\|.\|$ denotes Euclidean norm.   r is nonexpansive
and hence continuous (see for example [KS, Corollary 2.4]). By
construction g is also continuous and has a fixed point, which is

seen to be a stationary point of (f,K). //

Now let us demonstrate how the stationary point problem serves as a unifying model of various problems arising from the areas of mathematical economics, game theory and mathematical programming.

Exchange Economy

As Sonnenschein [S1] showed the crucial point whether a given function f from the n-dimensional unit simplex $S^n = \{ x : x \in R_+^{n+1}; \langle e,x \rangle = 1 \}$ into R^{n+1} is an excess demand function generated by aggregating individual utility maximizing behavior in some market is Walras' Law:

(2.4) $\langle x,f(x) \rangle = 0$ for any $x \in S^n$,

where e is an (n+1)-dimensional vector of ones and R_+^{n+1} is the nonnegative orthant of R^{n+1}. Therefore we assume here only (2.4) besides the continuity. Under this condition there is a price vector $s \in S^n$ satisfying $f(s) \le 0$. This price is called an equilibrium price. The existence of s is usually shown by applying Brouwer's Fixed Point Theorem (2.3a) to the function

$$g(x) = (x+f^+(x))/(1+\langle e,f^+(x) \rangle)$$

from S^n into itself, where $f_i^+(x) = \max \{0,f_i(x)\}$ for $i=1,\ldots,n+1$. We show below that a stationary point of $(-f,S^n)$ is an equilibrium price and (2.3b) also guarantees its existence.

(2.5) Theorem: Let f be a continuous function from the n-dimensional unit simplex S^n into R^{n+1} satisfying Walras' Law (2.4). Then a stationary point of $(-f,S^n)$ is an equilibrium price and vice versa.

Proof: Suppose s is a stationary point of $(-f,S^n)$ and so $f(s) \in N_{S^n}(s)$. Then

$$f_i(s) = \mu - \lambda_i, \quad \lambda_i \ge 0, \quad \lambda_i s_i = 0, \quad i=1,\ldots,n+1$$

for some μ and λ_i. By (2.4) and $s \in S^n$, $\mu = \mu \sum_i s_i = \langle s,f(s) \rangle = 0$.

Therefore $f(s) \leq 0$, that is s is an equilibrium. When $f(s) \leq 0$, by (2.4) $f_i(s) = 0$ if $s_i > 0$ and $f_i(s) \leq 0$ if $s_i = 0$. It is readily seen that $f(s) \in N_{S^n}(s)$, namely s is a stationary point of $(-f, S^n)$. //

When the economy involves production, the equilibrium is defined as follows: Let $A(p) \in R^{(n+1) \times m}$ be the technology matrix of input-output coefficients at price p, y denote the activity level, $d(p)$ denote the demand function at price p and b denote the endowment. Then a pair $(p, y) \in S^n \times R^m_+$ is called an equilibrium of this economy if

(2.6a) $d(p) - b - A(p)y \leq 0$;

   b) $A(p)^t p \leq 0$;

   c) $\langle y, A(p)^t p \rangle = 0$;

   d) $\langle p, A(p)y - d(p) + b \rangle = 0$.

(2.7) Theorem: Let $f(p) = b - d(p)$ and $K = \{ p : p \in S^n, A^t(p)p \leq 0 \}$. If $d(p) - b$ satisfies Walras' Law (2.4), an equilibrium of this economy is a stationary point of (f, K) and vice versa.

For this problem Mathiesen [M] proposed solving a sequence of linear complementarity problems obtained by taking a linear approximation of the problem and Dirven and Talman [DT4] proposed a simplicial fixed point algorithm for the problem with a constant technology matrix.

Noncooperative n-Person Game

Player j has $m_j + 1$ pure strategies and his expected loss is given by $L_j(x^1, x^2, \ldots, x^n)$ when player i chooses mixed strategy x^i for $i = 1, \ldots, n$ from his strategy space being an m_i-dimensional unit simplex S^{m_i}. Let the marginal loss of player j when he chooses his h^{th} pure strategy while others keep their mixed strategy be defined as $L_j(x^{-j}, e^h)$, where (x^{-j}, e^h) abbreviates $(x^1, \ldots, x^{j-1}, e^h, x^{j+1}, \ldots, x^n)$ and e^h denotes the hth unit vector of R^{m_j+1}. An equilibrium strategy is a point $s \in \Pi_i S^{m_i}$ satisfying

(2.8) $L_i(s) \leq L_i(s^{-i}, e^h)$ for any $h=1,\ldots,m_i+1$ and $i=1,\ldots,n.$

If we define

$$f_{ih}(x) = L_i(x) - L_i(x^{-i}, e^h);$$
$$f_i(x) = (f_{i1}(x),\ldots,f_{im_i+1}(x));$$
$$f(x) = (f_1(x),\ldots,f_n(x)),$$

the condition (2.8) is simply rewritten as

$$f(s) \leq 0.$$

By the definition of L_i we readily see that $\langle x^i, f_i(x) \rangle = 0$ for $i=1,\ldots,n$, and obtain the following theorem.

(2.9) Theorem: A stationary point of $(-f, \Pi_i S^{m_i})$ is an equilibrium strategy of the above game and vice versa.

In this example K is a cross product of several unit simplices. The same region appears in the competitive equilibrium problem of international trade market [L2] (see also [P1]). When the set K is a simplex or the cross product of several simplices as above, several path-following algorithms have been developed for (f, K) in [DET], [DLT], [DT1,2,3], [ELT], [ET], [LT5,9]. See also a survey [LT8]. Note that the normal cone $N_K(x)$ is known a priori for this kinds of K.

Nonlinear Program
Consider the following nonlinear program:

(2.10)   minimize $g_0(x)$ subject to $g_i(x) \leq 0$ for $i=1,\ldots,m.$

Under some appropriate constraints qualification (see for example [BS]) we see that a local minimizer x of this problem satisfies

(2.11a) $\nabla g_0(x) + Dg(x)u = 0;$
  b) $g_i(x) \leq 0$ for $i=1,\ldots,m;$
  c) $u \geq 0;$ $\langle u, g(x) \rangle = 0$

for Karush–Kuhn–Tucker multiplier vector $u = (u_1,\ldots,u_m)$, where $g(x)$
$= (g_1(x),\ldots, g_m(x))$ and $Dg(x)$ is its Jacobian matrix at x.

(2.12) Theorem: Suppose g_i is a convex function for $i=1,\ldots,m$.
Then under some appropriate constraint qualification a point
satisfying (2.11) is a stationary point of $(\nabla g_0,K)$ and vice versa,
where K is the feasible region of (2.10).

Robinson [R] proposed another formulation.

(2.13) Theorem: Let $f(x,u) = (\nabla g_0(x) + Dg(x)u, g(x))$ and K
$= R^n \times R^m_+$. Then (2.11) holds if and only if (x,u) is a stationary
point of $(-f,K)$.

Phan-huy-Hao [P2] also proposed an interesting method for reducing
nonliear programs to stationary point problems on the unit simplex of
multipliers (see also [D2]). We consider the problem (2.10) with a
constraint $x \in X$ added for some compact subset X of R^n. Given a
multiplier vector $\bar{u} = (u_0,u) = (u_0,u_1,\ldots,u_m)$ of the m–dimensional
unit simplex S^m of R^{1+m}, consider the problem with the aggregated
objective function:

(2.14)   minimize $\langle\bar{u},\bar{g}(x)\rangle$ subject to $x \in X$,

where $\bar{g}(x) = (g_0(x),g(x)) = (g_0(x),g_1(x),\ldots,g_m(x))$. Let us denote
by $x(\bar{u})$ the optimal solution which we assume for simplicity is
unique. Let

$$f_0(\bar{u}) = \langle u,g(x(\bar{u}))\rangle$$
$$f_i(\bar{u}) = - g_i(x(\bar{u})) \quad \text{for} \quad i=1,\ldots,m$$

and $f(\bar{u}) = (f_0(\bar{u}),f_1(\bar{u}),\ldots,f_m(\bar{u}))$.

(2.15) Theorem: Suppose there is a point z of X such that $g(z)$
< 0. Let $\bar{s} = (s_0,s) = (s_0,s_1,\ldots,s_m)$ be a stationary point of
(f,S^m). Then the optimal solution $x(\bar{s})$ of (2.14) with multiplier
$\bar{s}$ is an optimal solution of the original nonlinear program.

Proof: In the same way as in the proof of (2.5) there are μ
and $\lambda_0, \lambda_1, \ldots, \lambda_m$ such that

$$f_i(s) = \mu + \lambda_i, \quad \lambda_i \geq 0, \quad \lambda_i s_i = 0 \quad \text{for} \quad i=0,1,\ldots,m.$$

Therefore

$$(2.16) \qquad \mu + \lambda_0 = f_0(\bar{s}) = \langle s, g(x(\bar{s})) \rangle = -\sum_{i=1}^m s_i(\mu + \lambda_i) = \mu(s_0 - 1).$$

Suppose $s_0 > 0$.   Then $\lambda_0 = 0$ and $(2-s_0)\mu = 0$, which implies μ
$= 0$ and hence $f_i(\bar{s}) \geq 0$ for all i.   Thus

$$g(x(\bar{s})) \leq 0 \quad \text{and} \quad \langle s, g(x(\bar{s})) \rangle = 0.$$

Now let x be an arbitrary feasible point of the nonliear program.
Then

$$s_0 g_0(x(\bar{s})) = \langle \bar{s}, \bar{g}(x(\bar{s})) \rangle \leq \langle \bar{s}, \bar{g}(x) \rangle$$
$$= s_0 g_0(x) + \langle s, g(x) \rangle \leq s_0 g_0(x),$$

which yields the optimality of $x(\bar{s})$.   It remains to show that s_0
> 0.   Suppose the contrary, then $\mu = -\lambda_0/2$ by (2.16) and hence $f_0(\bar{s})$
$= \mu + \lambda_0 = \lambda_0/2 \geq 0$.   On the other hand for the point z in the
condition

$$f_0(\bar{s}) = \langle s, g(x(\bar{s})) \rangle = \langle \bar{s}, \bar{g}(x(\bar{s})) \rangle \leq \langle \bar{s}, \bar{g}(z) \rangle = \langle s, g(z) \rangle < 0.$$

This is a contradiction.   //

Generalized Complementarity Problem ([HP1],[K1],[S2])
By replacing the usual nonnegativity constraints of the
complementarity problems with partial orderings generated by a pair of
cones we obtain the following generalized complementarity problem of
finding a point s such that

$$s \in K, \quad f(s) \in K^+ \quad \text{and} \quad \langle s, f(s) \rangle = 0,$$

where K is a cone of R^n and $K^+ = \{ y : \langle y, x \rangle \geq 0 \text{ for any } x$
$\in K \}$.

(2.17) Theorem ([K1, Lemma 3.1]): A solution of the generalized
complementarity problem is a stationary point of (f,K) and vice
versa.

A typical approach for solving the stationary point problem
(f,K) is an iterative method generating a sequence of points each of
which is a stationary point of a linearized problem: x^{k+1} is given
as a(n) (approximate) solution of

$$0 \in f^k(x) + N_K(x) \equiv f(x^k) + A(x^k)(x-x^k) + N_K(x).$$

There are various methods depending on the choice of the matrix
$A(x^k)$. If $Df(x^k)$ or its approximation is chosen, we have Newton,
quasi-Newton, successive overrelaxation and linearized Jacobi methods.
A fixed symmetric positive definite matrix yields the projection
method. Newton-type methods have the same drawback as their
correspondents for ordinary equations. Namely, their convergence is
not global but local. See [PC] for convergence conditions.
The projection method is based on the fixed point formulation of the
stationary point used in the proof of Theorem (2.2) and hence is not
globally convergent unless the function f has some monotonicity.
Though the function f is a gradient mapping of some scalar valued
function only if its Jacobian matrix is symmetric at every point,
there are several algorithms exploiting the idea of the nonlinear
program. See for example [HM] and [IFI]. Instead of solving the
linearized problem on the entire set K one may solve it on a subset
K^k of K and update the subset so that the stationary point of
(f^k,K^k) may converge to a solution. A convex hull of several and
possibly few vertices of K is usually taken as K^k and one adds and
deletes vertices to obtain a new subset K^{k+1}. This is called the
simplicial decomposition method. The prcedure of updating K^k is very
crucial for the convergence. See for example [LH] and [PY]. For
further information of algorithms of these kinds the reader is referred
to the paper by Dafermos [D1] and the recent survey by Harker and Pang
[HP2] and their references.
    In the following section we show that the idea of path-following,

which is common in the area of fixed point algorithms, yields an
algorithm which converges under the assumption that f is continuous
and K is compact convex. For simplicity of discussion we will
assume that f is continuously differentiable.

3.   PDM AND THE BASIC MODEL

In this section we will give a brief review of subdivided manifolds,
a basic theorem for fixed point algorithms and a primal-dual pair of
subdivided manifolds introduced by Kojima and Yamamoto [KY1] as a
unifying framework for variable dimension algorithms.

We call an m-dimensional convex polyhedral set in R^k an m-cell
or a cell when there is no confusion. If a cell B is a face of cell
C, we write $B \prec C$. We call a finite or countable collection M of
m-cells a subdivided manifold if

(3.1a)   the intersection of two cells of M is either empty or
           their common face;

  b)   each (m-1)-cell of $\bar{M}$ lies in at most two m-cells of M;

  c)   M is locally finite: each point $x \in |M|$ has a neighbor-
           hood which intersects only finitely many cells of M,

where

$$\bar{M} = \{ \ B : B \ \text{is a face of some m-cell of} \ M \ \}$$
$$|M| = \bigcup \{ \ C : C \in M \ \}.$$

We call the collection of (m-1)-cells of $\bar{M}$ each of which lies in
exactly one m-cell of M the boundary of M and denote it by ∂M. A
continuous mapping h from $|M|$ into R^n is piecewise continuously
differentiable (abbreviated by PC^1) on M if the restriction of h
to each cell of M has a continuously differentiable extension.
We denote the Jacobian matrix of h at point x of cell C by
$Dh(x;C)$. A point $c \in R^n$ is a regular value of the PC^1 mapping h :
$|M| \to R^n$ if

$x \in B \prec C \in M$ and $h(x) = c$ imply $\dim\{Dh(x;C)y : y \in B\} = n.$

We see by Sard's theorem that almost every vector of R^n is a regular value. In case that $m = n+1$, we obtain (3.2) (see [A],[K2] and [KY2]).

(3.2) Theorem: Let M be a subdivided $(n+1)$-manifold in R^k and h : $|M| \to R^n$ be a PC^1 mapping on M. Suppose that $c \in R^n$ is a regular value of h. Then $h^{-1}(c)$ is a disjoint union of paths and loops, where by path we mean a connected one-dimensional manifold homeomorphic to one of the unit intervals $(0,1)$, $(0,1]$ and $[0,1]$, and by loop a connected one-dimensional manifold homeomorphic to the one-dimensional sphere. Furthermore

(3.3a) $h^{-1}(c) \cap C$ is either empty or a disjoint union of smooth one-dimensional manifolds for each cell $C \in M$;

b)   loops do not intersect $|\partial M|$;

c) $x \in h^{-1}(c)$ is an endpoint of a path if and only if $x \in |\partial M|$;

d)   if $|M|$ is a closed subset of R^k , every path homeomorphic to $(0,1)$ or $(0,1]$ is unbounded.

We say that a pair of subdivided manifolds P and D of R^k is a primal-dual pair of subdivided manifolds (PDM in short) if there is an operator d from $\bar{P} \cup \bar{D}$ to $\bar{P} \cup \bar{D} \cup \{\emptyset\}$ and a positive integer ℓ such that

(3.4a) $X^d \in \bar{D} \cup \{\emptyset\}$ for each $X \in \bar{P}$, and $Y^d \in \bar{P} \cup \{\emptyset\}$ for each $Y \in \bar{D}$;

b)   if $Z^d \neq \emptyset$, $(Z^d)^d = Z$ and $\dim Z + \dim Z^d = \ell$;

c)   if $Z_1, Z_2 \in \bar{P}$ (or $\bar{D}$), $Z_1 \prec Z_2$ and $Z_1^d \neq \emptyset$, $Z_2^d \neq \emptyset$, then $Z_2^d \prec Z_1^d$.

We call Z^d the dual of Z and ℓ the degree of the PDM. (3.4b) means that the operator d pairs off the cells and (3.4c) is a kind of complementarity. Since the dimensions of a cell and its dual sum up to the constant ℓ, (3.4c) means that when the dimension of a primal cell increases, that of dual one decreases. Given a PDM with degree ℓ let

$$\langle P,D;d \rangle = \{ \ X \times X^d \ : \ X \in \overline{P}; \ X^d \neq \emptyset \ \},$$

which could be written as

$$\langle P,D;d \rangle = \{ \ Y^d \times Y \ : \ Y \in \overline{D}; \ Y^d \neq \emptyset \ \}.$$

(3.5)   Theorem ([KY1], Theorems 3.2 and 3.3):   Let   (P,D)   form a PDM
with degree ℓ together with the dual operator   d.   Then   M
= $\langle P,D;d \rangle$ is a subdivided ℓ-manifold and has the boundary

$$\partial M = \{ \ X \times Y \ : \ X \in \overline{P}; \ Y \in \overline{D};$$
$$X \times Y \ \text{is an} \ (\ell-1)\text{-cell of} \ \overline{M};$$
$$\text{either} \ X^d \ \text{or} \ Y^d \ \text{is empty} \ \}.$$

     The basic model in [KY1] for the variable dimension fixed point
algorithms is the system

(3.6) $h(x,y) \equiv f(x) + y = 0; \quad (x,y) \in |M|.$

Suppose   f   is a continuously differentiable mapping from $|P|$ into
R^n.   If   (P,D)   is a PDM with degree n+1 and $0 \in R^n$ is a regular
value of the PC^1 mapping   h(x,y), then we can apply Theorem (3.2) to
system (3.6) and obtain paths and loops of solutions.   The keystone of
the algorithms is now the PDM, namely how we furnish the subdivided
manifold   M   with the boundary such that we may easily find a starting
point there and obtain a stationary point after toiling along a path.

4.   VARIABLE DIMENSION ALGORITHMS FOR STATIONARY POINT PROBLEMS

As the first example we take the algorithm by Talman and Yamamoto [TY]
for the stationary point problem   (f,K)   on a compact convex
polyhedral set

$$K = \{ \ x \in R^n \ : \ \langle a^i,x \rangle - b_i = 0 \ \text{for} \ i=1,\ldots,k;$$
$$\langle c^i,x \rangle - d_i \leq 0 \ \text{for} \ i=1,\ldots,m \ \}$$

and show how the PDM and the basic model (3.6) yield an algorithm.

The normal cone $N_K(x)$ is

$$N_K(x) = \{ \ y \in R^n \ : \ y = \sum_i \mu_i a^i + \sum_i \nu_i c^i; $$
$$\nu_i \geq 0, \ (<c^i,x>-d_i)\nu_i = 0, \ i=1,\ldots,m \ \}.$$

This is identical for all points x on a face of K, and so we denote by $N_K(F)$ the normal cone corresponding to face F. Then a face F contains a stationary point if and only if

$$0 \in f(F) + N_K(F).$$

Let w be an arbitrary starting point of K and wF be the convex hull of w and face F not containing w. Replacing F by wF in the above system we obtain

$$0 \in f(wF) + N_K(F)$$

or

$$0 = f(x) + y; \quad x \in wF; \quad y \in N_K(F).$$

The dimensions of wF and $N_K(F)$ are dim F+1 and n−dim F, respectively and sum up to n+1. Then it is quite natural to define a PDM as follows:

$$P = \{ \ wF \ : \ F \ \text{is a facet of} \ K \ \text{not containing} \ w \ \};$$
$$D = \{ \ N_K(v) \ : \ v \ \text{is a vertex of} \ K \ \};$$
$$wF \leftarrow d \rightarrow N_K(F) \ \text{for each face} \ F \ \text{of} \ K \ \text{not containing} \ w.$$

Then $M = <P,D;d>$ is a subdivided (n+1)-manifold and the boundary consists of

(4.1a) $F \times N_K(F)$ for face F not containing w;

  b) $wG \times N_K(F)$ for face F and its facet G such that $w \notin G$, $w \in F$;

  c) $\{w\} \times N_K(v)$ for vertex v of K other than w.

According to the basic model (3.6) we consider the system

(4.2) $h(x,y) \equiv f(x) + y = 0; \quad (x,y) \in |M|.$

(4.3) Theorem: Suppose that $0 \in R^n$ is a regular value of h : $|M|$
$\rightarrow R^n$ and the starting point $w \in K$ is not a stationary point of
(f,K). Then there is a path of solutions to (4.2) leading from
$(w,-f(w))$ to a point $(s,-f(s))$ such that s is a stationary point
of (f,K).

Proof: Clearly $(w,-f(w))$ is a solution of (4.2). Since K
is compact, $N_K(v)$ covers R^n as v ranges over all vertices of K.
By the assumption that w is not a stationary point $-f(w)$ does not
lie in $N_K(w)$. Therefore

$$(w,-f(w)) \in \{w\} \times N_K(v) \subset |\partial M|$$

for some vertex v of K. Then by Theorem (3.2) we obtain a path of
solutions to (4.2) which is homeomorphic to either (0,1] or [0,1].
Since $|M|$ is the finite union of closed subsets $wF \times N_K(F)$, it is
still closed. Moreover, by the compactness of K and the continuity
of f the entire set of solutions to (4.2) is contained in the
compact set $K \times f(K)$. Therefore by (3.3d) the path is homeomorphic to
[0,1] and has two distinct end points in $|\partial M|$. Let (s,y) be the
end point other than $(w,-f(w))$. Then $y = -f(s)$, and $(s,-f(s))$
lies in a cell of (4.1a) or b). Since otherwise $(s,-f(s)) \in \{w\}$
$\times N_K(v)$ for some vertex v and hence s = w and $f(s) = f(w)$. This
contradicts that (s,y) is different from $(w,-f(w))$. In either case
it is contained in $F \times N_K(F)$ for some face F of K. This
completes the proof. //

Thus tracing the path of solutions of (4.2) from $(w,-f(w))$ one will
reach a stationary point. When projected on the x-space one leaves
the starting point w along one of the line segments connecting w
to vertices of K. In this sense it has as many directions as the
vertices of K. This algorithm reduces to the (n+1)-ray algorithm of
when applied to the equilibrium problem on the unit simplex (see
[LT9]). An example of the path projected on the x-space is shown in
Figure 1 below. The function is

(4.4a) $f_1(x_1,x_2) = x_1$ if $\eta \leq 0$;

b) $= (\cos \eta\pi)x_1 + (\sin \eta\pi)x_2$ if $0 < \eta < 1;$

c) $= - x_1$ if $\eta \geq 1;$

d) $f_2(x_1,x_2) = x_2$ if $\eta \leq 0;$

e) $= (\sin \eta\pi)x_1 + (\cos \eta\pi)x_2$ if $0 < \eta < 1;$

f) $= - x_2$ if $\eta \geq 1,$

a continuous function which is minus the identity within the radius of
1/2 and the identity outside the unit circle and $K = \{ x \in R^2 : -1 \leq x_1, x_2 \leq 1 \}$. The origin is the unique stationary point of this
problem. When one moves conducted by minus the function value
illustrated by small arrows, one will spiral into the circle with a
radius of $1/\sqrt{2}$ and will never reach the stationary point.

Next one is the $(2^{n+1}-2)$-ray algorithm by Doup, van der Laan and
Talman [DLT] for the equilibrium problem on the unit simplex S^n. The
problem would be reduced to the stationary point problem $(-f,S^n)$,

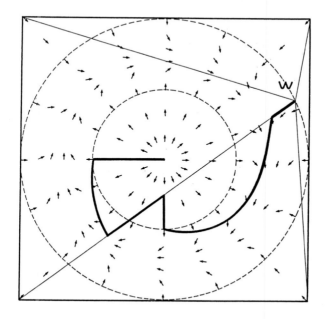

Fig.1.   Path projected on the x-space for the function of (4.4)

however the algorithm admits of economic interpretation of the trajectory it follows as a tatonnement process (see also [LT9]). Starting with an arbitrary point, say w, of S^n it increases the prices of the commodities in excess demand and decreases those of the commodities in excess supply proportionally to the starting point w. To make the complementarity between variables and function values more conspicuous we assume that $w > 0$. In general steps the algorithm traces a path of points x satisfying

$$
\begin{aligned}
x_i/w_i &= \max\{x_j/w_j : j=1,\ldots,n+1\} \quad \text{if} \quad f_i(x) > 0; \\
&= \min\{x_j/w_j : j=1,\ldots,n+1\} \quad \text{if} \quad f_i(x) < 0.
\end{aligned}
$$

We call a vector of -1, 0 and 1 a sign vector. To construct a PDM for this algorithm let $I^+(t)$, $I^0(t)$ and $I^-(t)$ be the index set of positive, zero and negative components of a sign vector t, respectively. For an $(n+1)$-dimensional sign vector t with nonempty $I^+(t)$ and $I^-(t)$ let

$$
\begin{aligned}
X(t) = \{ x \in S^n : \quad x_i/w_i &= \max\{x_j/w_j : j=1,\ldots,n+1\} \quad \text{if} \quad i \in I^+(t) \\
&= \min\{x_j/w_j : j=1,\ldots,n+1\} \quad \text{if} \quad i \in I^-(t) \}.
\end{aligned}
$$

$X(t)$ is the convex hull of the starting point w and a subset

$$
\begin{aligned}
F(t) = \{ x \in S^n : \quad x_i/w_i &= \max\{x_j/w_j : j=1,\ldots,n+1\} \quad \text{if} \quad i \in I^+(t) \\
&= 0 \qquad\qquad\qquad\qquad\qquad\quad \text{if} \quad i \in I^-(t) \}
\end{aligned}
$$

of face $F = \{ x \in S^n : x_i = 0 \text{ if } i \in I^-(t) \}$ and is of $|I^0(t)|+1$ dimension. Figure 2 illustrates $X(t)$'s of S^2. The dual of $X(t)$ is given by

$$
\begin{aligned}
Y(t) = \{ y \in R^{n+1} : \quad & y_i \geq 0 \quad \text{if} \quad i \in I^+(t) \\
& y_i = 0 \quad \text{if} \quad i \in I^0(t) \\
& y_i \leq 0 \quad \text{if} \quad i \in I^-(t) \},
\end{aligned}
$$

which is of $|I^+(t)|+|I^-(t)|$ dimension. Then

$$
\begin{aligned}
P = \{ X(t) : t \in R^{n+1} &\text{ is a sign vector with nonempty} \\
&I^+(t) \text{ and } I^-(t) \} \quad \text{and} \\
D = \{ Y(t) : t \in R^{n+1} &\text{ is a sign vector} \}
\end{aligned}
$$

form a PDM with degree n+2. Then by Theorem (3.5) we see that M
= <P,D;d> is an (n+2)-subdivided manifold and the boundary consists
of the following three kinds of (n+1)-cells:

(4.5a) $F(t) \times Y(t)$ for t with nonempty $I^+(t)$ and $I^-(t)$;

   b) $X(t') \times Y(t)$ for $t \geq 0$ (or $t \leq 0$) and t' obtained by
         replacing one of zeros of t by −1 (or +1);

   c) $\{w\} \times Y(t)$ for t with empty $I^0(t)$ and nonempty $I^+(t)$
         and $I^-(t)$.

According to the basic model (3.6) we consider the set of solutions to
the system

$$-f(x) + y = 0; \quad (x,y) \in |M|.$$

For a given starting point w of S^{n+1} $(x,y) = (w,f(w))$ is a

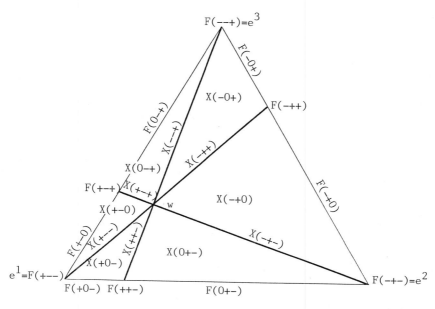

Fig.2.   Two-dimensional unit simplex and $X(t)$'s

trivial solution of this system and it lies in a cell of (4.5c) and
hence on the boundary of M. In fact it does not occur that $f(w)$
≥ 0 and $\neq 0$ since f satisfies Walras' Law, and w is already an
equilibrium when $f(w) \leq 0$. Therefore there is a path from $(w,f(w))$
to another end point, say $(s,f(s))$, which is in a cell of either
(4.5a) or b). When it is in a cell of (4.5a),

$$
\begin{aligned}
s_i &> 0, \quad f_i(s) \geq 0 \quad \text{for} \quad i \in I^+(t); \\
&\geq 0, \qquad\quad = 0 \quad \text{for} \quad i \in I^0(t); \\
&= 0, \qquad\quad \leq 0 \quad \text{for} \quad i \in I^-(t).
\end{aligned}
$$

Therefore $f(s) \leq 0$ by Walras' Law and hence s is an equilibrium.
We see $f(s) = 0$ in the case of (4.5b).

   A possibility of applying the well-known predictor-corrector
method for tracing the path of solutions to (4.2) is shown in [Y5].
Since the cell wF is the convex hull of w and all the vertices of
F, instead of (4.2) the following system is considered:

$$
\begin{aligned}
&f(\lambda_w w + \textstyle\sum_v \lambda_v v) + \sum_i \mu_i a^i + \sum_I \nu_i c^i = 0; \\
&\lambda_w + \textstyle\sum_v \lambda_v = 1; \\
&\lambda_w \geq 0; \lambda_v \geq 0 \quad \text{for all} \quad v; \\
&\nu_i \geq 0 \quad \text{for} \quad i \in I,
\end{aligned}
$$

where I is the index set of binding inequality constraints of F.
In general neither wF nor the normal cone $N_K(F)$ is simplicial.
That makes a difficulty of nonuniqueness of combination coefficients
λ, μ and ν. An idea of simplicial decomposition is presented in
[Y5] and also in [Y4] for the case of an affine function f. The
algorithm of [TY] traces the path of solutions to the system (4.2)
with f replaced by its piecewise linear approximation $\bar{f}$ on a
triangulation T of K, namely

(4.6) $\bar{h}(x,y) \equiv \bar{f}(x) + y = 0; \quad (x,y) \in |M|.$

They generalized the V-triangulation by Doup and Talman [DT1] to
triangulate K so that for each face F of K the set wF is also

triangulated by the lower dimensional simplices. Collecting the
families { σ × N_K(F) : σ ∈ T̄; σ ⊂ wF; dim σ = dim wF } for all
faces F of K not containing w, one obtain a refinement of M.
This is not necessarily simplicial but h̄ is affine on each of its
cells. Therefore the regular value assumption for piecewise linear
functions (see for example [E]) will suffice to ensure the existence
of the path of solutions to (4.6) leading from $(w,\bar{f}(w)) = (w,f(w))$
to a stationary point of $(\bar{f},K)$, that is an approximate stationary
point of (f,K). Let σ be a simplex of T̄ restricted to wF which
contains x, $v^1,\ldots,v^{k+1}$ be its vertices and I be the index set of
binding inequality constraints at face F. Then the system is

(4.7a) $\sum_j \lambda_j f(v^j) + \sum_i \mu_i a^i + \sum_I \nu_i c^i = 0;$

   b) $\sum_j \lambda_j = 1; \lambda_j \geq 0$ for all j;

   c) $\nu_i \geq 0$ for $i \in I.$

At the start we solve the linear program

$$\min. \sum_i b_i \mu_i + \sum_i d_i \nu_i;$$
$$\text{s.t. } \lambda_1 f(w) + \sum_i \mu_i a^i + \sum_i \nu_i c^i = 0;$$
$$\nu_i \geq 0 \text{ for all } i; \lambda_1 = 1,$$

and obtain an initial basic solution corresponding to the zero-
dimensional starting simplex {w}. The optimal basic variables
determine the normal cone $N_K(v)$ containing $-f(w)$ and further give
the cell $w\{v\} \times N_K(v)$ of M which the path of solutions of (4.6)
passes through. Since this problem is the dual of

$$\min. \langle f(w), x\rangle \quad \text{s.t. } x \in K,$$

it follows that the algorithm begins to move toward the first
candidate for the statinary point. In general steps when some λ_j
vanishes we obtain a facet τ of σ, then check if τ lies in the
boundary of wF. If not, there must be a unique simplex σ' in wF
sharing τ with σ. So we make an ordinary replacement. If τ is
in the boundary of wF, which is wG for some face G of F, we

introduce into the system (4.7) the coefficient vectors c^i's of the inequality constraints which become newly active at G. If ν_i vanishes, we first see if $y = \sum_i \mu_i a^i + \sum_I \nu_i c^i$ hits the boundary of $N_K(F)$. This will be done by checking if there remains a coefficient vector c^k, $k \in I$ on the opposite side of the subspace spanned by a^i's and c^i's corresponding to basic variables to the dropped one. If there is such a c^k, we introduce it to the system (4.7). If not, we delete the inequality constraints corresponding to the nonbasic variables to obtain a new face E of K containing F as a facet.

ACKNOWLEDGEMENT

The author wishes to thank Frieda Dueben for her careful reading of the earlier version of this paper.

REFERENCES

[A]   J.C. Allexander, "The topological theory of an embedding
        method", in: H. Wacker ed., Continuation Method (Academic
        Press, New York, 1978) pp.36-67.

[AG]   E.L. Allgower and K. Georg, "Predictor-corrector and simplic-
        ial methods for approximating fixed points and zero points of
        nonlinear mappings", in: A.Bachem, M.Grötschel and B.Korte eds.,
        Mathematical Programming, The State of the Art, Bonn 1982
        (Springer-Verlag, Berlin, 1983) pp.15-56.

[B]   M. Broadie, "An introduction to the octahedral algorithm for
        the computation of economic equilibria", Mathematical Pro-
        gramming Study 23 (1985) 121-143.

[BS]   M.S. Bazaraa and C.M. Shetty, Foundation of Optimization,
        Lecture Notes in Economics and Mathematical Systems 122
        (Springer-Verlag, Berlin, 1976).

[D1]   S. Dafermos, "An iterative scheme for variational inequalities",
        Mathematical Programming 26 (1983) 40-47.

[D2]   T. Doup, Simplicial algorithm on the simplotope, Ph.D.Disser-
        tation, Tilburg University (Tilburg, The Netherlands, Oct.
        1987).

[DET]   T. Doup, A.H. van den Elzen and A.J.J. Talman, "Simplicial
        algorithms for solving the nonlinear complementarity problem on
        the simplotope", in: A.J.J.Talman and G.van der Laan eds., The
        Computation and Modelling of Economic Equilibria (North-Holland,
        Amsterdam, 1987) pp.125-154.

[DLT]   T. Doup, G. van der Laan and A.J.J. Talman, "The $(2^{n+1}-2)$-ray algorithm: A new simplicial algorithm to compute economic equilibria", Mathematical Programming 39 (1987) 241-252.

[DT1]   T. Doup and A.J.J. Talman, "A new simplicial variable dimension algorithm to find equilibria on the product space of unit simplices", Mathematical Programming 37 (1987) 319-355.

[DT2]   T. Doup and A.J.J. Talman, "A continuous deformation algorithm on the product space of unit simplices", Mathematics of Operation Research 12 (1987) 485-521.

[DT3]   T. Doup and A.J.J. Talman, "The 2-ray algorithm for solving equilibrium problems on the unit simplex", Methods of Operations Research 57 (1987) 269-285.

[DT4]   C.A.J.M. Dirven and A.J.J. Talman, "A simplicial algorithm for finding equilibria in economies with linear production technologies", FEW 217, Department of Economics, Tilburg University (Tilburg, The Netherlands, 1988).

[E]   B.C. Eaves, "A short course in solving equations with PL-homotopies", SIAM-AMS Proceeding 9 (1976) 73-143.

[ELT]   A.H. van den Elzen, G. van der Laan and A.J.J. Talman, "Adjustment processes for finding equilibria on the simplotope", FEW 196, Tilburg University (Tilburg, The Netherlands, 1985).

[ET]   A.H. van den Elzen and A.J.J. Talman, "A new strategy-adjustment process for computing a Nash equilibrium in a noncooperative more-person game", REEKS TER DISCUSSIE No.85.06, Tilburg University (Tilburg, The Netherlands, 1985).

[F1]   K. Fan, "Fixed-point and related theorems for non-compact convex sets", in: O.Moeschlin and D.Pallaschke (eds.), Game Theory and Related Topics (North-Holland, Amsterdam, 1979) 151-156.

[F2]   R.M. Freund, "Variable dimension complexes, Part I: basic theory", Mathematics of Operations Research 9 (1984) 479-497.

[F3]   R.M. Freund, "Variable dimension complexes, Part II: a unified approach to some combinatorial lemmas in topology", Mathematics of Operations Research 9 (1984) 498-509.

[F4]   R.M. Freund, "Combinatorial theorems on the simplotope that generalize results on the simplex and cube", Mathematics of Operations Research 11 (1986) 169-179.

[F5]   R.M. Freund, "Combinatorial analogs of Brouwer's fixed point theorem on a bounded polyhedron", Sloan W.P. No.1720-85, Massachusetts Institute of Technology (Cambridge, U.S.A., 1985, Revised 1986).

[GLT]   R. Glowinski, J.L. Lions and R. Tremolieres, Numerical Analysis of Variational Inequalities (North-Holland, Amsterdam, 1981).

[GZ]   C.B. Garcia and W.I. Zangwill, Pathways to Solutions, Fixed Points and Equilibria (Prentice-Hall, Inglewood Cliffs, U.S.A., 1981).

[HM] J.H. Hammond and T.L. Magnanti, "Generalized descent methods for asymmetric systems of equations", Mathematics of Operations Research 12 (1987) 678-699.

[HP1] G.J. Habetler and A.L. Price, "Existence theory for generalized nonlinear complementarity problem", Journal of Optimization Theory and Applications 7 (1971) 223-239.

[HP2] P.T. Harker and J.-S. Pang, "Finite-dimensional variational inequality and nonlinear complementarity problems: A survey of theory, algorithms and applications", 87-12-06, Department of Decision Sciences, University of Pennsylvania (Philadelphia, U.S.A. 1987).

[I] V.I. Istrățescu, Fixed Point Theory (D.Reidel, Dordrecht, The Netherlands, 1981).

[IFI] T. Ito, M. Fukushima and T. Ibaraki, "An iterative method for variational inequalities with application to traffic equilibrium problems", Journal of the Operations Research Society of Japan 31 (1988) 82-104.

[K1] S. Karamadian, "Generalized complementarity problem", Journal of Optimization Theory and Applications 8 (1971) 161-168.

[K2] M. Kojima, "An introduction to variable dimension algorithms for solving systems of equations", in: E.L.Allgower, K.Glashoff and H.-O.Peitgen eds., Numerical Solution of Nonlinear Equations (Springer-Verlag, Berlin, 1981) pp.199-237.

[K3] H.W. Kuhn, "Approximate search for fixed points", in: Computing Methods in Optimization Problems 2 (Academic Press, New York, 1969).

[KS] D. Kinderlehrer and G. Stampacchia, An Introduction to Variational Inequalities and Their Applications (Academic Press, New York, 1980).

[KY1] M. Kojima and Y. Yamamoto, "Variable dimension algorithms: basic theory, interpretations and extensions of some existing methods", Mathematical Programming 24 (1982) 177-215.

[KY2] M. Kojima and Y. Yamamoto, "Variable dimension algorithms and a class of primal-dual subdivided manifolds", in: Proceedings of the 3rd Mathematical Programming Symposium, Japan (Tokyo, Japan, 1982) pp.45-61.

[KY3] M. Kojima and Y. Yamamoto, "A unified approach to the implementation of several restart fixed point algorithms and a new variable dimension algorithm", Mathematical Programming 28 (1984) 288-328.

[L1] G. van der Laan, Simplicial Fixed Point Algorithms, Mathematical Centre Tracts 129 (Mathematisch Centrum, Amsterdam, 1980).

[L2] G. van der Laan, "The computation of general equilibrium in economies with a block diagonal pattern", Econometrica 53 (1985) 659-665.

[LH] S. Lawphongpanich and D.W. Hearn, "Simplicial decomposition of

the asymmetric traffic assignment problem", Transportation
Research 18B (1984) 123–133.

[LS]   G. van der Laan and L.P. Seelen, "Efficiency and implementation
         of simplicial zero point algorithms", Mathematical Programming
         30 (1984) 196–217.

[LT1]   G. van der Laan and A.J.J. Talman, "A restart algorithm for com-
         puting fixed points without an extra dimension", Mathematical
         Programming 17 (1979) 74–84.

[LT2]   G. van der Laan and A.J.J. Talman, "A restart algorithm without
         an artificial level for computing fixed points on unbounded
         regions", in: H.-O. Peitgen and H.-O. Walther eds., Functional
         Differential Equations and Approximation of Fixed Points
         (Springer-Verlag, Berlin, 1979) pp.247–256.

[LT3]   G. van der Laan and A.J.J. Talman, "Convergence and properties
         of recent variable dimension algorithms", in: W. Forster ed.,
         Numerical Solution of Highly Nonlinear Problems (North-Holland,
         Amsterdam, 1980) pp.3–36.

[LT4]   G. van der Laan and A.J.J. Talman, "A class of simplicial re-
         start fixed point algorithms without an extra dimension",
         Mathematical Programming 20 (1981) 33–48.

[LT5]   G. van der Laan and A.J.J. Talman, "On the computation of fixed
         points in the product space of unit simplices and an applica-
         tion to noncooperative N-person games", Mathematics of Oper-
         ations Research 7 (1982) 1–13.

[LT6]   G. van der Laan and A.J.J. Talman, "Note on the path following
         approach of equilibrium programming", Mathematical Programming
         25 (1983) 363–367.

[LT7]   G. van der Laan and A.J.J. Talman, "Simplicial algorithms for
         finding stationary points, a unifying description", Journal of
         Optimization Theory and Applications 50 (1986) 262–281.

[LT8]   G. van der Laan and A.J.J. Talman, "Computing economic equilib-
         ria by variable dimension algorithms: State of the art",FEW 270,
         Tilburg University (Tilburg, The Netherlands, 1987).

[LT9]   G. van der Laan and A.J.J. Talman, "Adjustment process for
         finding economic equilibria", in: A.J.J.Talman and G.van der
         Laan eds., The Computation and Modelling of Economic Equilibria
         (North-Holland, Amsterdam, 1987) pp.85–124.

[LTV1]   G. van der Laan, A.J.J. Talman and L. Van der Heyden, "Sim-
         plicial variable dimension algorithms for solving the nonlinear
         complementarity problem on a product of unit simplices using a
         general labelling", Mathematics of Operations Research 12
         (1987) 377–397.

[LTV2]   G. van der Laan, A.J.J. Talman and L. Van der Heyden, "Variable
         dimension algorithms for unproper labellings", Research Memo-
         randum FEW 147, Tilburg University (Tilburg, The Netherlands,
         April 1984).

[M]   L. Mathiesen, "Computation of economic equilibria by a sequence

of linear complementarity problems", Mathematical Programming
Study 23 (1985) 144–162.

[P1]   J.-S. Pang, "Asymmetric variational inequality problems over
       product sets: Applications and iterative methods", Mathematical
       Programming 31 (1985) 206–219.

[P2]   E. Phan-huy-Hao, "Quadratically constrained quadratic
       programming", Zeitschrift für Operations Research 26 (1982)
       105–119.

[PC]   J.-S. Pang and D. Chan, "Iterative methods for variational and
       complementarity problems", Mathematical Programming 24 (1982)
       284–313.

[PY]   J.-S. Pang and C.-S. Yu, "Linearized simplicial decomposition
       method for computing traffic equilibria on networks", Networks
       14 (1984) 427–438.

[R]   S.M. Robinson, "Generalized Equations", in: A.Bachem,
       M.Grötschel and B.Korte eds., Mathematical Programming, The
       Stete of the Art, Bonn 1982 (Springer-Verlag, Berlin, 1983)
       pp.346–367.

[S1]   H. Sonnenschein, "Market excess demand functions", Econometrica
       40 (1972) 549–563.

[S2]   R. Saigal, "Extension of the generalized complementarity prob-
       lem", Mathematics of Operations Research 1 (1976) 260–266.

[S3]   L.S. Shapley, "On balanced games without side payments", in:
       T.C.Hu and S.M.Robinson eds., Mathematical Progarmning (Aca-
       demic Press, New York, 1973) pp.261–290.

[T1]   A.J.J. Talman, Variable Dimension Fixed Point Algorithms and
       Triangulations, Mathematical Centre Tracts 128 (Mathematisch
       Centrum, Amsterdam, 1980).

[T2]   M.J. Todd, The Computation of Fixed Points and Applications,
       (Springer-Verlag, Berlin, 1976).

[T3]   M.J. Todd, "Global and local convergence and monotonicity re-
       sults for a recent variable dimension simplicial algorithm",
       in: W.Forster ed., Numerical Solution of Highly Nonlinear Prob-
       lems (North-Holland, Amsterdam, 1980) pp.43–69.

[T4]   M.J. Todd, "An introduction to piecewise-linear homotopy algo-
       rithms for solving systems of equations", in: P.R. Turner ed.,
       Topics in Numerical Analysis (Springer-Verlag, Berlin, 1982)
       pp.149–202.

[TV]   A.J.J. Talman and L. Van der Heyden, "Algorithms for the linear
       complementarity problem which allow an arbitrary starting point",
       in: B.C. Eaves et.al. eds., Homotopy Methods and Global Con-
       vergence (Plenum Press, New York, 1983) pp.267–286.

[TY]   A.J.J. Talman and Y. Yamamoto, "A globally convergent simplicial
       algorithm for stationary point problems", Report No.86422-OR,
       Institut für Öconometrie und Operations Research, Univer-
       sität Bonn (Bonn, West Germany, June 1986) (to appear in Mathe-
       matics of Operations Research).

[W]   A.H. Wright, "The octahedral algorithm, a new simplicial fixed
        point algorithm", Mathematical Programming 21 (1981) 47-69.

[Y1]   Y. Yamamoto, "A new variable dimension algorithm for the fixed
        point problem", Mathematical Programming 25 (1983) 329-342.

[Y2]   Y. Yamamoto, "A unifying model based on retraction for fixed
        point algorithms", Mathematical Programming 28 (1984) 192-197.

[Y3]   Y. Yamamoto,"A variable dimension fixed point algorithm and the
        orientation of simplices", Mathematical Programming 30 (1984)
        301-312.

[Y4]   Y. Yamamoto, "A path following algorithm for stationary point
        problems", Journal of Operations Research Society of Japan 30
        (1987) 181-198.

[Y5]   Y. Yamamoto, "Stationary point problems and a path-following
        algorithm", in: Proceedings of the 8th Mathematical Programming
        Symposium, Japan (Hiroshima, Japan, 1987) pp.153-170.

[Y6]   Y. Yamamoto, "Orientability of a pseudomanifold and general-
        ization of Sperner's lemma", Journal of Operations Research
        Society of Japan 31 (1988) 19-43.

Stochastic Dynamic Optimization Approaches and Computation[1]

Pravin VARAIYA and Roger J-B WETS

University of California, Berkeley-Davis, U.S.A.

Abstract

The description of stochastic dynamical optimization models that follows is intended to exhibit some of the connections between various formulations that have appeared in the literature, and indicate some of the difficulties that must be overcome when trying to adapt solution methods that have been successfully applied to one class of problems to an apparently related but different class of problems. The emphasis will be on *solvable* models.

We begin with the least dynamical versions of stochastic optimization models, one- and two-stage models then consider discrete time models, and conclude with continuous time models.

1 ONE-STAGE MODELS

We consider the following simple one-stage stochastic optimization problem:

$$\begin{array}{ll} \text{minimize} & E\{h_0(z,\xi)\} \\ \text{subject to} & h_i(z) \le 0, \quad i = 1,\ldots,s, \\ & h_i(z) = 0, \quad i = s+1,\ldots,m, \\ & z \in Z \subset R^n \end{array}$$

where ξ is a random vector with support $\Xi \subset R^N$ and distribution P. We are looking for a vector z^* that is feasible, i.e., belongs to

$$S = \{z \in Z | h_i(z) \le 0, \ i = 1,\ldots,s; h_i(z) = 0, \ i = s+1,\ldots,m\},$$

and minimizes $E\{h_0(\cdot,\xi)\}$ on S. Of course, this is just a special instance of a nonlinear programming problem. Indeed, after integration, the objective can be rewritten as

$$\text{minimize } Eh_0(z),$$

where for each z,

$$Eh_0(z) := \int_\Xi h_0(z,\xi)\,dP(\xi).$$

[1]Supported in part by grants of the National Science Foundation and the Air Force Office of Scientific Research

M. Iri and K. Tanabe (eds.), Mathematical Programming, 309–331.
© *1989 by KTK Scientific Publishers, Tokyo.*

Such a function is called an *expectation functional*; the study of its properties is a major theme of the theory of stochastic programming. However, even this "simple" stochastic optimization problem cannot be solved by standard nonlinear optimization algorithms. The problem is with the evaluation of Eh_0 or its (sub)gradient. There are a few cases that can be managed:

1. when the function $h_0(z,\cdot)$ is separable so that

$$\int h_0(z,\xi)\,dP = \sum_{j=1}^{N} \int_{\Xi_j} h_{0j}(z,\xi_j)\,dP_j(\xi_j)$$

   (with P_j the marginal distribution function),

2. when $\Xi = \{\xi^1,\ldots,\xi^L\}$ is finite and L is not too large, then

$$\int h_0(z,\xi)\,dP = \sum_{\ell=1}^{L} p_\ell h_0(z,\xi^\ell)$$

   (where $p_\ell = P[\xi = \xi^\ell]$),

3. if h_0 is convex, sufficiently smooth, easy enough to evaluate and P is a multidimensional normal, Gamma or Dirichlet distribution function.

   The first case simply reflects the fact that univariate calculus, as well as one-dimensional numerical integration routines, are well developed. That is definitely not the case for multivariate calculus and multidimensional numerical integration. In the second case, the evaluation of Eh_0, or its gradient, at a point z is reduced to evaluating $h_0(z,\xi^\ell)$, or its gradient $\nabla_z f_0(z,\xi^\ell)$, for each ξ^ℓ in Ξ. And, in the third case, there are specific subroutines (developed by Hungarian computer scientists for stochastic programming problems) that combine Monte-Carlo techniques with some of the specific properties of those distributions. Because sampling is involved, the evaluation of $h_0(z,\xi)$ at any point ξ in Ξ should be "cheap" enough; unfortunately that is seldom the case in the most important applications.

   Because of this state of affairs, the research in stochastic programming has been concerned with either identifying classes of models that fit in those "solvable" categories, designing reliable and efficient solutions procedures for such problems, or developing theories and procedures that would allows us to solve any problem by solving approximating problems that belongs to the "solvable" categories.

One version of the one-stage model that has received limited attention in the literature is the case when the probability distribution of ξ depends on z. In terms of the essential objective, the problem would take on the following form:

$$\text{find } z^* \text{ that minimizes } \int h(z,\xi)\, dP(\xi;z).$$

Again, this is *just* a nonlinear optimization problem and an evaluation of the objective at any point z is not more complicated than it was before. What has changed are the properties of the function:

$$z \mapsto Eh(z) = \int h(z,\xi)\, dP(\xi;z).$$

For example, when P does not depend on z, the convexity of Eh follows immediately from the convexity of $h(\cdot,\xi)$ for all ξ. That is no longer the case when P depends on z. Similarly, the (sub)gradients of Eh can no longer be obtained by the (relatively) simple formula:

$$\partial Eh = \int \partial h(\cdot,\xi)\, dP(\xi).$$

The stochastic approximation-like techniques, e.g., stochastic quasi-gradient methods, can no longer be used to find (almost surely) a solution, at least not in the form in which these techniques have been used up to now. In fact, in this situation, the properties of Eh may very well have nothing in common with those of $h(\cdot,\xi)$.

The challenge would not be so much to design general solution procedures for this (richer, but ungainly) class of problems, but to identify those that possess properties that would still allow us to use "classical" solution procedures. Clearly, it all has to do with the type of dependence of P on z. For example, if P is defined on R^N, and $P(\xi;z) = Q(\xi + Hz)$, where Q is a probability distribution function and H is a (given) matrix of the appropriate size, the problem takes on the following form (after a simple change of variables):

$$\text{find } z^* \text{ that minimizes } \int h(z,\zeta - Hz)\, dQ(\zeta).$$

The properties of Eh will thus depend on the properties of $h(z,\xi)$ viewed as a function of (z,ξ) jointly.

2 TWO-STAGE MODELS

In addition to a (first stage) decision z_1, this model allows for a second stage or *recourse* decision z_2 that is taken *after* full or partial information is obtained about the values of the random components of the problem. The problem can be formulated as follows:

$$
\begin{aligned}
\text{minimize} \quad & f_{10}(z_1) + E\{f_{20}(z_2(\xi),\xi)\} \\
\text{subject to} \quad & f_{1i}(z_1) \le 0, && \text{for } i = 1,\ldots,m_1, \\
& f_{2i}(z_1,z_2(\xi),\xi) \le 0 \ a.s., && \text{for } i = 1,\ldots,m_2,
\end{aligned}
$$

where the function z_2 can depend (measurably) on ξ in a way that is consistent with the information that will be available in the second stage, i.e., when taking the recourse decision. A much more detailed discussion of the the modeling of the information process will follow; for the time being let us assume that full information is available before choosing z_2.

If we define

$$
f_0(z,\xi) := f_{10}(z) + \inf_{z_2}[f_{20}(z_2,\xi)|f_{2i}(z,z_2,\xi) \le 0, i = 1,\ldots,m_2],
$$

and

$$
f_i(z) := f_{1i}(z) \text{ for } i = 1,\ldots,m_1,
$$

we see that, at least from a theoretical viewpoint, the two-stage model can be analyzed in the framework provided by the one-stage model as long as we allow for a sufficiently general class of functions f_0, viz., infinite-valued (to account for the cases when for given z_1 and some ξ there is no z_2 that satisfies the second-stage constraints) and nondifferentiable (the infimal value of a mathematical program is seldom a differentiable function). Because, it covers a large number of applications, and because it is in some sense the first hurdle that must be mastered when considering dynamical optimization models, much of the algorithmic research in stochastic programming has been oriented at solving two-stage (recourse) models.

At first sight, the two-stage model may appear very restricted in its dynamical aspects. However, it is important to keep in mind that "stages" do not necessarily refer to time units. They correspond to stages in the decision process. The variable z_1 refers to all the decisions that must be taken *before* there will be any information about the values to be assigned to the random elements of the problem. The variables z_2 model *all* the decisions that will be made *after* the available information about these values will be collected. For example, z_1 could represent a sequence of decisions (control actions) to be made over a given time horizon, say $z_{11}, ..., z_{1t}, ..., z_{1T}$, and $z_2 = (z_{21}, ..., z_{2t}, ..., z_{2T})$, representing a similar sequence of decisions used to correct the basic trend set by the z_1-variables. Each one of the z_{2t} refers to a decision to be made at time t in response to the situation that would result from choosing z_1 and obtaining information about the random events that can be observed up to time t. Such models could be called *dynamical two-stage models*. As a special case, we could have $z_1 = z_{11}, ..., z_{1t}$, and $z_2 = z_{2,t+1}, ..., z_{2T}$, which would correspond to a mid-course correction. And, of course, there is no need to restrict oneself to discrete time.

Let us now turn to the case when the recourse decision must be made under less than full information. Before we start, let us stress the fact that although one may not observe ξ, there are many cases when the *observations* made allow us to recover enough information about the values of ξ that one can still refer to it as *full* information. This has sometimes been the source of some confusion between the "stochastic programming" formulation and the "stochastic optimal control" formulation. A typical, and simple, example could go as follows: instead of ξ, we observe the "state" x_1 of the system, with the state defined by a relation of the form:

$$x_1 = \varphi(z_1) + \xi.$$

In such cases, instead of viewing the recourse decision as a function of ξ, we could equally well think of it as a function of the "state" of the system.

If only partial information will be available, let $\mathcal{G}$ be the (sub)field of events that could be observed before taking the recourse decision; let $\mathcal{A}$ be the field of all events generated by ξ. In these terms, partial information would mean that $\mathcal{G}$ is a proper subcollection of $\mathcal{A}$. Since the recourse decision z_2 can only depend on the information that will become available, it must be $\mathcal{G}$-adapted or, equivalently, $\mathcal{G}$-measurable. Moreover, in evaluating the performance of a particular decision, only those events that lie in $\mathcal{G}$ can be taken into account, thus rather than using $f_{20}(z_2, \xi)$ as the objective function of the recourse problem, we would replace it by

$$E\{f_{20}(z_2, .) \,|\, \mathcal{G}\}(\xi).$$

Also, feasibility of a recourse function z_2 can only be checked up to events that lie in $\mathcal{G}$. Thus, a feasible first stage decision is one that satisfies the first stage constraints $f_{1i}(z_1) \leq 0, i = 1, .., m_1$ and to which one can associate a $\mathcal{G}$-measurable function z_2 such that almost surely satisfies:

$$f_{2i}(z_1, z_2(\xi), \xi) \leq 0, i = 1, ..., m_2.$$

This latter condition, may or may not impose restrictions on the choice of z_1 beyond those already imposed by the first stage constraints. If it does, one refers to them as *induced constraints*. Otherwise, the problem is said to have *relatively complete recourse*. This can also be expressed in terms of a certain property ($\mathcal{G}$-nonanticipativity) of the multifunction determined by the constraints; we shall return to this in the context of the multistage models.

Although the observations may very well depend on z_1, so far, we have only dealt with the case when the information available about the values taken on by the random quantities of the problem do *not* depend on the first stage decision. The solution of the two-stage model, defined at the beginning of this section, can be found by first finding z_1^* the optimal solution of the (finite dimensional) nonlinear program:

$$\text{minimize} \ \ f_0(z_1) \ \ \text{subject to} \ \ f_i(z_1) \leq 0, \ i = 1, \ldots, m_1,$$

with the functions $f_i, i = 0, \ldots, m_1$, as defined above, and then solving for each ξ (in the support of the probability measure), the deterministic nonlinear program:

$$\text{minimize} \ \ E\{f_{20}(z_2, \cdot) \,|\, \mathcal{G}\}(\xi) \ \ \text{subject to} \ \ f_{2i}(z_1^*, z_2, \xi) \leq 0, i = 1, \ldots, m_2.$$

As long as as there is a consistent rule for choosing the optimal solution when there are multiple (optimal) solutions, this will define an optimal $\mathcal{G}$-measurable function z_2^*. In most applications, only the here-and-now decision, i.e., the first stage decision, is of interest, and then there is no need to explicitly calculate the optimal z_2 function.

In general, all of this is no longer possible if the probability distribution of the random quantities depends on the first stage decision, or if the information (derived from the observations) depends on z_1.

To indicate that the (sub)field of events depends on z_1, let us denote it by $\mathcal{G}(z_1)$. The two-stage problem is then to find a pair (z_1, z_2) in

$$Z_\mathcal{G} := \{z_1 \in R^{n_1},\ z_2\ \mathcal{G}(z_1)-\text{measurable}\}$$

that satisfies the constraints and minimize the objective function as defined above. The space $Z_\mathcal{G}$ is no longer a linear space (as was the case when the field of information did not depend on z_1), in general it is neither convex (not even connected), nor closed. The nonlinearities introduced by the dependence of the information field on z_1 have changed the essence of the problem, and usually, it is a much more difficult problem to solve. The solution cannot be found, as before, by solving (in sequence) finite dimensional optimization problems. The optimal first stage decision cannot be found without finding an explicit description of the associate (optimal) second-stage decision function. There are examples in the literature (not exactly formulated in these terms), beginning with one due to Witsenhausen, that illustrate all of these difficulties. The fact that the problem becomes so complicated may suggest that there is a need to consider more carefully its formulation.

We shall return to this in the context of stochastic control models.

For purposes of illustration, let us consider a simple example: let

$$f_{20}(z_2, \xi) = q \cdot z_2,$$

and for $i = 1, \ldots, m_2$,

$$f_{2i}(z_1, z_2, \xi) = T_i z_1 + W_i z_2 - h_i(\xi),$$

where T_i, W_i are (fixed) vectors, and h_i is a random variable. Assuming that we have observed h, to find the optimal recourse decision, the problem that needs to be solved is a linear program. And, from parametric programming, we know that there is a piecewise linear function of $h - T z_1$ that yields the optimal recourse decision. If we do not observe h, or equivalently the "state" $h - T z_1$, but instead information is some (nonlinear) function of $h - T z_1$, then, in general, we loose the piecewise linearity of the optimal recourse decision with respect to the state. In order to be able to deal with such problems, we may very well want to restrict the class of acceptable second-stage decision functions to those that that depend on a finite number of parameters.

There is also the question of the dependence of the probability measure on the first-stage decision. We already discussed this in the framework of

the one-stage model. The situation is not any different here. There are no
new conceptual or theoretical difficulties, beyond those that we mentioned
in Section 1, except that we may have to deal with complications generated
by the dependencies of P on z_1 and by the restriction of z_2 to the class of
functions that are $\mathcal{G}(z_1)$ measurable.

3   MULTISTAGE MODELS

Conceptually, multistage models are straightforward extensions of two-stage
models. There are a few technical details that need to be taken care off, but
most assertions one can make about such models follow from those that have
been established for two-stage models. However, it does pay to analyze in
more detail the dynamical aspects of the problem. The real challenge comes
from having to deal with what has been called "the curse of dimensionality"
in the design of solution procedures. We shall begin with a rather general
formulation whose main virtue is that it is simple from a notational and
conceptual viewpoint. As in the previous section, we start with the case
when the information (inferred from the observations) and the distribution
of the random quantities do not depend on past decisions. Once more, let us
stress the fact that we do not exclude the possibility of having the observation
values depend on earlier decisions (controls).

  Although stages of a multistage stochastic optimization problem do not
necessarily correspond to time periods, let us use $t = \{1,\dots,T\}$ to denote
the stage-index and refer to it, by abuse of language, as "time". Let ξ_t denote
the random quantities that are observed at stage t *before* we have to make
our decision, i.e., the t-th stage decision function z_t can depend on all past
observations $\xi^t := \{\xi_s, s = 1,\dots,t\}$.

  With $T = 2$ and ξ_1 a degenerate random vector (i.e., whose distribution
is concentrated at one point), we recover the two-stage model; the variables
denoted ξ then, are now called ξ_2. We are now allowing for the possibility
that the problem considered in Section 2, was actually one of a possible
collection of problems obtained after observing ξ_1. This slight generalization
of the model comes from a shift in the type of questions that we like to
see answered. In the two-stage model, the emphasis was on calculating an
optimal first-stage decision, and this is still the case for many multistage
problems, but for another wide range of models the accent will be on finding

an optimal decision (control) rule that could be applied at *all* stages.

The random quantities of the problem will again be denoted by ξ with $\xi = (\xi_1, \ldots, \xi_T)$. The dependence of the (recourse) decision on past observations can be expressed in the following terms: let $(\Xi, \mathcal{A}, P)$ be the underlying probability space and let $\mathcal{B}_t$ be the $(\sigma\text{-})$field of events generated by the observations up to time t; this corresponds to the σ-field generated by the random vector ξ^t. The dependence of z_t on the past observation can thus be expressed in terms of the measurability of z_t with respect to $\mathcal{B}_t$, in other words, z_t must be $\mathcal{B}_t$-adaptable.

The constraints that are explicitly included in the formulation of the problem, will be represented by a multifunction:

$$\Gamma(t, \xi) := \{z^t = (z_1, \ldots, z_t) \text{ that satisfy the } t-\text{th stage constraints}\}.$$

(We use, somewhat indiscriminately, z_t to designate a function from Ξ into the decision space, say R^{n_t}, and a point in its range.)

Thus the *multistage recourse* problem, is to find

$$\begin{array}{ll} \text{for } t = 1, \ldots, T, & z_t \; \mathcal{B}_t - \text{measurable}, \\ \text{for } t = 1, \ldots, T, & z^t \in \Gamma(t, \xi),, \\ \text{that minimizes} & E\{h_0((z_1(\xi), \ldots, z_T(\xi)), \xi)\}. \end{array}$$

Most of the theory developed for one- and two-stage models can be applied to the multistage problem to obtain the basic properties of the deterministic equivalent problem, a number of useful characterizations of the optimal solutions (linearity, piecewise linearity, etc.), as well as necessary and sufficient optimality conditions. However, as already mentioned earlier, one is also interested in the dynamical properties of the solution, in particular in the role played by the dynamical restrictions on the z_t that comes from the $\mathcal{B}_t$-measurability condition.

Let Z be the space of all ($\mathcal{A}$-measurable) functions $z := (z_1, \ldots, z_T)$ defined on Ξ such that for all t, z_t is $\mathcal{B}_t$-measurable; such functions will be called *nonanticipative*. It is a linear subspace of the space of $\mathcal{A}$-measurable functions. From this simple observation follows an important optimality criterion: assuming that the problem at hand satisfies a "standard" constraint qualification, and the constraint-multifunction is nonanticipative, a necessary condition for optimality of z^*, that is also sufficient in the convex case, is that there exist multipliers $p = (p_1, \ldots, p_T)$ defined on Ξ, orthogonal to Z, i.e., such that

$$E\{p_t(\cdot)|\mathcal{B}_t\} = 0 \text{ } a.s., \text{ for } t = 1, \ldots, T,$$

and for almost all ξ:

$$z^*(\xi) \in \operatorname{argmin}\{h_0(z, \xi) - p(\xi) \cdot z|z^t \in \Gamma(t, \cdot) \text{ } a.s., \text{ for } t = 1. \ldots, T\}.$$

Knowledge of these multipliers would reduce the problem to one of point-wise minimization. One can interpret these multipliers as a price system associated with the *nonanticipativity* restrictions; a beautiful economic interpretation of these multipliers in terms of insurance prices has been sketched out by I. Evstigneev from C.E.M.I.(Moscow).

To state the optimality condition, we mentioned the concept of *nonanticipativity* of the constraint multifunction. By this one means the following: at any time time t there are no constraints induced on z^t beyond those already imposed by $\Gamma_s, s = 1, \ldots, t$; i.e., there are no constraints induced by potential future infeasibilities. This means: if z^t satisfies all the constraints up to time t, there exist functions $z_{t+1}, \ldots, z_T$, such that the resulting z is feasible for the multistage recourse problem. We referred to this, in Section 2, as *relatively complete recourse*. By deriving the induced constraints and including them explicitly in the formulation of the problem, any multistage recourse problem can be reduced to one with relatively complete recourse. However, deriving the induced constraint is not necessarily an easy task, and thus the general optimality theory must (and does) make provisions for the case when Γ is not necessarily nonanticipative, and the solution procedures must (and do) cope with the presence of these induced constraints (by introducing feasibility cuts).

In the choice of a solution technique, we have at our disposal all the experience gained from the study of one- and two-stage models, but all the difficulties that we have encountered so far are compounded by the fact that the number of possible realizations is exponentially increasing with the number of stages, the so-called "curse of dimensionality". The only possible remedy is *decomposition*. Decomposition not only with respect to possible realizations, but also, whenever possible, with respect to time (i.e., stages).

We have seen that introducing the multipliers associated with the nonanticipativity constraints, suggests a potential decomposition with respect to the sample (realization) space. This and the notion of an *average problem* have lead to the *aggregation principle* which allows us to solve any multistage recourse problem, by solving (repeatedly) deterministic versions of the original problem for particular realizations of ξ, sometimes called *scenarios*. The basic idea is captured in the *hedging* algorithm.

3.1   Partial Information, etc.

If instead of observing, or being able to infer, the values assumed by ξ^t, the information to which we have access determines a field $\mathcal{G}_t$, a strict subset of σ-field $\mathcal{B}_t$ of possible events, the (recourse) decision must now be $\mathcal{G}_t$-measurable. Let $Z_{\mathcal{G}}$ be the subspace of Z consisting of all $\mathcal{A}$-measurable functions z so that for all t, z_t is $\mathcal{G}_t$-measurable. This is a linear subspace of Z. The same arguments, and the same conditions as before, except for $\mathcal{B}_t$-nonanticipativity of the constraint-multifunction replaced by $\mathcal{G}_t$-nonanticipativity, will yield the following optimality criterion: if z^* solves the multistage recourse problem, there exist multipliers $q = (q_1, \ldots, q_T)$ defined on Ξ such that

$$E\{q_t(\cdot) \,|\, \mathcal{G}_t\} = 0 \ a.s., \ \text{for } t = 1, \ldots, T,$$

and for almost all ξ:

$$z^*(\xi) \in \operatorname{argmin}\{h_0(z, \xi) - q(\xi) \cdot z \,|\, z^t \in \Gamma(t, \xi), \text{for } t = 1. \ldots, T\}.$$

These conditions are of the same nature as those we already know for the full information case, the only differences are the stronger constraint qualification (nonanticipativity of $\Gamma(t, \cdot)$) with respect to $\mathcal{G}_t$, and the fact that now conditional expectation of q_t is taken with respect to a coarser σ-field. Again there is a rich economic interpretation that can be attached to these multipliers. If p corresponds to the multipliers associated with full information, then $q - p$ yields a price system that could be used to determine if it would be desirable or not, to seek full information; one could think of these multipliers as an information price-system.

As for the two-stage model, it is not always possible to express the information collected (from observations) independently of past decisions. We need to consider also the case when the information fields $\mathcal{G}_t$ depend on $z^{t-1} = (z_1, \ldots, z_T)$; we then write $\mathcal{G}_t(z^{t-1})$. And all the difficulties mentioned in connections with the two-stage model are still all present, except more so. The mathematical complexity generated by asking even the simplest of questions about such models is mind-boggling.

Because the search for an optimal solution will necessarily require, at each iteration a total description of $\xi \mapsto z_t(\xi)$ for all t, the challenge created by *this* formulation of the multistage recourse model may be, for ever, beyond our computational capabilities, unless one replaces the decision space and the sample space by a discrete set. In this discrete case, finding the optimal solution becomes a question of enumerating all possibilities, and this can be organized via dynamic programming techniques. And even that is only possible if the number of decisions in each time step (stage) is rather limited. One other approach is to replace the search for an optimal z^* with the search for the best z in a given class. We return to this in the context of the the stochastic optimal control model.

Finally, we could also have to deal with the dependency of the probability distribution on past decisions z^{t-1}. The added complexity is a function of the form of the relationship between P and z and the properties of h_0 and Γ, when viewed as functions of (z, ξ), just like for the one-stage model.

3.2   Stochastic Optimal Control Models

As we shall see, the formulation of the discrete-time stochastic optimal control model is very similar in nature to that of the multistage recourse models. However, the relationship between these models has not always been very well understood. The basic reason is motivation: the concept of solution is somewhat different in both models. The multistage recourse model is, in many cases, only concerned with z_1, the other decisions are of little interest. The stages 2 to T are only included in the problem to help evaluate the costs that may result from a particular choice of z_1. To the contrary, most of the motivation for the research on stochastic control problems comes from a class of applications where it is the decision *rule* (to be used in all time periods) that is of interest, i.e., the rule that will allow us to pass from observations to decisions. Hence, the insistence of finding a rule that depends on the observed (or estimated) state and not on the information we may infer about the underlying stochastic phenomena. This is only possible if there is a certain similarity between the stages. From a theoretical viewpoint, neither the multistage recourse model nor the stochastic optimal control model is a special case of the other, but there are fundamental differences when it comes to what practitioners will identify as "solvable" problems. Algorithmic research on multistage recourse models is oriented towards mathematical programming techniques, whereas the solution technique favored in the stochastic control literature is dynamic programming. This places natural limitations on the type of problems that can be approached in either way.

We consider the following formulation of a *discrete time, finite horizon, stochastic optimal control* problem:

$$x_t = f_t(x_{t-1}, u_t, \zeta_t^1), \quad t = 1, \ldots, T$$

with initial state x_0 about which we may only have probabilistic information. The variables x_t denotes the *state* of the system, u_t is the *control*, and ζ_t^1 models the system's disturbances (with given probability distribution). The *observations* $y^t = (y_1, \ldots y_t)$ that are available to the controller at time t are related to the state of the system by:

$$y_t = k_t(x_{t-1}, \zeta_t^2), \quad t = 1, \ldots, T$$

where ζ_t^2 are disturbances that affect the observations (again with known probability distribution). The choice of a control law is subject to *system* constraints (*state-space* constraints and *control* constraints):

$$x_t \in X_t, \quad u_t \in U_t, \quad t = 1, \ldots, T,$$

and *information* constraints:

$$\text{for } t = 1, \ldots T, \ u_t \text{ is } \mathcal{Y}_t - \text{measurable,}$$

where $\mathcal{Y}_t$ is the σ-field generated by the observations, i.e., $\mathcal{Y}_t = \sigma\{y_s \mid s \leq t\}$.

The choice of the control u_t must be a (measurable) function of the observations, let us denote it g_t,

$$u_t := g_t(y^t) = g_t(y_1, \ldots, y_t) \in U_t.$$

The vector-valued function

$$g = \{g_1, \ldots, g_T\}$$

is called the *feedback law*. Given g, we can define stochastic processes $\{x_t^g\}, \{y_t^g\}, \{u_t^g\}$ with

$$x_t^g = f_t(x_{t-1}^g, u_t^g, \zeta^1),$$

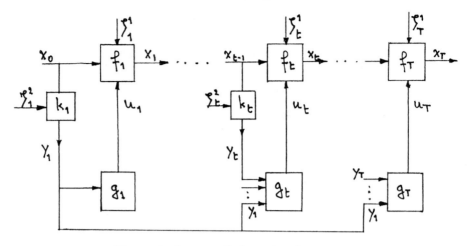

Figure 1: A controlled stochastic system

$$y_t^g = k_t(x_{t-1}^g, \zeta_t^2),$$
$$u_t^g = g_t(y_1^g, \ldots, y_t^g).$$

In the ensuing development, we usually drop the reference to g when referring to u, x or y but it is implicitly always there. Figure 1 gives a block diagram representation of the dynamics of the system.

The objective is to choose a feedback control law g^* that minimizes costs (or maximize performance):

$$J(g) := E^g\{\sum_{t=1}^T c_t(x_{t-1}, u_t, \zeta_1^t) + \Phi(x_T)\} := E\{\sum_{t=1}^T c_t(x_{t-1}^g, u_t^g, \zeta_t^1) + \Phi(x_T)\}.$$

The function Φ plays the role of a terminal condition.

The relation between this model and the multistage recourse model is immediate. Indeed, simply set $z_t := (x_t, u_t)$, $\xi_t := (\zeta_t^1, \zeta_t^2)$, $\xi_0 := x_0$, $\xi := (\xi_0, \ldots, \xi_T)$, $Z_t = X_t \times U_t$,

$$\Gamma(t, \xi) := \{z^t \in Z_t \mid x_t = f_t(x_{t-1}, u_t, \zeta_t^1)\},$$

and

$$h_0(z, \xi) := \sum_{t=1}^T c_t(x_{t-1}, u_t, \zeta_t^1) + \Phi(x_T).$$

The information constraint, which in the case of the stochastic optimal control model is explicitly included in the model in terms of a feedback law, would in the case of the multistage recourse model take the form: z_t must be $\mathcal{B}_t(z^t)$-measurable, where $\mathcal{B}_t(z^t) := \mathcal{Y}_t$.

There are thus no significant differences between these two models, at least as far as formulation goes. Certainly, any general theoretical result known about any one of these models, has a counterpart for the other one. To cite just a couple of examples, the optimality conditions mentioned earlier can easily be reformulated so that they apply to the stochastic control model. Similarly, qualitative results obtained about the value function of stochastic control problems could be applied to the corresponding class of multistage recourse problems. There are a few results that admit easy translation, whereas others are not so readily adaptable. There are two major features of stochastic control models that are not explicitly included in the recourse models. However, the differences are more a matter of perception (and formulation) than factual. First, the stochastic control model includes an explicit expression for the observation process, and second we are to use a feedback law based directly on the actual observations (rather than on the information gathered about "nature": ξ).

As for multistage recourse problems, the major classifications for stochastic control models is based on the type of feedback that will be called for, or/and the level of information that will be available to the controller.

OPEN LOOP : No information is collected that would enable us to adjust earlier decisions. This corresponds to having $y_t \equiv h_t \equiv 0$ for $t = 1, \ldots, T$. The selection of u_t, can as well be made from the very outset. We could extend this model to include those cases that allow for "local" adjustments, i.e., adjustment that are made at time t that do not affect the selected trajectory but try to remedy local deviations from a desired state. This latter case is then of the same nature as the dynamical two-stage model mentioned in Section 2. Such models are sometimes used with a *rolling horizon*, however the use of such an approach cannot always be recommended, since it arbitrarily ignores feedback (or recourse) possibilities that are inherent to all stochastic optimization problems. One further restriction would be to insist on *myopic* controls.

COMPLETE INFORMATION : Full information is available about the

state, i.e.
$$y_t = k_t(x_{t-1}, \zeta_t^2) = x_{t-1};$$

we refer to this case as *full state-information*. This should not be confused with what we have called full information in the framework of the multistage recourse model. In fact, full state-information, may or may not correspond to the full information case. A *nice* case when one can identify full state-information with full information, is when (ζ^1 and ζ^2 are strongly correlated):

$$
\begin{aligned}
y_t &= k_t(x_{t-1}, \zeta_t^2) = \zeta_t^1, \\
x_t &= f_t(x_{t-1}, u_t) + \zeta_t^1.
\end{aligned}
$$

If in addition, the random variables ζ_t are time-independent, then dynamic programming techniques can be used as a solution technique. This is the first time that we encounter in our discussion, this independence condition. This is *not* a modeling choice, but one dictated by the solution technique; more about this later.

PARTIAL INFORMATION : This is the general case. Let us stress once more that this does not correspond to what we have been calling partial information in the context of the multistage recourse model; to make sure that this distinction is not lost, we shall refer to this case as *partial state-information*. Here again is it is possible to appeal to dynamic programming techniques for finding the optimal feedback law. Instead of using the state of the system we rely on on an extended notion of state, viz., conditional distributions (on the state-space) will play the role of the state. These conditional distributions are sometimes called *hyperstates* or *information states*.

FEEDFORWARD : In this case the information available at time t, is either ζ_t^1 or a function of ζ_t^1, in other words the information is a random variable strongly correlated with ζ_t^1. If we take ζ_t^2 to be such a variable, then in terms of the stochastic optimal control problem, we could think of it as the case when

$$y_t = \zeta_t^2.$$

We receive direct information about the underlying stochastic phenomena. Without any need to adjust the information collected, we are in the framework of the multistage recourse model with full or partial information.

RESTRICTED FEEDBACK : Rather than allowing for g to be just any measurable function of the observations, we may want to restrict the class of admissible feedback laws to a particular (parametrized) class of functions. We already discussed this option in the context of the multistage recourse model. From a computational view point, this looks very attractive. But, before we really can use this approach, there are many unresolved theoretical questions that deserve serious investigation. More precisely, we need to characterize, as well as possible, the properties of optimal feedback and obtain error bounds when restrictions are placed on the class of admissible controls. Note that there are some models for which the optimal law is known and can be characterized in terms of a finite number of parameters, e.g., (s, S)-policies, impulse controls, certain bang-bang situations, etc..

The stochastic optimal control model may also include a *filtering* equation, i.e. a process used to analyze the observations in order to obtain an estimate of the state of the system. Instead of using the data that comes from the observations, we are to use the filtered data. If the filter is known *a priori*, then our formulation already allows for such a possibility, we simply define k_t appropriately and take y_t to be the filtered data. If, we are allowed to choose both an optimal control and an optimal filter, the problem is not so simple. In a few cases, one can appeal to the *Separation Lemma* which allows us to first calculate an optimal filter, and use it (redefining k_t) to calculate the optimal feedback law. In general, the situation is unfortunately much more complex. Although this is an important issue, we shall not be concerned with it here; we implicitly assume that we are using raw data (observations) or if it is filtered data (state estimates) the function k_t has been defined so as to include the filtering process.

There is a substantial literature devoted to the characterization of optimality centered around the Hamilton-Jacobi-Bellman equation (discrete or continuous time versions). The suggested solution methods for stochastic control problems are mostly based on solving that equation. They range from discretization (of state-space, controls and possible realizations) to Monte-Carlo simulations passing through finite element approximations of the Hamilton-Jacobi-Bellman equation. We shall only discuss the "discrete" case, and this in the setting of full or partial state-information; for simplicity's sake, we also assume that there no state-space constraints, i.e., no constraints of the type $x_t \in X_t$.

This approach relies on a crucial assumption that has not been needed up to now:

Assumption: The random variables $x_0, \zeta_t^1, \zeta_t^2, \ldots, \zeta_T^1, \zeta_T^2$ are mutually *independent*.

This has the following implication: for all g,

$$P^g\{x_t \in D \mid x_{t-1}, \ldots, x_0, u_t, \ldots, u_1\},$$
$$= P\{x_t \in D \mid x_{t-1}, u_t\} \quad \text{independent of } g,$$
$$= P\{\zeta_t^1 \in Q(x_{t-1}, u_t)\}$$

where

$$Q(x_{t-1}, u_t) := \{\zeta \mid f_t(x_{t-1}, u_t, \zeta) \in D\}.$$

We can reformulate the problem in terms of the following equivalent *Markov Decision Problem*: given the "controlled transition probabilities"

$$P(d\, x_{t-1} \mid x_{t-1}, u_t)$$

and the observation channel transition probabilities,

$$P(d\, y_t \mid x_{t-1}),$$

find $g = (g_1, \ldots, g_T)$, that minimizes

$$E^g \sum_{t=1}^{T} \hat{c}_t(x_{t-1}, u_t),$$

where

$$\hat{c}_t(x, u) := \int c_t(x, u, \zeta_t^1) P(d\, \zeta_t^1).$$

3.2.1   Full state-information

Now, if for all $t = 1, \ldots, T$, full state-information is available, i.e., $y_t \equiv x_t$, we define recursively the real-valued functions:

$$V_T(x_T), \ldots, V_0(x_0),$$

by

$$V_T(x_T) := \Phi(x_T),$$
$$\vdots$$
$$V_t(x_t) := \min_{u \in U_t}\{\hat{c}_t(x_t, u) + \int V_{t+1}(x_{t+1}) P(d\, x_{t+1} \mid x_t, u)\}$$

with $\hat{c}_0 \equiv 0$. Then

$$V_t(x) = \min_g\{E^g \sum_{s=t}^{T} \hat{c}_s(x_s.u_s) \mid x_t = x\}.$$

If

$$g_t^*(x) \in \operatorname*{argmin}_{u \in U_t}\{\hat{c}_t(x, u) + \int V_{t+1}(x_{t+1}) P(d\, x_{t+1} \mid x, u)\},$$

then

$$u_t = g_t^*(x_{t-t}), \text{ for } t = 1, \ldots, T.$$

is the optimal feedback law. In particular, note that u_t is Markovian, in that it only depends on $x_{t-1} = y_t$ and not on earlier observations $y_{t-1}, \ldots, y_1$.

3.2.2   Partial state-information

When only partial information is available, i.e., $y_t \neq x_{t-1}$, let

$$v^t := (y^t, u^{t-1}),$$

denote the information available when choosing u_t. Fix the feedback law g, and define

$$\pi_t^g(d\, x \mid v^t) := P\{x_t^g \in d\, x \mid v^t\}.$$

A fact which is of crucial importance to the development that follows is that π_t^g does not depend on g. It can be shown that there exists an operator S_t, sometimes called a 'filter', such that for $t = 1, \ldots, T$,

$$\pi_{t+1}(\cdot \mid v^{t+1}) = S_t[\pi_t(\cdot \mid v^t), y_{t+1}, u_t]$$

and

$$\pi_1(d\, x \mid v^1) = P\{x_0 \in d\, x \mid y_1\}.$$

Let Π be the space of all probability distributions on the state-space. For example, if $x_t \in \{1, \ldots, I\}$, then

$$\Pi = \{\pi_1, \ldots, \pi_I \mid \sum_1^I \pi_i = 1, \pi_i \geq 0\}.$$

In a manner similar to that used in the full state-information case, we define real-valued functions, but on Π, the hyperstate-space:.

$$V_T(\pi) \quad := \quad E\{\Phi(x) \mid \pi_T(\cdot \mid v^T) = \pi\},$$

$$\vdots$$

$$V_t(\pi) \quad = \quad \min_{u \in U_t} E\{\hat{c}_t(x, u) + V_{t+1}(S_t[\pi, y_{t+1}, u]) \mid \pi_t(\cdot \mid v^t) = \pi\}.$$

Then, for all g,

$$V_t(\pi_t(\cdot \mid v^t)) \leq E^g\{\sum_{s=t}^T c_s(x_{s-1}, u_s) \mid (y^t, u^{t-1})\},$$

and

$$u_t = g_t^*(\pi_t(\cdot \mid v^t))$$

is the optimal feedback law, where g_t^* is the argument that yields the minimum in the expression that defines V_t.

3.2.3   Computational implications

We have given a rather detailed description of the theoretical underpinnings of the methods used in practice to solve discrete-time stochastic optimal control problems. The reason is that we want to stress the differences between this approach and that favored for multistage recourse models. In both cases, full or partial state-information, the strategy has been to reduce the control problem to a Markov decision problem. To achieve this and to be able to solve the problem, we had to impose two unwelcome restrictions:

1. time-independence of the random variables plus independence between the disturbances that affect state and observations (although this latter restriction is inessential),

2. finite state-space, which in turn implies finitely distributed random variables and discrete control space.

These limitations are not always easy to justify in applications. At our present stage of development, that seems to be the price that needs to be paid to build a feedback control law based on information obtained about the state of the system rather than information about the underlying stochastic process.

Unless the state-space is actually discrete and the underlying stochastic process $\{\zeta_t\}_{t=1}^T$ consists of independent random variables, the solution obtained by solving the Markov decision model is, at best, an approximation of the problem at hand.

4   CONTINUOUS-TIME MODELS

We shall be very brief: there is not much to report from a (practical) computational viewpoint. Although the discrete time model did allow for a wide variety of stochastic disturbances, the only case that has really been studied in continuous-time is when the disturbances can be modeled by white noise (although, now, there are also martingale techniques). Defining the variables as the obvious continuous-time analogues of those of the discrete-time models, the *continuous-time* recourse model takes the form:

$$\text{minimize} \quad E\{\int h_0(z_t(\xi), \xi_t)dt\}$$
$$\text{subject to} \quad z_t \in \Gamma(t, \xi) \qquad \text{for all } t,$$
$$z_t \, \mathcal{B}_t - \text{measurable} \quad \text{for all } t,$$

where $\xi = (\xi_t)$ is a (continuous-time) stochastic process, $\mathcal{B}_t$ is a σ-field generated by earlier observations that may or may not depend on past decisions. Again the question of the nonanticipativity of the constraint-multifunction needs to be broached, and it plays a role in the type of conditions that can be used to characterize optimal solutions, etc..

The continuous-time version of the stochastic control problem that has received most of the attention in the literature is:

$$\text{minimize} \quad E\{\int c_t(x_t, u_t, \zeta_t^1)dt\}$$
$$\text{such that} \quad dx_t = f_t(x_t, u_t)dt + \sigma_1(x_t)d\zeta_t^1, \quad \text{for } t \in [0, T],$$
$$dy_t = k_t(x_t)dt + \sigma_2(x_t)d\zeta_t^2,$$

where ζ_t^1 and ζ_t^2 are Wiener processes (or more generally semi-martingales) that model disturbances that affect system and observations. The variable u_t is the control that is subject to the information constraint:

$$u_t \text{ is } \mathcal{Y}_t - \text{measurable,}$$

with $\mathcal{Y}_t$, as before, the σ-field generated by the observations $\{y_s \,|\, s \leq t\}$. There are some technical difficulties with giving a precise meaning to this constraint. To do so, one usually relies on a measure transformation (Girsanov's Lemma).

The continuous-time versions of the multistage recourse model as well as the stochastic optimal control model are (mathematical) analyst's delight. As soon as one goes beyond the quadratic regulator problem (a linear-quadratic model), there are essentially no closed-form solutions and most of the theory has been oriented at finding qualitative characterizations of optimal solutions. One could consult the work of Back and Pliska for the continuous-time recourse model, and that of Krasovskii, Fleming, Rishel, Kushner, Varaiya, Bensoussan, Evans, Lions (père & fils), Davis, Krylov and many others, for the continuous-time stochastic control model. The most computationally oriented work is probably that of Haussmann (Univ. British Columbia), beginning with his work on the stochastic maximum principle. However, very little success can be reported about the passage from theory to computationally implementable techniques; we exclude here, for obvious reasons, methods based on Monte-Carlo simulations and stochastic approximation techniques (that have a limited range of applicability).

Certain continuous time models have equivalent discrete-time (or discrete state-space) formulation, and sometime this can be exploited to solve (by successive approximations) more complicated problems. Let us give two examples. If the dynamics of the system are described by a continuous-time Markov chain (finite state-space), i.e.,

$$P\{x_{t+dt} = j \mid x_t = i, u_t = u\} = p_{ij}(u)dt,$$

it is usually possible to convert to problem to one in discrete time by a technique know as *uniformization*. The second example is a little bit more involved. It is a class of problems studied first by Vermes (Hungary), and at present, under further investigation by Davis (Imperial College). The state at any time t is the sum of a jump process (Markov jumps that occur at random times) and a dynamical system described by an ordinary differential equation that can be controlled. Certain maintenance problems and capacity expansion problems are easy to cast in this mold. Problems of this type can be converted to multistage recourse problems (possibly with an infinite number of stages), where each stage corresponds to the evolution that takes places between jumps and the (recourse) costs are random variables whose values depend on the length of time between jumps.

5 REFERENCES

[1] Marc H.A. Davis, "Martingale methods in stochastic control," in *Stochastic Control Theory and Stochastic Differential Systems*, M. Kohlmann & W. Vogel, eds., Springer Lecture Notes in Control and Information Sciences 16, Berlin, 1979, 85–117.

[2] Marc H.A. Davis, "Piecewise-deterministic Markov Processes: a general class of non-diffusion stochastic models," *J. Royal Statistical Society, Series B* **46(3)** (1984).

[3] Wendell H. Fleming & Raymond W. Rishel,, *Deterministic and Stochastic Optimal Control*, Springer Verlag, Berlin, 1975.

[4] P. R. Kumar & Pravin Varaiya, *Stochastic Systems. Estimation, Identification and Adaptive Control*, Prentice-Hall, Englewood Cliffs, 1986.

[5] Harold J. Kushner,, *Probability Methods for Approximations in Stochastic Control and for Elliptic Equations*, Academic Press, New York, 1977.

[6] Hans S. Witsenhausen, "A counterexample in stochastic control," *SIAM. J. on Control* **6(1)** (1968.).

Optimization with Multiple Objectives

F. A. LOOTSMA

Faculty of Mathematics and Informatics, Delft University of Technology, P.O. Box 356, 2600 AJ Delft, Netherlands

Abstract

We are concerned with a typical multi–objective optimization problem: there is a continuum of alternatives (described by a finite number of variables under a finite number of constraints), a finite number of conflicting objectives, and a decision–making committee with diverging viewpoints. First, we introduce some basic concepts such as efficient solutions and dominance cones, and we consider the geometric properties of the efficient set. Next, we discuss several classes of methods for solving the problem: multi–objective simplex methods, constraint methods with trade–offs, and ideal–point methods. Because the potential of these methods depends critically on the information–processing capacity of human beings, we finally concentrate on the interface between the decision makers and the methods, that is, the nature and the number of items to the judged in the decision process.

1. Introduction

In many real–life problems, one is usually confronted with several objectives which are in mutual conflict. A production manager, for instance, who is responsible for the operations in a plant, does not always want to maximize his profits only. For strategic reasons, he may also pursue the goal of minimizing the utilization of scarce resources in order to avoid their exhaustion. In energy models, several objective functions have been incorporated

M. Iri and K. Tanabe (eds.), Mathematical Programming, 333–364.
© *1989 by KTK Scientific Publishers, Tokyo.*

reflecting the critical issues in the energy debate such as minimization of costs, minimization of air pollution (SO_2, NO_x and dust emissions), minimization of nuclear power generation, minimization of net oil imports, minimization of natural gas consumption from domestic fields, and maximization of profits on domestic natural gas.

Goal programming (fully described by Ignizio (1976)) is one of the earliest tools for solving a multi–objective linear programming (MOLP) problem. For each of the objectives, the decision maker sets certain goals which should be satisfied as nearly as possible. In order to remain within the framework of linear programming, we take the sum of the absolute deviations from the respective goals as a quality index for a feasible solution. Minimization of the sum will, it is hoped, yield a compromise solution which is acceptable for the decision maker.

The present paper delves deeper into linear–programming problems with multiple objectives. First, we characterize and explore the set of efficient (non–dominated) feasible solutions. Thereafter, we concern ourselves with techniques for finding a feasible solution where the conflicting objectives are in a proper balance (at least in the eyes of the decision maker). The underlying ideas can mostly be generalized to solve nonlinear–programming problems with multiple objectives. We shall accordingly concentrate on the linear case.

This paper cannot be complete. Today, there are many excellent textbooks covering a wide range of techniques and applications (Keeney and Raiffa (1976), Cohon (1978), Zeleny (1982), Chankong and Haimes (1983), Yu (1985) and Steuer (1986), to mention just a few), so that we can only give a concise description of the main stream of ideas. In order to do so, we focus our attention on the key theorems, with simplified proofs.

Finally, it is worth noting that multi–objective optimization is basically concerned with a continuum of alternatives: the feasible solutions in the constraint set. In multi–criteria decision analysis (MCDA) we have a finite (usually small) number of alternatives (see, for instance, Saaty (1980), Hwang and Yoon (1981), Roy (1985)). Although the two fields are unrelated, at least at first sight, many techniques in MCDA have much to offer when we fathom the preferences of the decision makers. In fact, multi–objective optimization has two aspects. On the one hand, there is the identification of the efficient solutions, which can be carried out with the familiar mathematical objectivity. On the other hand, a choice must be made from the set of efficient solutions: we have to support the decision maker in selecting an efficient solution such that the objectives are in a proper balance, at least in his own, subjective opinion. Regrettably enough, human subjectivity is somewhat ignored in the literature on MOLP. This will be discussed at the end of the present paper.

2. Efficient (Non–dominated) Solutions

We consider here the linear problem of maximizing the objective functions

$$(c^i)x; \ i=1, \ ..., \ p \ ,$$

subject to the constraints

$$Ax \le b$$
$$x \ge 0, x \in E_n.$$

(1)

Let $\bar{x}^i$ denote a feasible solution when the i–th objective function is maximized. Mostly, the $\bar{x}^i$ do not coincide, and we are forced to find some compromise. We are accordingly led to the concept of the so–called efficient solutions, that is, a feasible solution $\tilde{x}$ is referred to as an efficient (non–dominated) solution if there is no feasible x such that

$$(c^i)^T x \ge (c^i)^T \tilde{x}; \qquad\qquad i = 1, \ ..., \ p \ ,$$

$$(c^k)^T x > (c^k)^T \tilde{x} \qquad\qquad \text{for some k, } 1 \le k \le p.$$

(2)

Obviously, no objective function can be improved by moving away from $\tilde{x}$ without a reduction of at least one of the other objective functions. A clear, geometrical picture is obtained when we use the so–called dominance cone

$$\Delta = \{x \,|\, (c^i)^T x \ge 0; \ i = 1, \ ..., \ p\} \ .$$

(3)

Under the additional regularity condition that at least one of the objective functions must change when we move away from any point in E_n, a feasible point $\tilde{x}$ is efficient if, and

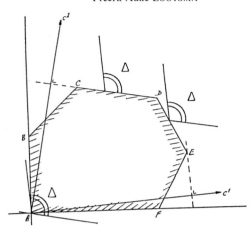

Figure 1. Linear Programming problem with two objective functions. The second objective has alternative solutions. Any point on the segment DE is efficient, any point on CD, with the exception of D, is weakly efficient (move the dominance cone Δ along the boundary of the constraint set ABCDEF).

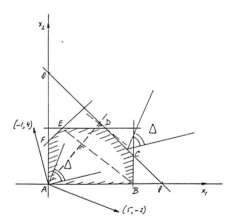

Figure 2. Linear-Programming example in decision space. A = (0,0), B = (6,0), C = (6,2), D = (4,4), E = (1,4), F = (0,3). The points on ED, DC, and CB are efficient (move the dominance cone Δ along the boundary of the constraint set.

only if, there is no other feasible point x such that $x - \bar{x} \in \Delta$. Moreover, the feasible point $\bar{x}$ is weakly efficient (weakly non–dominated) if there is no feasible point x such that $x - \bar{x}$ is in the interior of Δ. This implies that there is no feasible point x such that

$$(c^i)^T x > (c^i)^T \bar{x}, \ i = 1, ..., p. \tag{4}$$

In other words, it is impossible to improve all objective functions from a weakly efficient $\bar{x}$ simultaneously, but it may be possible to improve some objectives without a reduction of the other ones. In the two–dimensional example of Figure 1, with unique and alternative maximum solutions for the first and the second objective function respectively, the reader will find weakly efficient solutions (any point on the segment CD, except D) and efficient solutions (any point on the segment DE). This will be verified later in this section.

Frequently, the suggestion is made to solve the multi–objective linear–programming problem by maximizing a linear combination of the objective functions with positive weight coefficients $\lambda_1, ..., \lambda_p$. Thus, we consider the combination

$$\sum_{i=1}^{p} \lambda_i (c^i)^T x = \lambda^T C x \,,$$

and we are concerned with the problem

$$
\left.
\begin{aligned}
\text{maximize} \quad & \lambda^T C x \\
\text{subject to} \quad & Ax \leq b \\
& x \geq 0 \,.
\end{aligned}
\right\} \tag{5}
$$

The following theorem, which provides a justification for the idea to find an efficient solution of (1) by solving (5), is a key theorem in multi–objective programming.

Theorem 1 (Geoffrion (1968)). A feasible solution $\bar{x}$ of (1) is an efficient solution if, and only if, there exist positive weight coefficients $\lambda_1, ..., \lambda_p$ such that $\bar{x}$ is a solution of (5).

<u>Proof.</u> If $\bar{x}$ is a maximum solution of (5), then $\bar{x}$ is an efficient solution of the original multi–objective programming problem (1); if not, it would be possible to improve the objective function $\lambda^T Cx$ by moving away from $\bar{x}$. On the other hand, if $\bar{x}$ is an efficient solution of problem (1), then it is a maximum solution of the problem

$$
\left.
\begin{aligned}
&\text{maximize} \quad && (c^k)^T x \\[2mm]
&\text{subject to} \quad && (c^i)^T x \geq (c^i)^T \bar{x}, \ i \neq k , \\[2mm]
& && Ax \leq b \\
& && x \geq 0
\end{aligned}
\right\} \tag{6}
$$

for any $k = 1, ..., p$. Hence, there exist non–negative multipliers $\alpha_1^{(k)}, ..., \alpha_p^{(k)}, u_1^{(k)}, ..., u_m^{(k)}, v_1^{(k)}, ..., v_n^{(k)}$ such that

$$
c^k = - \sum_{\substack{i=1 \\ i \neq k}}^{p} \alpha_i^{(k)} c^i + \sum_{i=1}^{m} u_i^{(k)} a^i - \sum_{j=1}^{n} v_j^{(k)} e^j , \tag{7}
$$

where a^i stands for the i–th row of the matrix A, and e^j for the j–th unit vector. Furthermore, the multipliers satisfy the complementary slack relations at $\bar{x}$. We rewrite (7) as

$$
c^k + \sum_{\substack{i=1 \\ i \neq k}}^{p} \alpha_i^{(k)} c^i = \sum_{i=1}^{m} u_i^{(k)} a^i - \sum_{j=1}^{n} v_j^{(k)} e^j .
$$

Summing over k, and defining

$$
\lambda_i = 1 + \sum_{k \neq i} \alpha_i^{(k)} > 0 ,
$$

$$
\bar{u}_i = \sum_{k=1}^{p} u_i^{(k)} \geq 0 ,
$$

$$\overline{v}_j = \sum_{k=1}^{p} v_j^{(k)} \geq 0 \, ,$$

we obtain straightaway the relations

$$\sum_{i=1}^{p} \lambda_i c^i = \sum_{i=1}^{m} \overline{u}_i a^i - \sum_{j=1}^{n} \overline{v}_j e^j \, ,$$

whereas it is easy to verify that the multipliers $\overline{u}_1, \ldots, \overline{u}_m, \overline{v}_1, \ldots, \overline{v}_n$ satisfy the complementary slack relations

$$\overline{u}^T(A\overline{x} - b) = 0,$$
$$\overline{v}^T\overline{x} = 0.$$

These results imply that $\overline{x}$ maximizes the combined function $\lambda^T Cx$ over the feasible set. Hence, for any efficient solution there are positive weight coefficients (not necessarily unique) that could have been used to generate the solution in question. This proves Geoffrion's theorem, thereby characterizing the efficient (non–dominated) solutions of (1).

What happens if we maximize a linear combination $\lambda^T Cx$ with some zero weight coefficients? That depends on the uniqueness of the single–objective maximum solutions. Suppose, we maximize the i–th objective function regardless of the other ones, that is, $\lambda_k = 0$, $k \neq i$, $\lambda_i = 1$, and let $\overline{x}^i$ denote a maximum solution. If $\overline{x}^i$ is the unique solution, it is also efficient because whatever we do to improve the other objectives, as soon as we move away from $\overline{x}^i$ to another feasible solution, the i–th objective will decrease. So, <u>unique single–objective maximum solutions are efficient</u>. The two–dimensional example of Figure 1, however, shows that maximization of the linear combination

$$\lambda_1(c^1)^T x + \lambda_2(c^2)^T x$$

with $(\lambda_1,\lambda_2) = (0,1)$ does not necessarily yield an efficient solution if the second objective has <u>alternative maximum solutions</u>. Any point on the line segment CD maximizes the

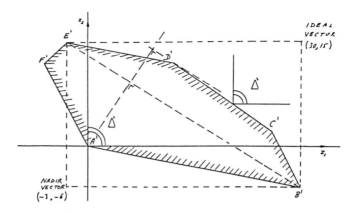

Figure 3. The same example in objective space. A' = (0,0),
B' = (30,-6), C' = (26,2), D' = (12,12), E' = (-3,15), F' = (-6,12).
The ideal vector (30,15) and the nadir vector (-3,-6) are both
unfeasible. The cross-effect matrix of this problem is given by

$$\begin{pmatrix} 30 & -3 \\ -6 & 15 \end{pmatrix}$$

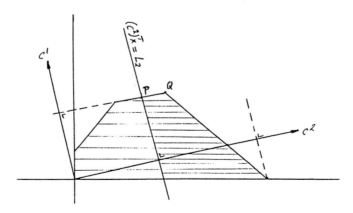

Figure 4. Any point on the line segment PQ maximizes $(c^1)^T x$ subject
to the constraints $(c^2)^T x \geq L_2$, $Ax \leq b$, $x \geq 0$, but with the exception
of Q they are all weakly efficient only (compare with Figure 1).

combination, but with the exception of D these points are weakly efficient only, not efficient. When we move along CD to D, the first objective improves without reducing the second objective. The points on the segment DE are indeed efficient. The reader will be able to verify this by "moving" the dominance cone along the boundary of the feasible set.

Decision space and objective space. It is easier to visualize the multi–objective programming problem (1), not in the n–dimensional decision space of vectors x, but in the p–dimensional objective space of vectors z = Cx. To illustrate matters, we consider the linear–programming problem

$$
\begin{aligned}
\text{maximize} \quad & z_1 = 5x_1 - 2x_2 \\
& z_2 = -x_1 + 4x_2 \\[1em]
\text{subject to} \quad & -x_1 + x_2 \le 3 \\
& x_1 + x_2 \le 8 \\
& x_1 \le 6 \\
& x_2 \le 4 \\
& x_1, x_2 \ge 0 .
\end{aligned}
\tag{8}
$$

In the two–dimensional decision space, the feasible set is the polygon ABCDEF exhibited by Figure 2. In the two–dimensional objective space the mapping z = Cx yields the feasible set A'B'C'D'E'F' which is shown in Figure 3. The efficient solutions can easily be identified here. Suppose we have two points z^1 and z^2 in the objective space, then z^2 dominates z^1 if, and only if, $z^2 \ge z^1$ and $z^2 \ne z^1$. So, the dominance cone is simply given by the set $\{z \,|\, z \ge 0\}$. By "moving" the cone along the boundary of the polygon one obtains that the efficient solutions of problem (8) are on the line segments B'C', C'D' and D'E' in the objective space or, equivalently, on the segments BC, CD, and DE in the decision space.

3. The Cross–Effect Matrix

A clear indication of the conflict between the respective objectives is given by the cross–effect matrix Q with components

$$q_{ij} = (c^i)^T \bar{x}^j, \tag{9}$$

where $\bar{x}^j$ stands for a single–objective maximum solution when the j–th objective function is optimized. The <u>ideal vector</u> z^* is the <u>unique</u> vector with components

$$z_i^* = (c^i)^T \bar{x}^i, \; i=1, ..., p \; . \tag{10}$$

It is obviously the main diagonal of the cross–effect matrix. The <u>nadir vector</u> n^* has as components the row minima

$$n_i^* = \min_{j=1,\,..,p} (c^i)^T \bar{x}^j, \; i = 1,..., p \; ,$$

which are not necessarily unique because some objective functions may have alternative maximum solutions. The ideal and the nadir vector of problem (8) are also shown in Figure 3. The ideal vector is usually unfeasible; otherwise there would be no <u>conflicting</u> objectives. The nadir vector may be feasible or unfeasible.

Even if the single–objective maximum solutions $\bar{x}^1, ..., \bar{x}^p$ are efficient, the components of the nadir vector satisfy the inequalities

$$n_i^* \geq \min_{x \in E} (c^i)^T x, \; i = 1,..., p, \tag{11}$$

where E stands for the set of efficient solutions in decision space (equality in (11) cannot be guaranteed, see also Isermann and Steuer (1985)). In many applications, however, we take the nadir vector as a pessimistic estimate for the outcome of the respective objective functions when decision makers balance their feelings, in contrast to the optimistic, ideal vector which is mostly not attainable.

In order to guarantee efficiency of the single–objective solutions, we may use <u>lexicographic maximization</u>. The point $\bar{x}^i$, for instance, is generated as follows. First, we maximize the i–th objective over the feasible set to obtain the maximum value $c_i^* = z_i^*$. If the maximum solution is not unique, we maximize the (i+1)–th objective over the feasible set <u>and</u> subject to the additional constraint

$$(c^j)^T x = c_j^*.$$

Let c_{i+1}^* stand for the maximum value, and note that $c_{i+1}^* \leq z_{i+1}^*$. If the maximum solution is not unique, we maximize the $(i+2)$-th objective over the feasible set <u>and</u> subject to the additional constraints

$$(c^j)^T x = c_i^*, (c^{i+1})^T x = c_{i+1}^*.$$

We continue the procedure until a unique maximum solution has been found (this is not possible in the unusual case that all objective functions have exactly the same set of alternative maximum solutions). The point $\bar{x}^i$ so obtained is efficient because a feasible solution dominating it does not exist.

4. Bi-criterion problems

In problems with two objective functions, the efficient set can easily be explored by the construction of a new objective function which is indifferent between the efficient solutions generated so far (the Non-Inferior Set Estimation method of Cohon (1978)). We describe the construction first in p dimensions. We let $\bar{x}^1$, .., $\bar{x}^p$ stand for the efficient single-objective maximum solutions of the respective objective functions, we suppose that the cross-effect matrix Q is also available, and we construct a linear combination

$$c^*(x) = \sum_{i=1}^{p} \lambda_i^* (c^i)^T x \tag{12}$$

with weight coefficient λ_i^*, $i = 1, ..., p$, such that

$$c^*(\bar{x}^1) = ... = c^*(\bar{x}^p),$$

that is, we compute weights λ_i^* so that the new objective function (12) is indifferent to the maximum solutions $\bar{x}^1, ..., \bar{x}^p$. The weights have to satisfy the equations

$$\lambda_1^* q_{11} + \cdots + \lambda_p^* q_{p1} = 1$$

$$\vdots \qquad\qquad \vdots$$

$$\lambda_1^* q_{1p} + \cdots + \lambda_p^* q_{pp} = 1$$

$$\left.\begin{array}{c} \\ \\ \\ \\ \\ \end{array}\right\}$$

or equivalently,

$$Q^T \lambda^* = e ,\tag{13}$$

where the right–hand side e stands for the p–vector with components 1 (note that the value 1 is not essential; what matters is equality of the right–hand side elements). If we find a <u>positive</u> solution to (13), we can further maximize the compromise objective function (12). The maximization will eventually yield a new efficient solution.

Applying this procedure to problem (5), we would start off with the single–objective maxima B and E in Figure 2. The new objective function would have contours parallel to BE, and maximization of this function would yield the point D. Starting off with B and D, we would obtain C, but continuation from B and C would show that the new objective function cannot be improved further.

If there are only two objective functions, and if the cross–effect matrix has elements q_{ij} such that $q_{11} > q_{12}$ and $q_{22} > q_{21}$, the weights λ_i^*, i = 1,2, computed by (13) will always be positive. This is not necessarily true if there are three or more objective functions. We show this by analyzing a three–dimensional example (due to Rietveld (1980)) written as

$$\begin{array}{lll} \text{maximize} & 100x_1 \\ & 75x_1 + 100x_2 \\ & 40x_1 \qquad\qquad + 100x_3 \\[1em] \text{subject to} & x_1 + x_2 + \quad x_3 = 1 \\[1em] & x_1, x_2, x_3 \geq 0 . \end{array} \right\} \tag{14}$$

The unit vectors e^1, e^2, and e^3 spanning the feasible set are the unique maximum solutions of the respective objective functions; hence, they are efficient. The cross–effect matrix is

$$\begin{pmatrix} 100 & 0 & 0 \\ 75 & 100 & 0 \\ 40 & 0 & 100 \end{pmatrix},$$

and the system (13) has the solution $\lambda_1^* = -0.0015$, $\lambda_2^* = \lambda_3^* = 0.01$, so that the construction of new efficient solutions breaks down.

5. Geometric Properties of the Efficient Set

We consider again problem (1) and the set E of efficient solutions, before we turn to methods for generating some points of E (generating all of them is mostly intractable).

Theorem 2 (Yu and Zeleny (1975)). Let x^1 and x^2 be feasible. If x^1 is not efficient then any point $x = \lambda x^1 + (1-\lambda)x^2$, $0 < \lambda \leq 1$, is not efficient.

Proof. There is a feasible point y dominating x^1, so we have

$$Cy \geq Cx^1, \ Cy \neq Cx^1 .$$

It is easy to show that $\lambda y + (1-\lambda)x^2$ dominates $\lambda x^1 + (1-\lambda) x^2$ for any $0 < \lambda \leq 1$ because

$$C(\lambda y + (1-\lambda)x^2) = \lambda Cy + (1-\lambda)Cx^2 \geq \lambda Cx^1 + (1-\lambda)Cx^2 = C(\lambda x^1 + (1-\lambda)x^2),$$

and

$$C(\lambda y + (1-\lambda)x^2) \neq C(\lambda x^1 + (1-\lambda)x^2).$$

<u>Corollary</u>. The set of non–efficient feasible solutions is convex.

This is not necessarily true for the efficient set E, as Figure 2 and Figure 3 show.

<u>Theorem 3 (Yu and Zeleny (1975))</u>. If the feasible set is bounded, so that it is the convex hull of a finite number of vertices x^1, ..., x^r, then any efficient solution $\tilde{x}$ is in the convex hull of the <u>efficient</u> vertices.

<u>Proof</u>. Assume the contrary, then there is a non–efficient vertex x^k, $1 \leq k \leq r$, such that

$$\tilde{x} = \sum_{j=1}^{r} \alpha_j x^j, \ \sum_{j=1}^{r} \alpha_j = 1 ,$$

$$\alpha_j \geq 0, j = 1, ..., r, \ \alpha_k > 0 .$$

If $\alpha_k = 1$, then $\tilde{x} = x^k$ and $\tilde{x}$ would not be efficient. Hence, $0 < \alpha_k < 1$. Now

$$\tilde{x} = \alpha_k x^k + (1-\alpha_k) \sum_{\substack{j=1 \\ j \neq k}}^{r} (\frac{\alpha_j}{1-\alpha_k}) x^j = \alpha_k x^k + (1-\alpha_k)y .$$

The previous theorem implies, since x^k and y are feasible and x^k is not efficient, that $\tilde{x}$ is not efficient. This contradiction proves the theorem.

Reversely, Figure 2 and Figure 3 show that a point in the convex hull of the efficient vertices is not necessarily efficient.

<u>Theorem 4</u>. Let $\tilde{x}$ and x^* be feasible solutions with exactly the same active constraints. Then $\tilde{x}$ is efficient if, and only if, x^* is efficient.

<u>Proof</u>. This theorem is a direct consequence of Geoffrion's theorem 1. Suppose that $\tilde{x}$ is efficient, then there is a positive weight vector $\tilde{\lambda}$ such that $\tilde{\lambda}^T Cx$ is maximized over the feasible set at $\tilde{x}$. The necessary and sufficient conditions for optimality are that there exist

non–negative vectors $\tilde{u}$ and $\tilde{v}$ of multipliers satisfying

$$\tilde{\lambda}^T C = \tilde{u}^T A - \tilde{v} ,$$

$$\tilde{u}^T(A\tilde{x} - b) = 0 ,$$

$$\tilde{v}^T \tilde{x} = 0 .$$

At the point x^* it must be true that

$$\tilde{u}^T(Ax^* - b) = 0$$

$$\tilde{v}^T x^* = 0 ,$$

implying that $\tilde{\lambda}^T Cx$ is also maximized at x^*. This proves the theorem.

6. Multi–Objective Simplex Tableaux

We still consider problem (1), but we now convert all constraints and objective functions into equalities by introducing slack variables x_{n+i} such that

$$\sum_{j=1}^{n} a_{ij} x_j + x_{n+i} = b_i; i = 1,..., m , \qquad (15)$$

and so–called goal variables z_i such that

$$z_i - \sum_{j=1}^{n} c_{ij} x_j = 0; i = 1,..., p . \qquad (16)$$

Moreover, we consider the maximization of $\lambda^T Cx$ for positive weights $\lambda_1, ..., \lambda_p$, and we introduce the equation

$$z_0 - \sum_{j=1}^{n} \sum_{i=1}^{p} \lambda_i c_{ij} x_j = 0 . \tag{17}$$

With the additional notation

$$w_{ij} = - c_{ij}; \; i = 1, \ldots \; p, \; j = 1, \ldots, n ,$$

$$w_{0j} = - \sum_{i=1}^{p} \lambda_i c_{ij} = \sum_{i=1}^{p} \lambda_i w_{ij}; \; j = 1, \ldots, n ,$$

we can form a multi–objective simplex tableau with the top row derived from (17), then p objective rows derived from (16), and finally m constraint rows with the coefficients taken from (15). Assuming that the right–hand side b is non–negative, we take the slack variables to be the basic variables. Let us now investigate the behaviour of the top row (the positive combination of objective functions) under the familiar pivot operation. Suppose $a_{rk} \neq 0$ is the pivot element, then w_{ij} $(1 \leq i \leq p, \; 1 \leq j \leq n)$ is transformed into w'_{ij} by

$$w'_{ij} = w_{ij} - \frac{w_{ik} \cdot a_{rj}}{a_{rk}} ,$$

and top–row elements w_{0j} yield

$$w'_{0j} = w_{0j} - \frac{w_{0k} \cdot a_{rj}}{a_{rk}} .$$

Hence,

$$w'_{0j} = \sum_{i=1}^{p} \lambda_i w_{ij} - \frac{\sum_{i=1}^{p} \lambda_i w_{ik} \cdot a_{rj}}{a_{rk}} =$$

$$= \sum_{i=1}^{p} \lambda_i (w_{ij} - \frac{w_{ik} \, a_{rj}}{a_{rk}}) = \sum_{i=1}^{p} \lambda_i w'_{ij} .$$

The top row remains clearly a linear combination of the objective rows, with exactly the same coefficients λ_1, ..., λ_p. We use this result to analyze Rietveld's example (14), in particular the possible efficiency of the vertices and their connecting edges, via the simplex tableau without the top row (17) explicitly added to it. Let us first take x_3 to be the basic variable, then the original system

$$
\begin{aligned}
z_1 && -100x_1 &&&& = 0 \\
& z_2 & -75x_1 - 100x_2 &&&& = 0 \\
&& z_3 & -40x_1 && -100x_3 & = 0 \\
&&& x_1 + x_2 + x_3 & = 1
\end{aligned}
\tag{18}
$$

is transformed into

$$
\begin{aligned}
z_1 && -100x_1 &&&& = 0 \\
& z_2 & -75x_1 - 100x_2 &&&& = 0 \\
&& z_3 & +60x_1 + 100x_2 &&& = 100 \\
&&& x_1 + x_2 + x_3 & = 1
\end{aligned}
\tag{19}
$$

Setting the non–basic variables to zero we find ourselves in the point $(0,0,1)$. Is it an efficient solution? Can we find a positive linear combination of the objective function satisfying the familiar optimality condition (non–negative coefficients in the transformed top row)? Then we have to check whether the system

$$
\begin{aligned}
-100\lambda_1 \quad -75\lambda_2 + 60\lambda_3 &\geq 0 \\
-100\lambda_2 + 100\lambda_3 &\geq 0
\end{aligned}
\tag{20}
$$

has a positive solution. This is indeed the case (we may take $\lambda_1 = \lambda_2 = 0.1$, $\lambda_3 = 0.8$). Incidentally, we see that the set of positive λ–vectors maximizing $\lambda^T Cx$ at $(0,0,1)$ is contained in a convex polyhedral cone, namely the cone defined by (20) and the non–negativity requirements.

Let us now traverse the edge connecting the current vertex $(0,0,1) = e^3$ and the vertex $(1,0,0) = e^1$; this is accomplished by a steady increase of x_1 from 0 to 1. Is it an efficient edge? To answer the question we imagine ourselves in the (x_1,x_2)–space of non–basic variables, where the problem under consideration (see system (19)) can be written as

$$
\left.
\begin{array}{ll}
\text{maximize} & 100x_1 \\
& 75x_1 + 100x_2 \\
& -60x_1 - 100x_2 \\
\\
\text{subject to} & x_1 + x_2 \leq 1 \\
& x_1 \geq 0, x_2 \geq 0.
\end{array}
\right\}
\tag{21}
$$

When x_1 increases, we traverse the edge between $(0,0)$ and $(1,0)$, and this is an efficient edge if we can find a positive linear combination of the objective functions which is maximized on the edge. Thus, we check whether the system

$$
\left.
\begin{array}{l}
-100\lambda_1 - \quad 75\lambda_2 \quad + 60\lambda_3 = 0 \\
\\
-100\lambda_2 + 100\lambda_3 \geq 0
\end{array}
\right\}
\tag{22}
$$

has a positive solution. This appears to be true ($\lambda_3 = 1$, $\lambda_1 = \lambda_2 = 60/175$).

We return to the original, three–dimensional decision space. The edge connecting e^3 and e^1 is efficient. Hence, the endpoint e^1 must be efficient, otherwise (see theorem 2) the edge connecting e^3 and e^1 would be inefficient. Obviously, <u>starting at an efficient vertex and proceeding via an efficient edge, we arrive at an efficient vertex.</u>

The reader can easily verify that the edge connecting e^3 and e^2 is inefficient because the system

$$
\left.
\begin{array}{l}
-100\lambda_1 - \quad 75\lambda_2 + \ 60\lambda_3 \geq 0 \\
\\
-100\lambda_2 + 100\lambda_3 = 0
\end{array}
\right\}
\tag{23}
$$

has no positive solution (it must be true that $\lambda_2 = \lambda_3$ whence $-100\lambda_1 - 15\lambda_2 \geq 0$, a contradiction). In order to check whether e^2 is efficient, we take x_2 as the basic variable so that the original system (18) is transformed into

$$
\left.
\begin{array}{llll}
z_1 & -100x_1 & & = 0 \\
\quad z_2 & +\ 25x_1 & +100x_3 & = 10 \\
\quad\quad z_3 & -\ 40x_1 & -100x_3 & = 0 \\
& x_1\ +\ x_2\ +\ x_3 & & = 1.
\end{array}
\right\}
\quad (24)
$$

It is easy to see that the system

$$
\left.
\begin{array}{l}
-100\lambda_1 + \quad\quad 25\lambda_2 - 40\lambda_3 \geq 0 \\[2mm]
\quad\quad\quad 100\lambda_2 - 100\lambda_3 \geq 0
\end{array}
\right\}
$$

has a positive solution. In summary, this example shows that the <u>edge connecting two efficient vertices is not necessarily efficient</u>.

Multi–objective simplex tableaux are extensively used in the method of Zionts and Wallenius (1976) which explores the efficient set as follows. Suppose that the current vertex (the current basic feasible solution) is efficient, then all efficient edges emanating from it are identified by inspection of the non–basic columns in the manner just sketched. The endpoints of these edges (again efficient vertices) are presented to a decision maker, and for each of them he is asked whether he prefers it to the current vertex, whether he prefers the current vertex to it, or whether he is indifferent between the two. The answers are used to generate a new positive combination of the objective functions which is maximized at a more acceptable, efficient vertex. The procedure continues until an efficient vertex is found such that no adjacent, efficient vertex is preferred. Even for medium–size problems with a few hundred variables and constraints, and with more than a few objectives, the method becomes unworkable.

7. Constraint Methods and Trade–offs

The intuitively appealing idea underlying constraint methods is to maximize one of the objective functions (the most important one if it can be identified) subject to the additional constraints that the remaining objective functions must be at or above certain lower bounds. Thus, the problem is

$$\text{maximize} \qquad (c^k)^T x$$

$$\text{subject to} \qquad -(c^i)^T x \leq -L_i, \ i \neq k, \tag{25}$$

$$Ax \leq b$$
$$x \geq 0.$$

Let $\bar{x}^k$ denote a maximum solution. As usual in multi–objective optimization, we now have to answer the following questions:

a) Is $\bar{x}^k$ an efficient solution of the original problem (1)?

b) Let x^* denote an arbitrary efficient solution of (1). Is x^* <u>attainable</u>, that is, can we find lower bounds L_i^* such that x^* solves (25)?

c) Which information can we supply to the decision maker about the quality of $\bar{x}^k$? How do we enable him to modify problem (25)?

Let us start with the first question. Because $\bar{x}^k$ is a maximum solution of (25), there exist non–negative multipliers $\alpha_i^{(k)}$, $i = 1, ..., p$, $i \neq k$, $u_i^{(k)}$, $i = 1, ..., m$, $v_j^{(k)}$, $j = 1 ,... , n$, such that

$$c^k = - \sum_{\substack{i=1 \\ i \neq k}}^{p} \alpha_i^{(k)} c^i + \sum_{i=1}^{m} u_i^{(k)} a^i - \sum_{j=1}^{n} v_j^{(k)} e^j , \tag{26a}$$

$$\alpha_i^{(k)} \{(c^i)^T \bar{x}^k - L_i\} = 0 , \ i=1, ..., p, \ i \neq k, \tag{26b}$$

$$u_i^{(k)} \{(a^i)^T \bar{x}^k - b_i\} = 0 , \ i=1, ..., m, \tag{26c}$$

$$v_j^{(k)} \bar{x}_j^{\,k} = 0 , \ j=1, ..., n, \tag{26d}$$

where a^i denotes the i–th row of A, and e^j the j–th unit vector. It is easy to see now that the condition $a_i^{(k)} > 0$ for any i = 1, ..., p, i ≠ k, is sufficient to guarantee efficiency of $\bar{x}^k$. For, formula (26a) can be written as

$$c^k + \sum_{\substack{i=1 \\ i \neq k}}^{p} a_i^{(k)} c^i = \sum_{i=1}^{m} u_i^{(k)} a^i - \sum_{j=1}^{p} v_j^{(k)} e^j .$$

Combining this with the complementary slack relations (26c) and (26d) we conclude that $\bar{x}^k$ maximizes the <u>positive</u> linear combination

$$(c^k)^T x + \sum_{\substack{i=1 \\ i \neq k}}^{p} a_i^{(k)} (c^i)^T x$$

over the feasible set of the original problem (1), which implies efficiency of $\bar{x}^k$. If at least one of the $a_i^{(k)}$ vanishes, the point $\bar{x}^k$ may be inefficient. This is illustrated in Figure 4.

For the second question, there is a straight affirmative answer. We substitute $L_i^* = (c^i)^T x^*$, i = 1, ..., p, i ≠ k, then x* is a feasible solution of (25). Suppose that $(c^k)^T x$ is not maximized at x*, then we would be able to find a point dominating x*, and this is a contradiction.

In order to answer the third question, we consider perturbations of the lower bounds in (25), and we investigate the behaviour of the objective functions under slight modifications of the bounds. So, the perturbed problem reads

maximize $\quad (c^k)^T x$

subject to $\quad -(c^i)^T x \geq -L_i + \epsilon_i,\ i \neq k,$ $\qquad\qquad\qquad\qquad$ (27)

$\qquad\qquad$ Ax ≤ b

$\qquad\qquad$ x ≥ 0,

and we take $x^k(\epsilon)$ to denote a maximum solution. Let us suppose that the multipliers $\alpha_i^{(k)}$ appearing in (26) are positive, so that the point $\bar{x}^k = x^k(0)$ is efficient. By (26b) it must be true that

$$(c^i)^T \bar{x}^k = L_i, \, i \neq k \, ,$$

and for sufficiently small perturbations we must accordingly have

$$(c^i)^T x^k(\epsilon) = L_i - \epsilon_i, \, i \neq k \, .$$

For the objective function of (25) and (27) we find (sensitivity analysis in linear programming)

$$(c^k)^T x^k(\epsilon) - (c^k)^T x^k(0) = \sum_{\substack{i=1 \\ i \neq k}}^{p} \epsilon_i \, \alpha_i^{(k)} \, .$$

Combination of the results yields the trade–off between the k–th and the i–th objective

$$\frac{(c^k)^T \{x^k(\epsilon) - x^k(0)\}}{(c^i)^T \{x^k(0) - x^k(\epsilon)\}} = \frac{\sum_{i \neq k} \epsilon_i \, \alpha_i^{(k)}}{\epsilon_i} \, . \tag{28}$$

In particular, if we only perturb the i–th lower bound L_i, the trade–off between the k–th and the i–th objective is simply $\alpha_i^{(k)}$. This is important information for a decision maker who is not satisfied with the current efficient solution $\bar{x}^k$. He may wish to modify certain bounds, and the multiplier $\alpha_i^{(k)}$ shows the increase (decrease) of the k–th objective per unit of decrease (increase) of the i–th objective.

This is the basis for the simple and popular versions in the class of constraint methods. In general, the decision maker increases (decreases) the lower bounds L_i, $i \neq k$, accordingly as he wishes to increase (decrease) the i–th objective function, at the expense of the preferred k–th objective function. The question of how to use the trade–off information more effectively is still unresolved. In the Surrogate Worth Trade–Off method (see Chankong and Haimes (1983)), the $\alpha_i^{(k)}$, $i \neq k$, are presented to a decision maker who is subsequently requested to assess how much he prefers $\alpha_i^{(k)}$ units of the i–th objective function in exchange for one unit of the k–th function. The surrogate worth is the numerical value $w_i^{(k)}$ assigned to the strength of his preference. It is a value on a numerical scale ranging from -10 to $+10$; a positive (negative) value means that he would accept (reject) the deal, with increasing intensity accordingly as he moves away from the value of zero which signifies indifference for the trade–off. Because both the trade–offs $\alpha_i^{(k)}$, $i \neq k$, and the surrogate worths $w_i^{(k)}$, $i \neq k$, depend on the lower bounds L_i, $i \neq k$, the above procedure of estimating surrogate worths is repeated for several sets of lower bounds (thus, problem (25) is repeatedly solved, and each time the decision maker is asked to assess the resulting trade–offs). The information so obtained is employed to estimate the behaviour of the surrogate worths as functions of the lower bounds. Finally, lower bounds are established (by solving a set of equations) where the surrogate worths are approximately zero. In the corresponding efficient solution, the conflicting views of the decision maker will hopefully be in equilibrium. It is worth noting that the scale sensitivity of the procedure (why the equidistant scale between -10 and $+10$?) has not been investigated, at least to our knowledge. Moreover, we still need a procedure for group decision making when a unanimously preferred objective function cannot be identified.

8. Ideal–Point Methods

In practical problems, the ideal vector (10) is always unfeasible (otherwise there would be no conflicts), but it is conceivable that the nearest feasible solution could be an acceptable compromise for the decision maker. Thus, we consider the problem of finding a feasible solution which minimizes

$$d_\alpha = \left\{ \sum_{i=1}^{p} |z_i^* - (c^i)^T x|^\alpha \right\}^{1/\alpha},$$

where the integer–valued parameter α, $1 \leq \alpha \leq \infty$, designates the norm in the objective space (we are clearly concerned with the distance between z^* and $z = Cx$).

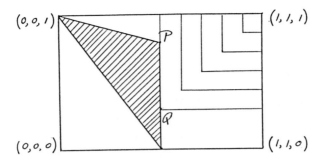

Figure 5. Example to show that a feasible solution, nearest to the ideal vector according to the d_∞ norm, is not necessarily unique. All points on the segment PQ are at the same distance from the ideal vector $(1,1,1)$.

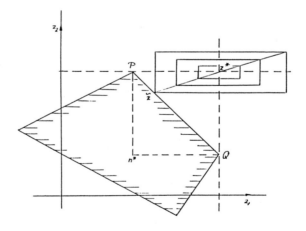

Figure 6. Exploration of the efficient set PQ by minimization of the distance from the ideal vector z^*, with varying weights in the Tchebycheff norm.

Let us, first, consider the choice of α. It will readily be clear that the familiar ℓ_1-norm (α=1) and ℓ_2-norm (α=2) are usually not interesting. For $\alpha = 1$, a nearest feasible solution is found by the minimization of

$$d_1 = \sum_{i=1}^{p} \{z_i^* - (c^i)^T x\}$$

subject to $Ax \leq b$, $x \geq 0$, and because the z_i^* are constants, this is equivalent to the maximization of

$$\sum_{i=1}^{p} (c^i)^T x$$

subject to the same constraints. The solution is clearly efficient (we maximize a positive linear combination of the objectives) but not necessarily unique, and when the simplex method is used, the solution will invariably be a vertex. For $\alpha = 2$, the problem of finding a nearest feasible solution reduces to the minimization of a quadratic function subject to linear constraints. There are many successful algorithms for solving this type of problems, but if the original problem (1) is large, it is not advisable to leave the area of linear programming.

There are several interesting methods based on the ℓ_∞-norm, where a nearest feasible solution is obtained by the minimization of

$$d_\infty = \max_{i=1,\ldots,p} \{z_i^* - (c^i)^T x\} ,$$

subject to $Ax \leq b$, $x \geq 0$. We can equivalently solve the linear-programming problem

minimize y subject to

$$\left.\begin{array}{l} y \geq z_i^* - (c^i)^T x, \quad i=1,\ldots, p , \\[2mm] Ax \leq b, x \geq 0 . \end{array}\right\} \tag{29}$$

We need a three–dimensional example to show that a nearest feasible solution is not necessarily efficient and not necessarily unique. Consider a problem with the objective functions x_1, x_2, and x_3 (then the decision space is also the objective space), and a feasible set spanned by four vectors: the unit vectors and a point P in the unit cube; then the ideal vector is (1,1,1). In Figure 5 we show the cross–section through the origin, the ideal vector, and the points (0,0,1) and (1,1,0) for a particular choice of P. Any point on PQ is a solution of (29), but only P is efficient.

An important step is the introduction of positive weight factors $w_1,..., w_p$ into the Tchebycheff norm. From now, we concern ourselves with the minimization of the distance function

$$\max_{i=1,\ldots,p} w_i\{z_i^* - (c^i)^T x\} .$$

In Figure 6, where the contours are sketched, the reader will see the ideas behind the method: when the distance to the constraint set is minimized with varying values of the weight factors, the point $\tilde{z}$ skratches the surface of the constraint set, hopefully the efficient set only. Moreover, in the point $\tilde{z} = C\tilde{x}$ we might have

$$w_i\{z_i^* - (c^i)^T \tilde{x}\} = w_j\{z_j^* - (c^j)^T \tilde{x}) . \tag{30}$$

Appealing ideas, but not always correct as Figure 5 shows in the case that the weight factors are equal! Nevertheless they are useful and they have been heavily tested (Kok and Lootsma (1985)) under the provision that an efficient solution is generated. This is accomplished by a perturbation of the distance function; see also the description in Lewandowski and Wierzbicki (1988) of the reference–point method underlying the DIDAS system for multi–objective optimization. We minimize

$$\left.\begin{array}{l} \max_{i=1,\ldots,p} w_i\{z_i^* - (c^i)^T x\} + \sum_{i=1}^{p} \epsilon_i\{z_i^* - (c^i)^T x\} \\ \\ \text{subject to } Ax \leq b,\ x \geq 0 , \end{array}\right\} \tag{31}$$

with small, positive numbers $\epsilon_1, ..., \epsilon_p$. Any minimum solution of (31) is now efficient, and it can reversely be shown that any efficient solution is attainable, in the sense that it can be found by solving (31) when the weight factors w_1, ..., w_p and the numbers ϵ_1, ..., ϵ_p are properly chosen.

The strength of the method is that the process for choosing the weight factors can be structured to reveal the preferences of the decision makers as well as the diverging opinions in a decision–making body. Using a nadir vector n* to indicate a possible range of objective–function values we write

$$w_i = \rho_i / (z_i^* - n_i^*), \; i = 1, ..., p \; . \tag{32}$$

We should now ask the decision makers to assign positive values to the ρ_i; for simplicity we take these values to be normalized so that $\Sigma \; \rho_i = 1$. Formula (30) suggests what the assignment would imply: a high (low) value for ρ_i designates a strong (weak) reluctance to deviate from the ideal value z_i^* of the i–th objective function. This is explained to the decision makers, whereafter the objective functions with their ideal and nadir values are presented in pairs. So, the basic experiment is a pairwise comparison by a single decision maker who explains in qualitative terms which ratio he is willing to accept for deviations from the ideal vector in the direction of the nadir vector. The subsequent mathematical analysis of the judgemental statements (note that the decision makers do not always agree, and that each of them may be inconsistent in his judgement) will produce individual weights as well as a set of group weights, which can be used to calculate the corresponding efficient solutions via (31) and (32). The results will hopefully enable the decisision makers to arrive at a concensus. This is typically the purpose of multi–objective optimization.

9. Concluding Remarks about Multiple Objectives

Multi–objective optimization methods are difficult to use because human subjectivity is an integral part of them. We cannot just formulate a model and leave it to an optimization expert to calculate an optimal solution. At various stages, we have to interrupt the calculations to fathom the preferences of the decision makers. Certain parameters (weights, targets, desired levels) are adjusted on the basis of current information (objective values, trade–offs), whereafter the computations proceed in a somewhat modified direction. So, interactive hardware and software support is highly desirable, but not an absolute necessity in this field. In designing a multi—objective optimization method, however, we are subject to several, severe limitations.

a) Decision makers with significant degrees of freedom do not really exist. Particularly in public administration, there are many so—called <u>actors</u> who prepare a decision via a complicated network of policy committees, advisory committees, pressure groups, lobbies, etc.

b) Decision makers (or actors) do not really understand mathematical models. On the contrary, they suspect that the analysts involved in the design of the model understand neither the decision problem nor the decision process within the managerial or official hierarchy, and this is frequently correct. Reversely, the analysts suspect that the problem under study is not the real problem. They feel that there are hidden objectives which cannot be discussed, and they are frequently right.

c) The information supplied at the subsequent stages is not sufficiently transparent. The decision makers cannot use the intermediate results (adjacent efficient vertices, trade—offs) to express their preferences because these items still have too many contradictory (attractive as well as repellent) properties.

d) Decision makers (or actors) cannot spend much <u>time</u> on a particular decision problem. It is an illusion to think that <u>many</u> interruptions are possible to elicit preference information from the decision makers. One or two sessions of the decision—making committee, communication via the mail and the telephone in the time intervals between sessions, and that's all!

Hence, it is not easy to evaluate methods for multi—objective optimization on the basis of their effectiveness in actual decision making. An attempt has nevertheless been made by Kok (1986). He introduced a number of highly relevant criteria to judge the available methods.

a) The information load, the amount of information to be digested by the decision maker, in particular the <u>number of stimuli</u> to be assessed <u>simultaneously</u>. Under this criterion, there are wide variations between the methods. In the multi—objective simplex method of Zionts and Wallenius (1976), the decision maker has to compare and to assess a number of p–dimensional vectors (the objective–function values at various efficient solutions), so that he is confronted with p stimuli at the same time. In the ideal–point methods, particularly in the version of Kok and Lootsma (1985), he only has to judge pairs of one–dimensional stimuli (deviations from ideal values). We note in passing that the ability of the decision maker to oversee a multiple of stimuli is somewhat overestimated in multi–objective optimization.

b) Requisite convergence, termination of the interactive decision process at an acceptable solution after a finite number of interruptions. The term requisite is due to Phillips (1982): the interactive procedure for solving a multi–objective optimization problem models the preferences of the decision maker, and this procedure continues "until no new intuitions emerge about the problem". The acceptable solution, a feasible point where the objectives are felt to be in balance, could accordingly be the somewhat arbitrary result of the procedure itself. This is in sharp contrast with the tacit assumptions which are usually made: the assumption that a genuine solution, the balance of power between the preferences, norms, and social values of the decision maker, exists at the start of the procedure (it may be deeply hidden in the back of his mind), and the assumption that the procedure generates a sequence of intermediate solutions converging towards that genuine solution. The acceptable solution just mentioned would accordingly be an approximation to it. Under the criterion of requisite convergence, these dubious assumptions are left out of consideration. Nevertheless, it remains very difficult to evaluate the multi–objective optimization methods from this viewpoint and to estimate their ability to bring the decision maker so far that no new intuitions about the problem emerge. In fact, we have an open area for research here.

c) Applicability in groups of decision makers, in particular the ability of the methods to suggest a compromise solution in accordance with the relative power position of the group members. In the field of multi–criteria analysis, where we consider a finite number of alternatives, such methods do exist (see also the author (1987, 1988a)), and they can immediately be applied in the ideal–point method when we are confronted with a continuum of alternatives. The key issue is the numerical scaling of judgemental statements, expressed in pairwise comparisons of simple, one–dimensional stimuli. The idea is also present in the Surrogate Worth Trade–Off method of Chankong and Haimes (1983). We expect further improvements here by the analysis of trade–offs, recently proposed by the author (1988b) and studied further by S. Wang (Chinese Academy of Sciences, Beijing, temporarily Delft University of Technology).

In summary, modelling with multiple objectives is a significant step forwards, because we are moving away from the preoccupation with the measurable aspects in a decision problem (costs, profits) towards an approach where also non–measurable aspects (conflicting aspirations, human judgement and expertise) are taken into account. We are concerned with the decision–making process itself, with individual and group decisions, with the preferences of the decision makers, the identification of diverging opinions, the analysis of conflicts, and the attempts to arrive at a joint conclusion supported by a maximum concensus. The result is that the solution of a multi–objective optimization problem may be remarkably <u>robust</u>. It is a common experience that solutions of single–objective problems are usually very sensitive to perturbations of the cost or profit coefficients. The compromise of several objectives, however, is not easily perturbed by uncertainty in the cost data.

<u>References</u>

[1]   V. Chankong and Y.Y. Haimes, <u>Multiobjective Decision Making</u>. North–Holland, Amsterdam, 1983.

[2]   J.L. Cohon, <u>Multi–Objective Programming and Planning</u>. Academic Press, New York, 1978.

[3]   A.M. Geoffrion, Proper Efficiency and Theory of Vector Maximization. <u>Journal of Mathematical Analysis and Applications</u> 22, 618–630, 1968

[4]   C.L. Hwang and K. Yoon, <u>Multiple Attribute Decision Making</u>. Springer, Berlin, 1981.

[5]   J.P. Ignizio, <u>Goal Programming and Extensions</u>. Lexington Books, Lexington, Mass., 1976.

[6]   H. Isermann and R.E. Steuer, Pay—off Tables and Minimum Criterion Values over the Efficient Set. Report 85–178, College of Business Administration, University of Georgia, Athens, Georgia, USA.

[7]   R. Keeney and H. Raiffa, Decisions with Multiple Objectives: Preferences and Value Trade—offs. Wiley, New York, 1976.

[8]   M. Kok, Conflict Analysis via Multiple Objective Programming, with Experiences in Energy Planning. Thesis, Delft University of Technology, The Netherlands, 1986.

[9]   M. Kok and F.A. Lootsma, Pairwise Comparison Methods in Multi—Objective Programming, with Applications in a Long—Term Energy—Planning Model. European Journal of Operational Research 22, 44–55, 1985.

[10]   A. Lewandowski and A. Wierzbicki, Aspiration—based Decision Analysis and Support. Part I, Theoretical and Methodological Backgrounds. Working Paper 88–03, IIASA, Laxenburg, Austria, 1988.

[11]   F.A. Lootsma, Modélisation du Jugement Humain dans l'Analyse Multicritère au moyen de Comparaisons par Paires. R.A.I.R.O. Recherche Opérationelle 21, 241–257, 1987.

[12]   F.A. Lootsma, Numerical Scaling of Human Judgement in Pairwise Comparison Methods for Fuzzy Multi—Criteria Decision Analysis. To appear in G. Mitra (ed.), Mathematical Models for Decision Support. Springer, Berlin, 1988a.

[13]   F.A. Lootsma, Conflict Resolution via Pairwise Comparison of Concessions. Manuscript, Delft University of Technology, The Netherlands, 1988b. To appear in the European Journal of Operational Research.

[14]   P. Nijkamp and P. Rietveld, Multi—Objective Programming Models. Regional Science and Urban Economics 6, 253–274, 1976.

[15]   L. Phillips, Requisite Decision Making, a Case study. Journal of the Operational Research Society 33, 303–311, 1982.

[16]　P. Rietveld, Multiple Objective Decision Methods and Regional Planning. Studies in Regional Science and Urban Economics 7, North–Holland, Amsterdam, 1980.

[17]　B. Roy, Méthodologie Multicritère d'Aide à la Décision. Economica, Paris, 1985.

[18]　Th.L. Saaty, The Analytical Hierarchy Process, Planning, Priority Setting, Resource Allocation. McGraw–Hill, New York, 1980.

[19]　R.E. Steuer, Multiple Criteria Optimization: Theory, Computation and Application. Wiley, New York, 1986.

[20]　P.L. Yu, Multiple Criteria Decision Making: Concepts, Techniques and Extensions. Plenum Press, New York, 1985.

[21]　P.L. Yu and M. Zeleny, The Set of all Nondominated Solutions in Linear Cases and a Multicriteria Simplex Method. Journal of Mathematical Analysis and Applications 49, 430–468, 1975.

[22]　M. Zeleny, Multiple Criteria Decision Making. McGraw–Hill, New York, 1982.

[23]　S. Zionts and J. Wallenius, An Interactive Programming Method for Solving the Multiple Criteria Problem. Management Science 22, 652–663, 1976.

L. V. Kantorovich's Works in Mathematical Programming

J. V. ROMANOVSKY

Leningrad University, U.S.S.R.

1   Introduction

We have lost now the possibility to listen to L.V.Kantorovich himself about his early work in Mathematical Programming. But in addition to the published papers we possess at least two of his own evidences. They are, firstly, his letter to Professor R.Thrall who edited *Management Science* in the sixties, and, secondly, his posthumous memoirs [26] which were recorded when he was mortally ill.

He displayed such courage at that hard time. Up to the last days of his life he was very active, we discussed some possible algorithms of Mathematical Programming, and he kept on planning to visit this beautiful country (which was too interesting to him) in connection with the Conference on Input/Output Analysis.

Formally Kantorovich[1] started his investigation in this field in 1939 with the well-know booklet [2] *Mathematical methods of organizing and planning of production* and finished in 1985 with our joint lecture on column generation [25] at the previous Conference, but, certainly, more attention should be paid to the early period.

I think that it would be difficult to follow in details an old- fashioned explanation of terms, algorithms and theorems, because the theory has developed for forty years in traditions started by George Dantzig, and these traditions seem more harmonious in the main. I intend to turn to the original explanation only sometimes, when it is necessary.

---

[1]It is difficult to say too often **Kantorovich** and psychologically quite impossible for me to use only his name Leonid. Following Russian manner I shall use sometimes letters **LV** which form abbreviation for **Leonid Vitalievich** - the full name of our author.

M. Iri and K. Tanabe (eds.), Mathematical Programming, 365–382.
© *1989 by KTK Scientific Publishers, Tokyo.*

Let us start with a concise description of the main results as it was made in the letter of Kantorovich to Professor Thrall:

> Perhaps it might be desirable to present my own view as to what I obtained in the pre-War years.
>
> 1. I indicated a wide class of practical problems of management leading to the same mathematical extremal problem. One property of this mathematical problem, to use current terminology, is a guarantee of the existence of a feasible solution. However, this circumstance is dictated by the very form of the practical problems which I considered...
>
> 2. I found necessary and sufficient conditions for a solution of this mathematical problem to be optimal. These necessary and sufficient conditions are based on the existence of a supporting hyperplane at the maximum point in the production space. The coefficients of this hyperplane (the *resolving multipliers*) appear as the solution of the dual extremal problem... Still, there is no need to take the last fact into account; it is sufficient to use the properties of a supporting hyperplane.
>
> 3. On the base of the above mentioned necessary and sufficient conditions I constructed (finite) methods of solution (called *the method of resolving multipliers*) which are on a par, in effectiveness and generality, with subsequently developed methods; for example, the simplex method. The method of multiplier correction which is one of the above methods coincides essentially with Primal-Dual methods, and is perhaps the closest to the method developed in 1957 by Dantzig- Ford-Fulkerson.
>
> A method of potentials was worked out for the transportation problem. This problem was then extended to other problems, under the name of the *method of successive improvements of plan.* It is closely akin to the so-called revised simplex method...
>
> 4. I clarified to some extent the economic meaning of the resolving multipliers...
>
> 5. The extremal problem mentioned above and its solution were generalized to the infinitely-dimensional case...
>
> Now I should like to discuss what I did **not** accomplish – of course, from the point of view of contemporary achievements and future needs.

1. The properties of duality were not studied, nor was the dual problem formulated with sufficient accuracy.

2. Questions of degeneracy and the practical implementation of methods of solution for this case were not investigated (except a little something).

3. I did not give the algorithms in definitive mathematical form, although the necessary stages of calculation, including possible complications, were given in a form covering the regular cases.

4. No generalizations of the extremal problems for the case where feasible solutions do not exist and for the case of the unbounded linear form.

5. I did not present a detailed exposition of the results that I obtained. This circumstance came about as follows: Whenever I would begin to think of this, I have to decide whether to devote myself to the mathematical side of the question, which is so simple in comparison with other mathematical problems (judging by my own experience) that any mathematician should be able to formulate and present it rigorously - or else to direct my attention to the applied, economic side, which was new, non-trivial, and thought- provoking. Naturally, I prefer the latter.

After this preliminary I'd like to give you some personal information about the author and his experience up to the time of his first works in Mathematical Programming and then to touch some special topics.

The first of my special topics will be the continuous version of the transportation problem and some its generalizations in LV papers and in forthcoming studies. I take the liberty to mention here our joint work that became unknown.

The second one will be Kantorovich's work in applications of linear programming. The selected applications are connected with optimization in iron industry. It was pleasant to mark that the conference schedule includes a special section of applications in steel industry.

After that let us touch some applications of optimization methods in the problems of Numerical Analysis where LV was also a known authority. So it is interesting to follow his point of view about connection of these three his preferable fields. Yes, three fields, because the third one is Functional Analysis.

And then let me say some words about Kantorovich's work in software design and in connection with this work about current development of software for mathematical programming.[2]

2   What was LV before his works in MP

I am sure that many of you have met Kantorovich before. But I think you would like to have an impression about Kantorovich of the time when he began working in Mathematical Programming. Please look at the picture of 1935, three years before that time. Kantorovich was born in 1912, so in 1939 he was 27.

He started his active mathematical life when he was 16 and he was still a student of the Leningrad University. At the time we are talking about he was an experienced mathematician with a good reputation, with about sixty published research papers and with a stable social position of a professor of the University. It was natural that he was an object of admiration of elder Leningrad mathematicians and had a team of pupils of about the same age. In previous years he had valuable results in various fields of abstract and applied mathematics. The list includes the theory of functions and descriptive set theory, complex variable theory, functional analysis, the theory of approximation with especial interest to Bernstein polynomials, some extremal problems arising in mathematical physics. He have some research papers and even a textbook on the Calculus of Variations.

It might to be especially significant that he had an experience in functional analysis and was interested in ideas of optimality, convexity and partial ordering. At that time functional analysis was the field of mathematics where the notions and technique of convex geometry were actively used. I can refer to Hahn-Banach theorem, to lattice theory of G.Birkhoff, to works of M.G.Krein. Von Neumann worked both in the theory of games and in functional analysis.

LV paid many attention to various topics of Numerical Analysis. His personal computational experience before the World War was not large from the point of view of current practice but he was very active in computational research. He considered a lot of computational schemes for analytical, mechanical and physical problems with a clear aim to reduce human labour. In 1936 together with V.I.Krylov he published a large book on methods of approximate solution of partial differential equations [1]. It was republished in significantly revised form in 1941 as *Approximate methods of Higher*

---

[2]In advance I beg your pardon for gaps in references to topics, persons and papers.

L. V. Kantorovich

Analysis and this version ran into many editions and translations.

After the War LV was at the head of a large computational group in the Leningrad Branch of the Academy of Science Mathematical Institute (LOMI), where his duties included organization of some responsible computational works with a very modest hardware. LV was proud of what he and his colleagues achieved in a calculation of higher orders Bessel functions with a so-called Hollerith set of business perfocard calculators ([6], [11]).

In his activity LV neither calculated anything by himself nor make a detailed computational scheme. But persons who worked with him that time made evidence that he always had a perfect understanding of pluses and minuses of various computational ideas and selected the best plan of computation (by quite unforeseen ways).

3   The first LV's works in MP

As it was told above, Kantorovich's work in Mathematical Programming was directed by his considerations of the economical inferences from the mathematical results. The great significance of improvement of economical relations in his country made him to return to the economical presentation of his ideas again and again.

He wrote the first version of his book *The economic calculation of the best use of resources* [17] in the War years and presented the manuscript for discussion in the academic Institute of Economy and in the State Planning Committee (Gosplan). But that time our economists considered the mathematics itself as a death threat for themselves (and frankly speaking they had some reasons). And their response for a new approach was so decisively negative that LV made a delay for some years. I refer you for details to his memoirs [26].

For a long time the mathematical part of his research was presented with the booklet of 1939 which was oriented on possible applicators of the methods and had only a few pages of a pure mathematical text and two small papers for mathematicians in the Doklady in 1940 [3] and 1942 [4].

Let us start from the booklet. As a foundation of his economic reasoning LV introduced a model of an enterprise planning with the task of proportional output of several kinds of production under some linear technological restrictions. He understood quite well that the original model should not include any prices of goods because at that time the administrative management of the Soviet economy eluded to orient its decisions on such "capitalistic" categories as prices and profits.

His initial model was like following (I choose a representative model from a few Kantorovich's models and made some changes of the statement):

Let us consider an enterprise with m kinds of tools and n goods to be produced. Let a_{ij} be the productivity of the tool i producing the j-th good, i.e. the output of the good when the tool produces all its time. Let p_j be the coefficient of proportionality of j-th good in the total output. Find the matrix of non-negative x_{ij} such that maximizes

$$\min\{z_j/p_j | j \in 1:n\}$$

where

$$z_j = \sum_i a_{ij} * x_{ij}$$

$$\sum_j x_{ij} \le 1, \qquad i \in 1 : m.$$

This minimax problem with linear restrictions can be reduced obviously to a linear program. It had no difficulties with the existence and construction of an initial feasible solution and some details of optimization process were simplified. LV even assumed directly that all z_j had the same value.

The dual problem and the duality theorem also did not appear yet in this explanation. But the dual variables were introduced under the name of *resolving multipliers* and the criterium of optimality was formulated in terms of existence of the set of non-negative multipliers $\lambda_1, \lambda_2, ..., \lambda_m$ such that $x > 0$ implies

$$\lambda_i = \max\{\lambda_k a_{kj} | k = 1, ..., m\}.$$

The main aim of the booklet was to show the practical importance and universal applicability of the multipliers in various situations and this aim dictated the level of explanation. The level of author's understanding of the material is obvious from his small paper in Doklady 1940 [3]. Let me read its theorems.

Theorem 1. *Let E be a compact normed space and A be a convex weak compact set in E. Let a functional $F(x)$ be defined and bounded on A with the following properties:*

1) F is upper weak-semicontinuous,

2) $F(x)$ doesn't reach its maximum inside A, i.e. for each internal point x in each its neighborhood there exists a point x' with a greater value of the functional: $F(x') > F(x)$.

Then there exist a point x_0 on the boundary of A where $F(x)$ reaches its maximum and a linear functional $f_0(x)$ reaching its maximum in the same point.

Theorem 2. *Let the set $\{x | F(x) > C\}$ be convex for each C under the same assumptions. If a point x_0 and a functional f_0 are such that x_0 maximizes the value of f_0 on A and the equality $F(x_0) = F(x)$ implies $f_0(x) > f_0(x_0)$ then $F(x)$ reaches its maximum in x_0. Moreover the existence of x_0 and f_0 is guarantied.*

As the main technique the author used various kinds of separation theorems for convex sets and supporting plane theorems. The algorithms of finding the extremal points were represented only by their ideas, but the ideas remains quite modern.

In this paper LV enumerated some important applications of the mentioned theorems. He included the classical problem of the best approximations and the approximate solving of inconsistent equations. We know that the both applications present now valuable directions of Numerical Analysis. But the very first of the applications was named an interesting continuous version of a linear program:

Let functions $\alpha_i(t), i \in 1 : m$ be defined and integrable on $[a, b]$. Find measurable non-negative functions $h_i(t)$ such that maximize

$$\min_i \int_{[a,b]} \alpha_i(t) h_i(t) dt$$

under condition that $\sum_i h_i(t) = 1$ in each point $t \in [a, b]$.

The author noted that the optimal solution could be obtained in class of 0-1 functions.

Note that here like in the finite-dimensional case the author considered the space of goods outputs rather than the space of variable intensities. It simplified the treatment of optimality condition and wide variation of the models with more stable system of resolving multipliers and led him to the idea of economical indices.

4   Translocation of masses

The last of the early papers was *On the Translocation of Masses* [4] where LV gave the statement of the transportation problem mentioned shortly in 1939 booklet and the criterium of optimality for the feasible solution (the plan of transportation) in terms of existence of a special *potential function* on a set of states. This study of transportation problems accompanied by a rather large paper which was written for the practical transport managers and demonstrated the potential method on a concrete transhipment problem with a fragment of a Soviet railway net. This second paper (with M.K.Gavurin) was published only in 1949 [9].

Only 10 lines were given in [4] for the statement of the transportation problem and for the remark about [9]. The main part of the paper dealt with a infinite-dimensional transportation problem which influenced a rather active theoretical contributions.

Namely, LV introduced a problem of the best transformation of one given mass on a metric compact space into another with a continuous translocation cost function. Later [5] he found that such a problem had been considered by Gaspar Monge in his memoir of 1781 [27] and by Appell in 1881 [28]. The condition of optimality of the plan of the mass translocation was formulated in terms of existence of a potential function $u(x)$ such that

$$u(x) - u(y) < cost(y, x)$$

for all possible points x and y and the relation holds as an equality in all pairs of points which are connected by an essential translocation (the word *essential* means that a positive mass moves between every two neighborhoods of the points). This condition gave him means to prove the Monge conjecture that the optimal translocation consisted of non-crossing paths. It should be remarked that the structure of the optimal translocation remains unstudied.

Note that in the Soviet researches on the transportation problem there was no essential difference of the matrix and network statement and the term *potential* appeared in the very first of his works on the transportation problem. It was natural for LV to take in his mind physical and mathematical relations of the new conceptions with classic ones.

In the paper 1942 LV paid attention to topics related with functional analyzis and proposed to consider a metric space of mass distributions with the minimal translocation cost as the metric. The continuation of these words ensued in 1957 and 1958 in joint papers by Kantorovich and G.Sh.Rubinstein [15], [16]. They considered the properties of the corresponding functional space in a classic tradition, describing the definition of a norm in the space, the convergence in the sense of this norm, the conjugate space. The modern explanation of this topic can be founded in [24].

Let us note that one important topic of the problem remained out of LV's interest and the interest of his pupils of that time. I mean the problems related with the structure of the set of feasible transportation plans. It seems that LV had absolutely no interest in the geometric side of the problematic both in the finite- and infinite-dimensional case. But let us touch it, because the problematic was too close to the Monge-Appell statement of problem and definitely many forthcoming researches in the field were influenced namely by Kantorovich's researchs.

For more details connecting the problem with many problems of measure theory see ([29], [30], [31]).

I think that it would not be right to consider the continuous transportation problem only as a field of pure mathematical research. It seems a useful tool for studies of more complicated transportation problems. For instance, if we want to consider a transportation model with stochastic volumes of supply and demand it can be more convenient to find the answer in terms of the Dirichlet sets of consumers given for all producers or vice versa. Such an approach seems to be useful in the computation of initial solution for usual matrix problems with many small consumers. I had a corresponding experience in this field. And certainly the initial Monge-Kantorovich problem arises in some agricultural and meliorative problems.

One more topic related with the continuous transportation problems is connected with extremal problems for measures preserved under transformation. Such kind of problems arises in the theory of controlled processes and particularly in the theory of Markov decision processes where the optimal stationary policy defines a transformation on the set of process state and the measure preserved under this transformation corresponds to the optimal stationary state of the process.

As an example of such considerations let me mention a model that was studied in our joint papers in 1965-1966 [19], [20] to support one economic idea of LV. It was a model for the development of optimal structure of machine park under variable load and for study the depreciation charges. LV assumed that the depreciation charge should consist of two parts: one part should be a payment for the load and has to be dependent on the load level in the system and the second part should be a payment for the keeping the unit of a given age.

We considered an equipment park consisting of separate units of the same kind. The units have various degrees of wear and maintenance costs depend on the wear of a given unit. At each point of time a whole spectrum of load appears. We can consider it as a limit of daily repeated periodical load structure when the period tends to zero. So we have here rather specific two-dimensional time, where one dimension describes the ordinary homogeneous time and the second is used only for introducing the periodical non-homogeneity.

5 Applications

It is necessary to say that Kantorovich always tend to applications, and the booklet of 1939 gives us a good example of his interest. He made many efforts to apply some concrete economic ideas, for example, in his archive we have found a letter with a proposition of different payments for power which was written in forties to Leningrad electric-power administration. I also remember his work for changing payment system for taxi-cars which was realised in 1961.

But certainly the linear programming applications were considered to be the main. The first attempt was to apply it to the problem of cutting of stocks which was mentioned in the booklet of 1939. LV came into contact with specialists of wood processing and wrote a paper for the magazine *Forest Industry* but it was published only after the War [10].

The real applications of the model were started only in 50-es when V.A.Zalgaller jointed to LV. It was a young mathematician who had just graduated as a geometer from the Leningrad University and possessed a hard experience of military signaller during the War (he is still working at LOMI and teaching Geometry at the University, and nevertheless he remains one of the most qualified experts in the problem of optimal stocks cutting). Young Zalgaller was ordered to go to wagon-building enterprise and to try to solve problems arising in practice. Certainly there were no computers. And moreover there was no interest of enterprise staff in results.

The results were published in their book *Rational cutting of industry stocks*, Leningrad, 1951 [12] (the second edition [23] was published in 1971 in Novosibirsk). It seems to me that this book is a masterpiece in Operations Research, and I can only regret that the Western world ignored it. It contains the presentation of problem, the case study with a lot of technical hints and versions. And it is shown how Zalgaller solved problems with several dozens of restrictions without any computers.

It was a problem of so-called serial cutting where there was given a constant demand on a given set of m details in fixed proportions. There are given positive numbers $k_1, k_2, ..., k_m$ and the task is to produce from a given stock the details in quantities $z_1, z_2, ..., z_m$ such that maximize the minimal ratio z_i/k_i.

In this application it was especially clear that the free-style approach to computational methods mentioned above had its preferences. While solving the problem Kantorovich and Zalgaller were not restricted by any quite formal scheme (such as, for instance, primal simplex-method). They knew that it was necessary to solve some linear systems, the matrices were sparse, and Zalgaller spared his strength on appropriate elimination of variables and choice of initial basic solution. They found and used very efficient technique of computations for the linear programs with a large number of variables.

This approach is well known now as the column generation technique.

The list of their novelties in optimization was not exhausted by linear programming. Each stage of this procedure includes an extremal problem: under given prices of details which are the dual variables find a cutting corresponding to maximal profit. Remember that for the case of linear stock it is the knapsack problem: to find a set of non-negative integers $t_1, ..., t_m$ such that maximizes the total income $\sum_i u_i t_i$ under restriction $\sum_i l_i t_i < L$. Here u_i are dual variables, l_i are lengths of details and L is the length of stock.

Zalgaller told me that it was Kantorovich who proposed to use there such recurrent relations which are used now in dynamic programming and Zalgaller himself used something like the branch-and-bound method. They succeeded in the both approaches.

Another large practical problem was that of production and transportation of several goods that seemed to have a lot of possible concrete applications in our economy. LV had chosen the problem of allocation and transportation of iron production [21], and it was studied by his team in Institute of Mathematics in Novosibirsk.

This problem deals with a given set of rolling mills with their resources of hot time and their capabilities to roll various assortments of steel. The set of orders is known also; each order includes a assortment type, an ordered quantity of the assortment and the delivery point. It is a linear program with trivial balance relations for the hot time of each mill and for each order, since the order could be formally divided between some mills. The concrete problems which were solved by Kantorovich's group dealt with all rolling mills of our country (that means more then a hundred mills and about 40 thousand orders).

Since their computer base being rather modest, the special versions of simplex-method which took into account the peculiarities of the matrix were made, and the problem was solved. The results of the calculations were very optimistic since the optimization gave about 6 percents of productivity increasing with the same transportation costs.

However, the real implementation of the results has met with a difficulty that turned to be very attractive for research. The managing body of the industry receives the orders not at a time but continually.

The integral properties of this flow of orders are well known, so it is possible to calculate the total optimal plan of production. But when the plan is known it is necessary to realise that plan or its main part in the sequence of operational decisions. It seems that we meet here a general need of schemes and approaches which permit to joint the finding of the main tendency from a static general model and the operational implementation of the tendency.[3]

The same difficulty appears now in the problem of stock cutting mentioned above. Sometimes a producer who is compounding a complex order neglects some lacks of stocks. My colleagues who are working with applications of the problem case across this aspect in concrete situations.

6 Computational aspects

It was mentioned that LV worked hardly in Numerical Analysis and in [3] he listed some applications of the new technique in this field. He returned to the theme twice.

For the first time it happened in 1948 when he discovered important relations between functional analysis and applied mathematics. He had written at that time two papers with the same title - [7] and [8]. The second paper in *Uspekhi* was the enlarged version of the first part of [7]. It was assumed that its continuation would enlarge the second part of [7], but unfortunately the intention was not realized. This part contained a unified explanation of computational methods for extremal problems both in classic and modern versions.

For the second time he returned to the computational questions in 1962 [18] when he worked in Novosibirsk and his activity in classical Numerical Analysis was rather weakened. But it was the time of active computational work in linear and non-linear programming, and LV paid some attention to possible mathematical applications. He said that it could be useful to take into account various kinds of a priori information in computational problems, such as boundaries for coefficients of approximating form, and to use the modern optimization technique to solve the improved problems.

I like very much one good idea of this paper which had been repeated by LV several times in his lectures. In Numerical Analysis it happens very often that it is necessary to find a solution of a difficult equation not for the solution itself but for to calculate some rather simple functions of the solution.

For instance, in seismic geology if one wants to estimate a volume of underground ore layer he tries usually to calculate the layer form and to

---

[3]I think that the paper by R.Johnston on this conference was iniciated by the same difficulties.

find its volume. After this the idea is clear.

LV proposed to estimate the needed information directly without intermediate computational unstable problems. Some applications of this idea appeared later in his papers together with Soviet seismologists [22].

7   Connections with software design

Since in the post-War time LV dealt with computations and had some interests in hardware and software. Concerning hardware I can add that in his work with Gavurin a hydraulic model to solve transportation problems was proposed. His post-War ideas in hardware were quite right, they have been reopened and realized in various forms. (For example, LV proposed to add to computer various coprocessors for special duties).In software he started researches for using computers in analytical manipulations [13]. Certainly we can't follow Kantorovich's software ideas directly, because now there exists a giant experience of use various algorithmic languages, programming tools and systems. But something remains still.

As to me, the most productive idea of Kantorovich in software was the idea of *prorab*. In Russian the word is an abbreviated form of "PROisvoditel' RABot" or "Jobs manager", at that time it was a common name for a low-level manager in building industry. LV proposed to consider a program or a computational plan as a field of interconnection of different prorabs acting each in its own field of computation (in modern situation we can consider input/output prorab, memory manager, solver of linear systems and so on). The whole computational plan should be divided into separate jobs that are transmitted to corresponding prorabs. It was assumed that the prorab programs could be prepared separately and could be more stable and universal units.

In contemporary software design this idea becomes widely used in rather different implementations. The simplest reference can be made to the popular idea of program interpretation, and to refresh in memory a lot of programming tools based on this idea.Even in software for linear programming we can name some modern packages which give to user a set of regimes executable by their LP-prorabs.

Another prorab treatment is related with languages supporting special modular structures of programming systems. I mean languages such as "Modula-2" and Ada with their concept of module and C language with its construction of procedure file and the concept of abstract data types so widely treated in American software literature a few years ago. This second approach is close to my current interests. In the last decade I hadan experience in designing software for optimization, so it permits me to finish my lecture with some words on this topic.

Following the spirit of "prorab" I defined a software *service* [32] as a set of procedures or procedure-like possibilities (functions, entries etc.) jointed together by their common information that is not known directly by their software environment. The idea of the service was principally clear and close to many professional programmers and it received its implementation (however, rather limited) in various languages and software systems. Only in the last decade the idea has drown the attention in connection with the development of modularity principles, particularly with ideas of hiding information and abstract data types.

The idea happens to be fruitful in designing the optimization software. Let me demonstrate that on the structure of simplex method with possibilities of column generation.

In the main part of the simplex-method we can select two important services, namely, the service of the basic linear system and the service of data. Certainly, there are possible and useful special interface services for dialogue and other forms of input/output. It is convenient to have simultaneously a few types and copies of the data service, because even in quite usual problems it is better to consider separately the ordinary problem variables and the system variables, artificial, complementary and so on. The set of ordinary variables itself can be partied onto subsets of the same and of different nature.

Such a structure gives us a possibility to separate completely the service of data and to make it free exchangeable in future applications. In last decade we have designed a few LP packages with the possibilities of including the user-made modules.

I want to thank the Organizing Committee for this double privelege - to read a lecture about Kantorovich and to take part in the Conference. I am grateful to all persons who helped me in this matter and especially to Professor George Dantzig who chaired the Memorial Session.

References

[1] Kantorovich L.V., Krylov V.I. *Methods of approximate solution of partial differential equations*, Leningrad-Moscow, 1936.

[2] Kantorovich L.V. *Mathematical methods for organizing and planning of production*, Leningrad State University, 1939.

[3] Kantorovich L.V. *An efficient method for solving some classes of extremal problems*, Doklady Akademii Nauk SSSR, vol. 28, # 3, pp. 212-215

[4] Kantorovich L.V. *On the translocation of masses*, Doklady Akademii Nauk SSSR, vol. 37, # 1, pp. 227-230

[5] Kantorovich L.V., *On a problem of Monge*, Uspekhi Matematicheskih Nauk, 1948, vol. 3, # 2, pp. 225-226.

[6] Kantorovich L.V., Gavurin M.K. *On some hits of calculation the sums of products using binary representations*, Uspekhi Matematicheskih Nauk, 1948, vol 3, # 4, pp.160-162

[7] Kantorovich L.V., *Functional analysis and applied mathematics*, Vestnik Leningradskogo Universiteta, 1948, # 6, pp. 3-18

[8] Kantorovich L.V., *Functional analysis and applied mathematics*, Uspekhi Matematicheskih Nauk, 1948, vol. 3, # 6, pp.91-185.

[9] Kantorovich L.V., Gavurin M.K., *The application of mathematical methods in problems of cargo-flow analysis*, In: Problemy povyshenija effectivnosyi raboty transporta , Moscow-Leningrad, 1949, pp. 110-138.

[10] Kantorovich L.V., *Selection of deliveries giving the maximal output within a given assortment*, Lesnaja Promyshlennost', 1949, # 7, pp.15-17, # 8, pp. 17-19.

[11] Gavurin M.K., Faddeeva V.N., *The tables of Bessel functions $J_m(x)$ of integer orders from 0 to 120*, Moscow-Leningrad, 1950.[4]

[12] Kantorovich L.V., Zalgaller V.A. *Calculation of rational cutting of industrial stocks*, Leningrad, 1951.

[13] Kantorovich L.V., *On a mathematical symbolic which is convenient for computer-oriented calculations*, Doklady Akademii Nauk SSSR, 1957, vol. 113, # 4, pp. 738-741.

---

[4]Kantorovich was editor of this work.

[14] Kantorovich L.V., *On methods of analysis of some extremal problems of industrial planning*, Doklady Akademii Nauk SSSR, 1957, vol. 115, # 3, pp. 441-444.

[15] Kantorovich L.V., Rubinstein G.Sh. *On a functional space and some extremal problems*, Doklady Akademii Nauk SSSR, 1957, vol. 115, # 6, pp. 1058-1061.

[16] Kantorovich L.V., Rubinstein G.Sh., *On a space of completely additive functions*, Vestnik Leningradskogo Universiteta, 1958, # 7, pp. 52-59.

[17] Kantorovich L.V., *Economic Calculation of the Best Use of Resources*, Moscow, 1959.

[18] Kantorovich L.V., *On some new approaches to computational methods and treatment of observations*, Sibirskij Matematicheskij Jurnal, 1962, vol. 3, # 3, pp.701-709.

[19] Kantorovich L.V., Romanovsky J.V. *Depreciation charges under optimal use of tools*, Doklady Akademii Nauk SSSR, 1965, vol. 162, # 5, pp. 1115-1118.

[20] Kantorovich L.V., Romanovsky J.V. *The structure of depreciation charges under stationary load of machinery*, Doklady Akademii Nauk SSSR, 1966, vol. 166, # 2,pp. 309-312.

[21] Kantorovich L.V. *Mathematical methods of optimal loading of rolling mills*. Preprint of Institute of Mathematics, Novosibirsk, 1966.

[22] Kantorovich L.V., Keilis-Borok V.I., *Algorithms of interpretation of seismic data*, Izvestija Akademii Nauk SSSR, ser. Fizika zemli, 1970.

[23] Kantorovich L.V., Zalgaller V.A., *Rational cutting of industrial stocks*, Novosibirsk, 1971.

[24] Kantorovich L.V., Akilov G.P. *Functional Analysis*, 2-nd ed., Moscow, 1977.

[25] Kantorovich L.V., Romanovsky J.V. *Column generation in simplex method*, Economika i matematicheskie metody, 1985, vol. 21, # 1, pp. 128-138.

[26] Kantorovich L.V., *My way in science (An intended lecture in Moscow Mathematical Society)*, Uspekhi Matematicheskih Nauk, 1987, vol. 42, # 2, pp. 183-213.

[27] Monge G, *Memoire sur la théorie les déblais et remblais*, Histoire de l'Académie des Sciences de Paris avec les Memoirs de mathematique et de physique pour la meme annee, 1781, pp. 666-704.

[28] Appell P., *Memoire sur les déblais et remblais des systemes continues et discontinues*, Memoirs des sav. etr., 2 ser., 1884, t. 29, 3 memoire.

[29] Sudakov V.N., *The geometric problems of the theory of infinite-dimension probability distributions*, Trudy Matematicheskogo Instituta im. Steklova AN SSSR, 1976, vol. 141

[30] Levin V.L., *The Monge-Kantorovich problem on the translocation of masses*, Metody funkchional'nogo analiza v matematicheskoj economike, Moscow, 1978, pp. 23-55.

[31] Rachev S.T., *The Monge-Kantorovich problem on the translocation of masses and its applications in stochastics*, Teorija verojatnostej i ee primenenija, 1984, vol. 29, #4, pp. 625-653.

[32] Romanovsky J.V., *A package version of simplex-method. An evolutionary description of the main constructions*, Issledovanie Operacij i Matematicheskoe modelirovanie, iss. 5, Leningrad, 1979, pp. 55-71.